Arabic Dictionary of Civil Engineering

ENGLISH - ARABIC
ARABIC - ENGLISH

Compiled by Multi-Lingual International Publishers Ltd
under the general editorship of
Ernest Kay

Routledge & Kegan Paul
London, Boston and Henley

Words which the authors, editors and publishers have reason to believe constitute registered trade marks have been labelled as such, using the abbreviation TM. However, neither the presence nor the absence of such designation should be regarded as affecting in any way the legal status of any trademark.

First published in 1986
by Routledge & Kegan Paul plc

14 Leicester Square, London WC2H 7PH, England

9 Park Street, Boston, Mass. 02108, USA and

Broadway House, Newtown Road,
Henley-on-Thames, Oxon RG9 1EN, England

Printed in Great Britain
by T.J. Press (Padstow) Ltd
Padstow, Cornwall

ISBN 0-7102-0429-9

CONTENTS

INTRODUCTION

THIS translation dictionary of Civil Engineering words and terms has been compiled with the needs of the on-site engineer in mind. It is essentially a practical dictionary, avoiding definitions, because those using a word know the meaning of that word.

Technical coverage of the dictionary extends throughout the sphere of Civil Engineering and includes airport construction, bridges, building, demolition, docks and harbours, tunnels, utility supplies and water. Those activities associated with the work of the civil engineer, such as design, contract management, structure of materials, surveying and environmental engineering, were also taken into consideration during the selection of the word list. Even the latest electronic techniques, especially computer technology, have been taken into account, although on a carefully selected basis.

Not only have translation requirements been taken into account during the compilation of this dictionary, but careful consideration has been given to the problems associated with the pronunciation of Arabic script so frequently encountered by English-speaking people. Because of these difficulties, each word in Arabic script is accompanied by its own pronunciation aid — a transliteration which has been proved successful over the years. A short guide to the use of this transliteration follows this introduction.

As the first dictionary of its kind in Civil Engineering, it will be an essential tool for civil engineers, whether on-site or based in head office. Because of its exclusive subject coverage, those associated with civil engineering in any way, including those from other engineering fields, plant and equipment manufacturers and suppliers, will find the dictionary useful. Teachers and students of engineering both in high school, technical colleges and universities will find the dictionary very helpful, as will translators, interpreters and anyone charged with the task of providing an equivalent Arabic word or term.

In order to keep the dictionary to a manageable size words have been selected carefully — and in many cases words have been presented as part of a term or phrase which is in frequent use — but some omissions will have occurred. Certain omissions have been deliberate because the words are in such common use that they occur in most general dictionaries. Others are not intended and constructive comments are welcome to help improve subsequent editions.

NOTES ON TRANSLITERATION

Arabic is written from right to left using an alphabet of 29 letters. There are no capitals, but some letters (when joined together) change their shape slightly, as shown in brackets. The alphabet is set out below with the corresponding transliteration for each letter.

أ		*a*	ط		*ṭ*
ب	(بـ)	*b*	ظ		*ẓ*
ت	(تـ)	*t*	ع	(عـ)	ʿ
ث	(ثـ)	*ṯ*	غ	(غـ)	*ḡ*
ج	(جـ)	*j*	ف	(فـ)	*f*
ح	(حـ)	*ḥ*	ق	(قـ)	*q*
خ	(خـ)	*ḵ*	ك	(كـ)	*k*
د		*d*	ل		*l*
ذ		*ḏ*	م	(مـ)	*m*
ر		*r*	ن	(نـ)	*n*
ز		*z*	هـ	(ـه)	*h*
س	(سـ)	*s*	و		*w*
ش	(شـ)	*š*	ي	(يـ)	*y*
ص	(صـ)	*ṣ*	ء		ʾ
ض	(ضـ)	*ḍ*			

The following letters are pronounced like their English equivalents:

بـ = *b*, تـ = *t*, هـ = *h*, و = *w*, يـ = *y*,
د = *d*, ر = *r*, ز = *z*, سـ = *s*, جـ = *j*,
نـ = *n*, فـ = *f*, كـ = *k*, ل = *l*, مـ = *m*.

شـ transliterated (*š*) is pronounced as the English sound *sh* in *should*
ثـ transliterated (*ṯ*) is pronounced as the voiceless English sound *th* in *thin*
ذ transliterated (*ḏ*) is pronounced as the voiced English sound *th* in *the.*

The following sounds need special attention:

- حـ transliterated *(ḥ)*, is a throaty sound like *h* in a loud whisper;
- خـ transliterated *(ḵ)* is another throaty sound as in the Scottish word *loch*;
- ء transliterated *(ʾ)*. This is a glottal stop. It can be heard in Cockney English, e.g. *"bread and bu-ʾer";*
- عـ transliterated *(ʿ)*. This is a throaty sound. To produce it, make a long sound *aaa* while pressing on the throat;
- غـ transliterated *(ḡ)* is a gargling sound like a long *r*;
- قـ transliterated *(q)*. It sounds like a *k* pronounced in the back of the throat.

صـ (*ṣ*), ضـ (*ḍ*), ط (*ṭ*), and ظ (*ẓ*) are emphatic counterparts of
سـ (*s*), د (*d*), تـ (*t*), and ذ (*ḏ*).

Pronounce each strongly as if it was followed by the Arabic vowel *u*, thus *ṣu* for ضـ. (The *u* part of the sound is not included in the transliteration).

There are three short vowels in Arabic: *a* as in *fat, i* as in *fit,* and *u* as in *pull.* These are not written in the Arabic but they are transliterated. Corresponding to the short vowels are three long vowels indicated by a dash placed above them: *ā* like the American *'a'* in *path, ī* as in *feet, ū* as in *pool.*

There are two diphthongs in Arabic: *ai* as in *may* and *au* as in *how.*

Short vowels must be pronounced short, and long ones pronounced long; double consonants must be emphasised as double. If this is not done the meaning may change, e.g.:
malik = king, *mālik* = owner,
darasa = to learn, *darrasa* = to teach.

ENGLISH - ARABIC

انكليزي - عربي

WHAT THE FUCK IS ALL THIS SHIT?

THIS IS BRITAIN NOT SAUDI ARABIA

A

Abney level ميزان تسوية يدوي
mīzān taswiya yadawī

Abrams' law قانون «أبرامز»
qānūn abrāmz

abrasives مواد حاكة
mawādd ḥākka

abscissa الإحداثي السيني
al-iḥdāṯī as-sīnī

absolute addressing أمر التحديد المطلق
amr at-taḥdīd al-muṭlaq

absolute humidity الرطوبة المطلقة
ar-ruṭūba al-muṭlaqa

absolute permissive block signalling الاشارة الهادية الجائزة المطلقة
al-išāra al-hāwiya al-jā'iza al-muṭlaqa

absorbing well بئر امتصاص
bi'r imtiṣāṣ

absorption امتصاص
imtiṣāṣ

absorption loss فقد امتصاصي
fuqud imtiṣāṣī

absorption pit حفرة امتصاص
ḥufrat imtiṣāṣ

abutment كتف
katif

abutment wall جدار ارتكاز
jidār irtikāz

Abyssinian well بئر حبشي
bi'r ḥabašī

accelerated curing انضاج مُعجل
inḍāj mu'ajal

accelerator جهاز تسارع
jihāz tasāru'

access منفذ. مسلك
manfaḏ, maslak

access ladder سلم وصول
sullam wuṣūl

access road طريق موصل
ṭarīq muwaṣṣil

access shaft بئر الوصول
bi'r al-wuṣūl

accident حادث
ḥādiṯ

accidental damage تلف عرضي
talaf 'araḍī

accidental error خطأ عفوي
ḵaṭa' 'afawī

accident prevention منع الحوادث
man' al-ḥawādiṯ

accident statistics احصاءات الحوادث
iḥṣā'āt al-ḥawādiṯ

accuracy دقة
diqqa

acid goggles نظارات واقية من الحوامض
naẓārāt wāqiya min al-ḥawāmiḍ

acid-proof صامد للحوامض
ṣāmid lil-ḥawāmiḍ

acid-resisting مقاوم للحوامض
muqāwim lil-ḥawāmiḍ

acid-resisting brickwork بناء بالطوب مقاوم للحوامض
binā' biṭ-ṭaub muqāwim lil-ḥawāmiḍ

acid steel فولاذ حامضي
fūlāḏ ḥāmiḍī

acoustical property الخاصية الصوتية
al-ḵāṣṣīya aṣ-ṣauṭīya

acoustic insulation عزل الصوت
'azl aṣ-ṣauṭ

acoustics علم الصوت
'ilm aṣ-ṣauṭ

acoustic strain gauge مقياس حدة الصوت
miqyās ḥiddat aṣ-ṣauṭ

acre فدان
faddān

acre foot فدان قدم
faddān qadam

acrylics أكريليك
akrīlīk

Ac system نظام تصنيف التربة
niẓām taṣnīf at-turba

activated-biofilter process عملية الترشيح البيولوجي المنشطة
'amalīyat at-taršīḥ al-bīyūlījī al-munašiṭa

activated-sludge process عملية الحمأة المنشطة
'amalīyat al-ḥama' al-munašiṭa

activated-sludge system نظام الحمأة المنشطة
niẓām al-ḥama' al-munašiṭa

active earth pressure ضغط التربة النشط
ḍaġṭ at-turba an-našiṭ

active-effluent treatment plant محطة معالجة النفاية النشطة
maḥaṭṭat mu'ālajat an-nifāya an-našiṭa

active layer طبقة نشطة
ṭabaqa našiṭa

active pressure ضغط فعال
ḍaġṭ fa'āl

A

active waste disposal التخلص من النفاية النشطة
at-takalluṣ min an-nifāya an-našiṭa

acute angle crossing تقاطع الزاوية الحادة
taqāṭuʿ az-zāwīya al-ḥādda

adaptor مهايئ
muhāyiʾ

addenda ملاحق
mulāḥiq

additional site information معلومات ميدانية إضافية
maʿlūmāt maidānīyn iḍāfīya

additive مادة مضافة
mādda muḍāfa

additive constant ثابت جمعي
ṯābit jamʿī

addressing mode نمط الأمر لايجاد المعلومات
mamaṭ al-amr li-ījād al-maʿlūmāt

adhesion التصاق
iltiṣāq

adjusting screw برغي معايرة
burǧī muʿāyara

adjustment معايرة . تعديل
muʿāyara, taʿdīl

admissible load الحمل المسموح
al-ḥaml al-masmūḥ

admissible stress الاجهاد المسموح
al-ijhād al-masmūḥ

admixture مزيج . خليط
mazīj, ḵalīṭ

adopted street طريق تحت الاشراف
ṭariq taḥt al-isrāf

adsorption استجذاب
istijḏāb

advanced gas cooled reactor مفاعل متقدم يبرد بالغاز
mufāʿil ǧāzī mutaqaddim yubarrad bil-ǧāz

advanced gas reactor مفاعل غازي متقدم
mufāʿil ǧāzī ,itaqaddo,

advanced waste treatment المعالجة المتقدمة للنفايات
al-muʿālaja al-mutaqaddima lin-nifāyāt

advanced waste water treatment المعالجة المتقدمة للماء المهدر
al-muʿālaja al-mutaqaddima lil-māʾ al-muhdar

advertisement for bids إعلان عن العطاءات
iʿlān ʿan al-ʿaṭāʾāt

aeolian ريحي
rīḥī

aerated concrete خرسانة مهواة
ḵarasāna muhawāt

aeration مزج بالهواء
mazj bil-hawāʾ

aeration tank خزان تهوية
ḵazzān tahwīya

aeration-tank effluent دفق خزان التهوية
dafq ḵazzān at-tahwīya

aerial cableway طريق الكبل الهوائي
ṭarīq al-kabl al-hawāʾī

aerial photograph صورة جوية
ṣūra jawwīya

aerial ropeway طريق حبلي هوائي
ṭarīq ḥablī hawāʾī

aerial survey مسح جوي
masḥ jawwī

aerial surveying المساحة الجوية
al-misāḥa al-jawwīya

aerial tramway ناقل هوائي
naqil hawāʾī

aerodrome مطار
maṭār

aerodynamic instability عدم الاستقرار الايروديناامي
ʿadam al-istiqrār al-īrūdīnāmī

A-frame هيكل على شكل حرف A
haikl ʿala šakl ḥarf ayy

aftercooler مبرد لاحق
mvbarrid laḥiq

age factor عامل الزمن
ʿāmil az-zaman

ageing التصلد بمرور الزمن
at-taṣallud bi-murūr az-zaman

agent عامل . وسيط
ʿāmil wasīṭ

aggregate ركام
rukām

aggregate batch meter عداد دفعات الركام
ʿaddād dafaʿāt ar-rukām

aggregate/cement ratio نسبة الركام الى الاسمنت
nisbat ar-rukām ilal-ismant

aggregate grading curve منحنى تصنيف الركام
munḥana taṣnīf ar-rukām

aggregate interlock تشابك الركام
tašābuk ar-rukām

aggregate weigh batcher skip قادوس وزن دفعة الركام
qādūs wazn dufʿat ar-rukām

A

agitating truck شاحنة خلاطة
šāhina kallāṭa
Aglite (TM) ركام خفيف
rukām kafīf
agonic line خط اللا إنحراف
katt al-lā-inḥirāf
agreement اتفاقية
ittifāqīya
agricultural drain صرف زراعي
ṣarf zirāʿī
agriculturalist خبير زراعي
kabīr zirāʿī
agricultural requirement المتطلبات الزراعية
al-mutaṭallabāt az-zirāʿīya
A-horizon أعلى طبقات التربة
aʿlā ṭabaqāt at-turba
air base قاعدة جوية
qāʿida jawwīya
airborne contaminants التلوثات العالقة بالهواء
at-talawwuṯāt al-ʿāliqa bil-hawā'
air bricks طوب مفرغ
ṭaub mufarrag
air compressor ضاغط الهواء
ḍāgiṭ al-hawā'
air conditioning system نظام تكييف الهواء
niẓām takyīf al-hawā'
air content of fresh concrete المحتوى الهوائي للخرسانة الجديدة
al-muḥtawa al-hawā'ī lil-karasāna al-jadīda
air-cooled condenser مكثف يبرد بالهواء
mukaṯṯif yubarrad bil-hawā'
air cooler مبرد هوائي
mubarrid hawā'ī
air core قلب هوائي
qalb hawā'ī
aircraft طائرة
ṭā'ira
aircraft catering building مبنى تموين الطائرات بالطعام
mabnā tamwīn aṭ-ṭā'irāt biṭ-ṭaʿām
aircraft movements تحركات الطائرة
taḥarrukāt aṭ-ṭā'ira
air duct مسلك الهواء
maslak al-hawā'
air embolism إنسداد هوائي
insidād hawā'ī
air-entrained concrete خرسانة مسحوبة الهواء
karasāna masḥūbat al-hawā'
air-entraining agent عنصر سحب الهواء
ʿunṣur saḥb al-hawā'
airfield lighting إضاءة المطار
iḍā'at al-maṭār
airfield soil classification تصنيف تربة المطار
taṣnīf turbat al-maṭār
air flow مجرى هواء
majrā hawā'
air flue مصرف هواء
maṣraf hawā'
air-flush drilling الحفر بضغط الهواء
al-ḥafr bi-ḍagṭ al-hawā'
air grate شبكة تصريف الهواء
šabakat taṣrīf al-hawā'
air-lift pump مضخة رافعة بالهواء
miḍakka rāfiʿa bil-hawā'
air lock غلق هوائي
galq hawā'ī
airplane mapping رسم خرائط بالطائرات
rasm karā'iṭ biṭ-ṭā'irāt
air-pollution control مكافحة تلوث الهواء
mukafaḥat talawwuṯ al-hawā'
airport مطار
maṭār
airport beacon منارة المطار
manārat al-maṭār
airport design تصميم المطار
taṣmīm al-maṭār
airport layout مخطط المطار
mukaṭṭaṭ al-maṭār
airport location موقع المطار
mauqiʿ al-maṭār
airport reference point نقطة الاسناد في المطار
nuqṭaṭ al-isnād fil-maṭār
airproof صامد للهواء
ṣāmid lil-hawā'
air pump مضخة هواء
miḍakkat hawā'
air-relief shaft بئر تنفيس الهواء
bi'r tanfīs al-hawā'
air-space فضاء جوي
faḍā' jawwī
air-space clearance خلوص فضائي
kulūṣ faḍā'ī
air survey مسح جوي
meseh jawwī
air temperature درجة حرارة الهواء
darajat ḥarārat al-hawā'

air traffic control — مراقبة حركة المرور الجوي
murāqabat ḥarakat al-murūr al-jawwī

air turbulence — إضطراب جوي
iḍṭirāb jawwī

air valve — صمام الهواء
ṣimām al-hawā'

air vessel — وعاء هوائي
wi'ā' hawā'ī

air-water system — نظام الهواء والماء
niẓām al-hawā' wal-mā'

alarm — إنذار
inḏār

Alclad (TM) — تصفيح بالألومنيوم
tasfīḥ bil-alūmīnyūm

alcove — فجوة
fajwa

alidade — عضادة
'uḍāda

align — تسوية
taswīya

alignment — محاذاة
muḥāḏāt

alignment chart — مخطط بياني
mukaṭṭaṭ bayānī

aline — يحاذي
yuḥāḏī

all-air system — نظام هوائي كامل
niẓām hawā'ī kāmil

alley — زقاق
zuqāq

all-in contract — عقد شامل
'aqd šāmil

allowable bearing pressure — ضغط التحميل المسموح
ḍaġṭ at-taḥmīl al-masmūḥ

allowable stress — الاجهاد المسموح
al-ijhād al-masmūḥ

alloy — سبيكة
sabīka

alloy steel — فولاذ سبيكي
fūlāḏ sabīkī

all-weather road — طريق مناسب لجميع الفصول
ṭarīq munāsib li-jamī' al-fuṣūl

altar — كتيفة حوض جاف
katīfat ḥauḍ jāf

alternator — منوب
munawwib

altimetry — قياس الارتفاعات
qiyās al-irtifā'āt

altitude — إرتفاع
irtifā'

altitude level — مستوى الارتفاع
mustawa al-irtifā'

altitude valve — صمام الارتفاع
ṣimām al-irtifā'

alumina — ألومينا
alūmīnā

aluminium — ألومنيوم
alūmīnyūm

aluminium alloy — سبيكة ألومنيوم
sabīkat alūmīnyūm

aluminium oxide — أكسيد الألومنيوم
aksīd al-alūmīnyūm

aluminium pipe — أنبوب ألومنيوم
unbūb alūmīnyūm

alumino-thermic reaction — تفاعل حراري ألوميني
tafā'ul ḥarārī alūmīnī

aluminous cement — إسمنت ألوميني
ismant alūmīnī

amendment — تعديل
ta'dīl

American caisson — قيسون أمريكي
kaisūn amrīkī

American Ephemeris and Nautical Almanac — التقويم الفلكي والبحري الأمريكي
at-taqwīm al-falakī wal-baḥrī al-amrīkī

American wire gauge — مقياس أسلاك أمريكي
miqyās aslāk amrīkī

ammonium nitrate — نترات الأمونيوم
nītrāt al-amūnyūm

ampere — أمبير
ambīr

amplifier output — خرج المضخم
ḵarj al-muḍakkim

amplitude — سعة الذبذبة
si'at aḏ-ḏabḏaba

anallatic lens — عدسة التركيز البؤري
'adasat at-tarkīz al-bu'rī

anallatic telescope — تلسكوب لمسح الأبعاد
talskūb li-masḥ al-ab'ād

analogue computer — حاسبة بالقياس
ḥāsiba bil-qiyās

analogue data — بيانات نسبية
bayānāt nisbīya

analogue-to-digital conversion — تحويل النسب الى أرقام
taḥwīl an-nisab ila arqām
analogy — تناظر
tanāḍur
analysis — تحليل
taḥlīl
analysis of requirements — تحليل المتلطبات
taḥlīl al-mutaṭalabāt
anchor — مرساة
mirsāh
anchorage area — منطقة الارساء
minṭaqat al-irsā'
anchorage distance — مسافة الارساء
masāfat al-irsā'
anchor and collar — مفصلة معدنية لغلق البوابة
mifṣala ma'danīya li-ġalq al-bawwāba
anchor block — بكرة التثبيت
bakarat at-taṯbīt
anchor bolt — مسمار ملولب للتثبيت
mismār mulaulab lit-taṯbīt
anchor chain — سلسلة الارساء
silsilat al-irsā'
anchor gate — بوابة الارساء
bawwābat al-irsā'
anchoring spud — عمود تثبيت فولاذي
'amūd taṯbīt fūlāḏī
anchor pile — ركيزة ارساء
rakīzat irsā'
anchor plate — لوح الارساء
lauḥ al-irsā'
anchor tower — برج الارساء
burj al-irsā'
anchor wall — جدار الارساء
jidār al-irsā'
ancillary buildings — مبان فرعية
mabāni far'īya
ancillary frame — هيكل فرعي
haikl far'ī
aneroid barometer — بارومتر معدني
bārūmītar ma'danī
angle bar — قضيب زاوي
qaḍīb zāwī
angle cleat — كتيفة من زاوية حديدية
katīfa min zāwīya ḥadīdīya
angledozer — جرافة لتسوية الطرق
jarāfa li-taswīyat aṭ-ṭuruq
angle float — مسطرين زاوي
masṭarīn zāwī
angle iron — حديد زاوي
ḥadīd zāwī
angle of friction — زاوية الاحتكاك
zāwīyat al-iḥtikāk
angle of inclination — زاوية الميل
zāwīyat al-mail
angle of incline — زاوية الانحدار
zāwīyat al-inḥidār
angle of internal friction — زاوية الاحتكاك الداخلي
zāwīyat al-iḥtikāk ad-dāḵilī
angle of repose — زاوية الاستقرار
zāwīyat al-istiqrār
angle of shearing resistance — زاوية مقاومة القص
zāwīyat muqāwamat al-qaṣṣ
angle section — مقطع زاوي
maqṭa' zāwī
anion — أنيون
anyūn
anisotropic — متباين الخواص
mutabāyin al-ḵawāṣ
annealed wire — سلك ملدن
silk muladdan
annealing — تلدين
taldīn
annex to a building — ملحق للمبنى
mulḥaq il-mabna
annual run-off — الصرف السطحي السنوي
aṣ-ṣarf as-saṭḥī as-sanawī
annual storage — تخزين سنوي
taḵzīn sanawī
annual variation — التغير السنوي
at-taġayur as-sanawī
annulment of contract — إبطال العقد
ibṭāl al-'aqd
anode — أنود . مَصْعَدْ
anūd, maṣ'ad
anodizing — معالجة بالطريقة الأنودية
mu'ālaja biṭ-ṭarīqa al-anūdīya
antechamber — غرفة الاحتراق المتقدم
ġurfat al-iḥtirāq al-mutaqaddim
anticlastic — تضادّ
taḍād
anti-crack — مضاد للتصدع
muḍād lit-taṣaddu'
anti-crack reinforcement — تسليح مضاد للتصدع
taslīḥ muḍād lit-taṣaddu'

anti-flood and tidal valve — صمام منع الفيضان والمد والجزر
sammām manʿ al-fayaḍān wal-madd wal-jazr

anti-flood device — وسيلة منع الفيضان
wasīlat manʿ al-fayaḍān

anti-friction metal — معدن مقاوم للاحتكاك
maʿdan muqāwim lil-iḥtikāk

anti-sag bar — شدادة منع الارتخاء
šaddādat manʿ al-irtikāʾ

anti-slip paint — طلاء مانع للانزلاق
ṭalāʾ māniʿ lil-inzilāq

apparent horizon — الافق الجغرافي
al-ufuq al-juġrāfī

appearance — ظهور . مظهر
ẓuhūr, maẓhar

apprentice — متدرب
mutadarrib

approach channel — قناة الاقتراب
qanāt al-iqtirāb

approach lighting — إضاءة الاقتراب
iḍāʾat al-iqtirāb

approach surface — سطح الاقتراب
saṭḥ al-iqtirāb

approach-zone clearance — خلوص منطقة الاقتراب
kulūṣ minṭaqat al-iqtirāb

approval of plans — اعتماد الخطط
iʿtimād al-kuṭaṭ

apron — مدرج طيران
madraj ṭayarān

apron capacity — سعة المدرج
siʿat al-madraj

apron control — مراقبة المدرج
murāqabat al-madraj

apron control cabin — مقصورة مراقبة المدرج
maqsūrat murāqabat al-madraj

aquaplaning — إنزلاق مائي
inzilāq māʾī

aqueduct — قناة اصطناعية
qanā iṣṭināʿīya

aquiclude — تكوين جيولوجي منخفض الانفاذية
takwīn jūlūji munkafiḍ an-nafādīya

aquifuge — أرض قاحلة
arḍ qāḥila

aquitard — تكوين صخري منخفض الانفاذية
takwīn ṣakrī munkafiḍ al-infādīya

arbor — شياق
šīyāq

arc — قوس كهربائي
qaus kahrabāʾī

arcade — سلسلة قناطر
silsilat qanāṭir

arch — قنطرة . عقد
qanṭara, ʿaqd

arch and catenary construction — إنشاء ذو عقد وقناطر
inšāʾ ḏū ʿaqd wa qanaṭir

arch bridge — جسر قنطري
jisr qanṭarī

arch dam — سد قنطري
sadd qanṭarī

arch design — تصميم قنطري
taṣmīm qanṭarī

Archimedean screw — شادوف «ارخميدس»
šādūf arkamīdīs

Archimedes' principle — قاعدة «ارخميدس»
qāʿidat arkamīdīs

architect — مهندس معماري
muhandis miʿmārī

architecture — هندسة البناء
handasat al-bināʾ

architrave — رأس العمود
raʾs al-ʿamūd

arch rib — ضلع القنطرة
ḍilʿ al-qanṭara

arch ring — طوق القنطرة
ṭauq al-qanṭara

arc welding — لحام قوسي
liḥām qausī

area available — المنطقة المتوفرة
al-minṭaqa al-mutawafira

argon-arc welding — لحام قوسي أرجوني
liḥām qausī arjūnī

arithmetic instruction — تعليم الحساب
taʿlīm al-ḥisāb

arm — ذراع
ḏirāʿ

armour — درع . تدريع
durʿ, tadrīʿ

armoured cable — كبل مدرع
kabl mudarraʿ

armoured pipe — أنبوب مدرع
unbūb mudarraʿ

armoured slopes — منحدرات مدرعة
munḥadarāt mudarraʿa

Armstrong scale — مقياس أرمسترونغ
miqyās armstrunġ

A

arpent أربنت: وحدة قياس فرنسية
arbant, wiḥdat qiyās fransīya
arrest point نقطة الكبح
nuqṭat al-kabḥ
arrow سهم
sahm
arterial road طريق رئيسي
ṭarīq ra'īsī
artesian well بئر ارتوازية
bi'r irtiwāzīya
artificial اصطناعي
iṣṭinā'ī
artificial cooling تبريد اصطناعي
tabrīd iṣṭinā'ī
artificial harbour مرفأ اصطناعي
marfa' iṣṭinā'ī
artificial horizon أفق اصطناعي
ufuq iṣṭinā'ī
artificial islands جزر اصطناعية
juzur iṣṭinā'īya
artificial lake بحيرة اصطناعية
buḥaira iṣṭinā'īya
artificial recharge إعادة شحن اصطناعي
i'ādat šaḥn iṣṭinā'ī
asbestos أسبستوس
isbastus
asbestos pipe أنبوب الأسبستوس
unbūb min al-isbastus
as-built survey مسح نهائي للبناء
masḥ nihā'ī lil-binā'
ASCE الاتحاد الأمريكي للمهندسين المدنيين
al-ittiḥād al-amrīkī lil-muhandisīn al-madanīyīn
ash content المحتوى الرمادي
al-muḥtawa ar-ramādi
ashlar حجر منحوت للبناء
ḥajar manḥūt lil-binā'
ashlar bricks طوب من الحجر المنحوت
ṭaub min al-ḥajar al-manḥūt
ashlar facing تلبيس من الحجر المنحوت
talbīs min al-ḥajar al-manḥūt
ashlar masonry بناء من الحجر المنحوت
binā' min al-ḥajar al-manḥūt
ashlar pieces قطع من الحجر المنحوت
qiṭa' min al-ḥajar al-manḥūt
asphalt أسفلت . قار
isfalt, qār
asphalt cement لصاق زفتي
liṣāq ziftī
asphaltic concrete خرسانة أسفلتية
karasāna isfaltīya
asphalt tanking أسفلت مانع للتسرب
isfalt māna' lit-tasarrub
asphalt tile آجر أسفلتي
ājur isfaltī
assembling تجميع . تركيب
tajmī', tarkīb
astronomical eyepiece عدسة عينية
'adasa 'ainīya
astronomy علم الفلك
'ilm al-falak
asynchronous serial interface توصيل متتابع لا متزامن
tauṣīl mutatābi' lā-mutazāmin
atomic-hydrogen welding لحام في جو من الهيدروجين
liḥām fi-jaww min al-hīdrūjīn
attenuator مُوهن
mu'ahhin
Atterberg limits حدود التماسك
ḥudūd at-tamāsuk
attic سقيفة
saqīfa
audio amplifier مضخم سمعي
muḍakkim sam'ī
audio-frequency overlay circuit دائرة تغطية الترددات الصوتية
dā'irat taġṭiyat at-taraddudāt aṣ-ṣautīya
audio measurement قياس الترددات الصوتية
qiyās at-taraddudāt aṣ-ṣautīya
audit تدقيق
tadqīq
auditorium قاعة اجتماع
qā'at ijtimā'
auger مثقاب
miṯqāb
auger drills مثاقب حفر
maṯāqib ḥafr
autoclaving تعقيم بالبخار المحمى وبالضغط
ta'qīm bil-bukār al-maḥmī wa biḍ-ḍaġṭ
auto-decrementing تناقص ذاتي
tanāquṣ ḏātī

A

autogenous healing إلتئام ذاتي
ilti'ām ḏātī

autogenous welding لحام ذاتي
liḥām ḏātī

auto-incrementing تزايد ذاتي
tazāyud ḏātī

Automated Guideway Transit ترانزيت الممر الآلي
tranzīt al-mamarr al-ālī

automatic-block signal system نظام الإشارات الهادية الأوتوماتيكي
niẓām al-išārāt al-hādīya al-ūtūmātīkī

automatic control تحكم أوتوماتي
taḥakkum ūtūmātī

automatic data logging تسجيل أوتوماتي للبيانات
tasjīl ūtūmātī lil-bayānāt

automatic door باب أوتوماتي
bāb ūtūmātī

automatic electric pumping ضخ كهربائي أوتوماتي
ḍakk kahrabā'ī ūtūmātī

automatic fire damper مخمد حريق أوتوماتي
mikmad ḥarīq ūtūmātī

automatic level مستوى أوتوماتي
mustawa ūtūmāti

automatic process control تحكم أوتوماتي في العمليات
taḥakkum ūtūmātī fil-'amalīyāt

automatic reset إعادة ضبط أوتوماتي
i'ādat ḍabṭ ūtūmātī

automatic sewage pumping station محطة ضخ أوتوماتي للمجاري
maḥaṭṭat ḍakk ūtūmātī lil-majārī

automatic siphon spillway قناة أوتوماتية لتصريف الفائض بالطرد
qanāt ūtūmātīya li-taṣrīf al-fā'iḍ biṭ-ṭard

automatic sprinkler system نظام مرشات أوتوماتي
niẓām miraššāt ūtūmātī

automatic testing اختبار أوتوماتي
iktibār ūtūmātī

automatic train control تحكم أوتوماتي في القطارات
taḥakkum ūtūmātī fil-qiṭārāt

automatic train operation تشغيل أوتوماتي للقطارات
tašġīl ūtūmātī lil-qiṭārāt

automatic welding لحام أوتوماتي
liḥām ūtūmātī

automation شغيل آلي
tašġīl ālī

automobile liability سؤولية عن السيارات
al-mas'ūlīya 'an as-sayyārāt

autopatrol درجه آلية
madraja ālīya

autoset level ستوى الضبط الآلي
mustawa aḍ-ḍabṭ al-ālī

auxiliary cable بل مساعد
kabl musā'id

auxiliary exhaust fan روحة طرد مساعدة
mirwaḥat ṭard musā'ida

available power طاقة المتوفرة
aṭ-ṭāqa al-mutawafira

avenue ثارع عريض
šāri' 'arīḍ

average متوسط . معدل
mutawassiṭ, mu'addal

average annual evaporation معدل التبخر السنوي
mu'addal at-tabakkur as-sanawī

average annual rainfall معدل سقوط الأمطار السنوي
mu'addal suqūṭ al-amṭār as-sanawī

average daily consumption معدل الاستهلاك اليومي
mu'addal al-istihlāk al-yaumī

average level المنسوب المتوسط
al-mansūb al-mutawassiṭ

award of contract منح العقد
manḥ al-'aqd

axe فأس . بلطة
fa's, balṭa

axial-flow fan مروحة ذات دفق محوري
mirwaḥa ḏāt dafq miḥwarī

axial flow piston pump مضخة بكباس ذو دفق محوري
miḍakka bi-kabbās ḏū dafq miḥwarī

axial load حمل محوري
ḥiml miḥwari

axial stress إجهاد محوري
ijhād miḥwari

axis محور
miḥwar

axman عامل مساحة
'āmil misāḥa

axonometric اكسونومتري
aksūnūmitrī

B

axonometric projection إسقاط اكسونومتري
isqāṭ aksūnūmitrī
azimuth السمت
as-samt
azimuthal projection إسقاط سمتي
isqāṭ samtī
Bacillus coli جراثيم عضوية
jarāṯīm ʻuḍwīya
backacter مجرفة عكسية
mijrafa ʻaksīya
back cutting حفر إضافي
ḥafr iḍāfī
back gauge معيار المسافة الخلفية
miʻyār al-masāfa al-ḵalfīya
backhoe معزقة خلفية
miʻzaqa ḵalfīya
backing store ذاكرة مساندة
ḏākira musānida
back-inlet gulley أنبوبة مجرور مسيكة
unbūba majrūr masīka
back mark العلامة الخلفية
al-ʻalāma al-ḵalfīya
back observation رصد خلفي
raṣd ḵalfī
back prop دعامة خلفية
daʻāma ḵalfīya
back-up copy نسخة داعمة
nusḵa dāʻima
back water ماء محجوز
māʼ maḥjūz
backwater curve منحنى الماء المحجوز
munḥana al-māʼ al-maḥjūz
bacteria بكتريا . جراثيم
baktīrīya, jarāṯīm
bacteria bed طبقة بكتريا
ṭabaqat baktīrīya
bad drainage تصريف سيئ
taṣrīf sayyiʼ
bad mortar ملاط سيئ
milāṭ sayyiʼ
bad soil تربة سيئة
turba sayyiʼa
bad weather طقس رديئ
ṭaqs radīʼ
baffle pier مرطم أمواج
mirṭam amwāj
baffle plate لوح اعتراضي لتغيير اتجاه التيار
lauḥ iʻtirāḍī li-taḡyīr ittijāh at-tayār
baffle-type distribution توزيع اعتراضي
tauzīʻ iʻtirāḍī
bagwork جدار ساتر
jidār sātir
bail نزح
nazaḥ
bailer منزحة
minzaḥa
Bailey bridge جسر «بيلي»
jisr bailī
Bakelite (TM) باكليت
bāklīt
Baker bell dolphin مرسى جرسي الشكل
marsā jarasī aš-šakl
balance توازن
tawāzun
balance bar قضيب الموازنة
qaḍīb al-muwāzana
balance box صندوق الموازنة
ṣundūq al-muwāzana
balance bridge جسر قلاب متوازن
jisr qallāb mutawāzin
balanced earthworks أعمال ترابية متوازنة
aʻmāl turābīya mutawāzina
balanced foundation أساس متوازن
asās mutawāzin
balance point نقطة التوازن
nuqṭat at-tawāzun
balance window نافذة توازن
nāfiḏat tawāzun
balancing إتزان
ittizān
balancing reservoir خزان موازنة
ḵazzān muwāzana
balata بلاته . صمغ لزج جيد العزل
balāta, ṣamḡ lazj jayyid al-ʻazl
balcony شرفة
šurfa
balk ردم بين حفرتين . رافدة
radm bain ḥufratain, rāfida
ball and socket joint مفصل كروي حُقي
mifṣal kurawī ḥuqqī
ballast حصى رصف
ḥaṣā raṣf
ballast bed طبقة حصى الرصف
ṭabaqat ḥaṣā raṣf
ball bearing محمل كريات
maḥmal kurayāt

B

ball hammer مطرقة كروية
miṭraqa kurawīya
ball mill مطحنة كروية
miṭḥana kurawīya
balloon framing هيكل مقوى بالشكالات
haikal muqawwa biš-šakālāt
balsa خشب البلزا
k̲ašab al-bulza
baluster عمود درابزين
'amūd darābzīn
balustrade درابزين
darābzīn
band شريط
šarīṭ
band chain شريط قياس فولاذي
šarīṭ qiyās fūlādī
banderolle شاخص
šāk̲iṣ
band saw منشار شريطي
minšār šarīṭī
band screen مصفاه شريطية
miṣfā šarīṭīya
bank ضفة
ḍaffa
banking ميل
mail
bank of transformers مجموعة محولات
majmū'at muḥawwilāt
bank protection حماية الضفة
ḥimāyat aḍ-ḍaffa
banksman مساعد عامل المرفاع
musā'id 'āmil al-mirfā'
bank storage مخزون الضفة
mak̲zūn aḍ-ḍaffa
banquette رصيف ضيق
raṣīf ḍayyiq
bar قضيب رواسب رملية وحصباوية
qaḍīb, rawāsib ramlīya wa-ḥaṣbāwīya
barbed wire سلك شائك
silk šā'ik
bar bender مكنة لّي القضبان
makinat layy al-quḍbān
Barber Greene tamping levelling finisher مكنة رصف وتسوية الطرق
makinat raṣf wa-taswiyat aṭ-ṭuruq
bar chart مخطط بياني قضيبي
muk̲aṭṭaṭ bayānī qaḍībī
barge طوف . قارب مسطح
ṭauf, qārib musaṭaḥ
barge bed قاع طيني
qā' ṭīnī
barium plaster جص الباريوم
jiṣṣ al-bāryūm
barn-door hanger بكرات تعليق باب مخزن الحبوب
bakarāt ta'līq bāb mak̲zan al-ḥubūb
Barnes's formula قانون «بارنز»
qānun barnz
Barnes's formula for flow in slimy sewers قانون «بارنز» للجريان في المجاري الانسيابية
qanūn barnz lil-jarayān fil-majārī al-insiyābīya
barometer بارومتر. مقياس الضغط الجوي
bārūmitar, miqyās aḍ-ḍaġṭ al-jawwī
barometric pressure ضغط بارومتري
ḍaġṭ bārūmitrī
baroque style أسلوب باروكي
uslūb bārūkī
barrage حاجز اصطناعي
ḥajiz iṣṭinā'i
bar reinforcement التسليح بالأسياخ
at-taslīḥ bil-asyāk̲
barrel vault عقد قنطري
'aqd qanṭarī
barrette مشبك
mišbak
barricade متراس
mitrās
barrier railings درابزين حاجز
darabzīn ḥajiz
bar section مقطع القضيب
maqṭa' al-qaḍīb
barytes بارايت
bārait
basalt بازلت
bazalt
bascule bridge جسر قلاب
jisr qallāb
bascule gate بوابة قلابة
bawwāba qallāba
base قاعدة
qā'ida
base area مساحة القاعدة
masāḥat al-qā'ida
base board لوح القاعدة
lauḥ al-qā'ida

base course طبقة القاعدة التحتية
ṭabaqat al-qā'ida at-taḥtīya
base drain صرف القاعدة التحتية
ṣarf al-qā'ida at-taḥtīya
base exchange استبدال فلز القاعدة بآخر
istibdāl filliz al-qā'ida bi-āḵar
base flow تدفق الماء الباطني
tadaffuq al-mā' al-bāṭinī
base line خط القاعدة
ḵaṭṭ al-qā'ida
basement طابق سفلي
ṭābaq suflī
basement wall جدار أساسي
jidār asāsī
base plate لوح القاعدة
lauḥ al-qā'ida
basic refractory مادة جيرية مقاومة للحرارة
mādda jīrīya muqāwima lil-ḥarāra
basic steel فولاذ أساسي
fūlāḏ asāsī
batching plant وحدة قياس كميات الخلط
wiḥda qiyās kammīyāt al-ḵalṭ
batch mixer خلاطة خرسانة
ḵallātat ḵarasāna
batch mixing unit وحدة خلط على دفعات
wiḥdat ḵalṭ 'ala dufu'āt
bat faggot حزمة قضبان خشبية
ḥuzmat quḍbān ḵašabīya
bathotonic reagent مفاعل خافض للتوتر السطحي
mufā'il ḥāmiḍī lit-tawattur as-saṭḥī
batten عارضة
'āriḍa
batten door باب بعوارض خشبية
bāb bi-'awāriḍ ḵašabīya
batten plate لوح خشبي
lauḥ ḵašabī
batter منحدر . ميل
munḥadar, mail
batter level مسوّاة المنحدر
musawwāt al-munḥadar
batter pile ركيزة مائلة
rakīza mā'ila
battery بطارية
baṭārīya
battery terminal مربط وصل البطارية
mirbaṭ waṣl al-baṭṭāriya
baud rate معدل سرعة الارسال البرقي (بود)
mu'addal sur'at al-irsāl al-barqī būd
baulk رافدة . جائز
rāfida, jā'iz
Bauschinger effect ظاهرة «باوشينجر»
ẓāhira būšingar
bauxite بوكسيت
būksīt
bay فسحة ما بين عمودين
fusḥa mā bain 'amūdain
beaching إنزال على الشاطئ
inzāl 'ala aš-šāṭi
beach replenishment إستعاضة الماء الباطني
isti'āḍat al-mā' al-bāṭinī
beacon فنار
fanār
beaded section قطاع بطرف محدب
qiṭa' bi-ṭarf muḥaddab
beam عارضة خشبية
'āriḍa ḵašabīya
beam-and-girder construction إنشاء ذو جائز وعارضة
inšā' ḏū jā'iz wa-'āriḍa
beam and girder framing هيكل ذو جائز وعارضة
haikal ḏū jā'iz wa-'ariḍa
beam-and-slab construction هيكل ذو جائز وبلاطة
haikal ḏū jā'iz wa-balāṭa
beam and slab floor أرضية ذات ركيزة وبلاطة
arḍīya ḏāt rakīza wa-balāṭa
Beaman stadia arc تاكومتر ذو قراءة مباشرة
tākūmitar ḏū qirā'a mubāšira
beam bender مكنة حني الروافد
makinat ḥanī ar-rawāfid
beam compasses فرجار ذو عاتق
firjār ḏū 'ātiq
beam deflection إنحراف الشعاع
inḥirāf aš-šu'ā'
beam engine محرك بخاري
muharrik buḵārī
beamless floor أرضية بدون عوارض
arḍīyat bidūn 'awāriḍ
beam test إختبار معامل التمزق
iḵtibār mu'āmil at-tamazzuq
bearing محمل
maḥmal
bearing capacity استطاعة المحمل
istiṭā'at al-maḥmal

B

bearing pile ركيزة تحميل
rakīzat taḥmīl
bearing pressure ضغط التحميل
ḍagṭ at-taḥmīl
bearings إتجاهات زاوية
ittijāhāt zāwīya
bearing stratum طبقة تحميل
ṭabaqat taḥmīl
bearing stress إجهاد التحميل
ijhād at-taḥmīl
bearing test إختبار التحميل
ikṭibār at-taḥmīl
bearing wall جدار داعم
jidār dā'im
bear trap gate بوابة بمحبس دعم
bawwāba bi-miḥbas da'm
Beaufort scale مقياس «بوفورت» لسرعة الريح
miqyās būfūrt li-sur'at ar-rīḥ
becquerel أشعة «بيكريل»
aši''at bīkīrīl
bed قاع . قاعدة
qa', qā'ida
bedding فرشة من الملاط
farša min al-milāṭ
bed joint إتصال أفقي
ittiṣāl ufuqī
bed load حمل المواد المترسبة
ḥiml al-mawād al-mutarassiba
bed plate لوح الأساس
lauḥ al-asās
bedrock صخر القاعدة
ṣakr al-qā'ida
beech خشب الزان
kašab az-zān
beetle head رأس المطرقة
ra's al-miṭraqa
Belanger's critical velocity السرعة الحرجة لـ «بلانجر»
as-sur'a al-ḥarija li-bilānjīr
Belgian truss جُملون بلجيكي
jamlūn baljīkī
bell dolphin مرسى جرسي الشكل
marsa jarasī aš-šakl
bellmouth overflow طفح الخزان
ṭafḥ al-kazzān
bellmouth spillway قناة تصريف ناقوسية الفم
qanāt taṣrīf nāqūsīyat al-famm
belly rod قضيب مقوّس
qaḍīb muqawwas
belt conveyor ناقلة بالسير
nāqila bis-sair
bench منضدة عمل
minḍadat 'amal
benched foundation أساس مدرج
asās mudarraj
benching مصطبة
maṣṭaba
benching iron لوح فولاذي مثلث
lauḥ fūlāḏī muṯallaṯ
bench mark علامة المنسوب
'alāmat al-mansūb
bench saw منشار منضدي
minšār minḍadī
bend كوع
kū'
bending إنحناء
inḥinā'
bending formula معادلة الانحناء
mu'ādalat al-inḥinā'
bending moment عزم الثني
'azm aṯ-ṯanī
bending-moment diagram رسم بياني لعزم الثني
rasm bayāni li-'azm aṯ-ṯanī
bending-moment envelope منحنى التغير لعزم الثني
munḥana at-tagayyur li-'azm aṯ-ṯanī
bending schedule جدول الانحناء
jadwal al-inḥinā'
bending strength مقاومة الانحناء
muqāwamat al-inḥinā'
bending stress إجهاد الانحناء
ijhād al-inḥinā'
bends مرض التحني
maraḍ at-taḥannī
bend test إختبار الانحناء
ikṭibār al-inḥinā'
Benoto caisson قيسون «بنتون»
qaisūn binūtū
bent منحني
munḥani
bentonite صلصال مركب
ṣalṣāl murakkab
bentonite mud طين صلصالي
ṭīn ṣalṣālī
berm حافة ناتئة
ḥāffa nāti'a

B

Bernoulli's assumption افتراض «برنولي»
iftirāḍ birnūlī
Bernoulli's theorem نظرية «برنولي»
naẓarīyat birnūlī
berth مرسى
marsa
berthing impact تأثير الإرساء
ta'ṯīr al-irsā'
bevelled washer فلكة مائلة
falaka mā'ila
B-horizon طبقة التربة الوسيطة (ب)
ṭabaqat at-turba al-wasīṭa bī
bi-cable ropeway طريق حبلي ثنائي الكبل
ṭarīq ḥablī ṯunā'ī al-kabl
bid عطاء
'aṭā'
bidding documents مستندات العطاء
mustanadāt al-'aṭā'
bidding requirements متطلبات العطاء
mutaṭallabāt al-'atā'
billet كتلة خشبية
kutla ḵašabīya
bill of quantities قائمة الكميات
qā'imat al-kammīyāt
bi-metal strip شريحة من معدنين
šarīḥa min ma'danain
binary ثنائي العنصر
ṯunā'ī al-'unṣur
binary program برنامج ثنائي
barnāmij ṯunā'ī
binder رباط
ribāṭ
binding material مواد ربط
mawwād rabṭ
binding wire سلك ربط
silk rabṭ
bio-chemical oxygen demand طلب أحيائي كيماوي للأوكسجين
ṭalab al-bīyūkīmāwī lil-uksijīn
biodisk قرص بيولوجي
qurṣ bīyūlūjī
biological-contactor process عملية التلامس البيولوجي
'amalīyat at-talāmus al-bīyūlūjī
biological filter مرشح بيولوجي
muraššiḥ bīyūlūji
biological shield درع واق من الاشعاع
dur' wāqi min al-iš'ā'
BIPOLAR ثنائي القطب
ṯunā'ī al-quṭb
bipolar transistor ترانزستور ثنائي القطب
tranzistūr ṯunā'ī al-quṭb
birch خشب البتولا
ḵašab al-batūlā
bird's eye view منظر عام
manẓar 'āmm
bird's mouth joint وصلة فم العصفور
waṣlat famm al-'aṣfūr
Birmingham wire gauge مقياس الأسلاك البرمنجهامي
miqyās al-aslāk al-birminġham
BISTRO بيسترو
bīstrū
bit لقمة حفر
luqmat ḥafr
bitumen قار . زفت
qār, zift
bitumen lining تبطين بالقار
tabṭīn bil-qār
bitumen macadam قار لرصف الطرق
qār liraṣf aṭ-ṭuruq
bitumen sheathing تغطية بالقار
taġṭīya bil-qār
bituminous carpet فرشة قارية
farša qārrīya
bituminous emulsion مستحلب قاري
mustaḥlab qārrī
bituminous felt لباد قاري
lubbād qārrī
bituminous mixing plant وحدة خلط القار
wiḥdat ḵalṭ al-qār
bituminous paint دهان قاري
duhān qārrī
bituminous pavement رصيف قاري
raṣīf qārrī
bituminous plastics لدائن قارية
ladā'in qārrīya
bituminous surfacing رصف قاري
rasf qārrī
black أسود
aswad
blackboard سبورة
sabbūra
black bolt مسمار ملولب أسود
mismār mulaulab aswad

B

black diamond ألماس أسود
almās aswad
black japan ورنيش اليابان الأسود
warnīš al-yābān al-aswad
black mortar ملاط أسود
milāṭ aswad
black oxide أوكسيد أسود
aksīd aswad
black powder بارود أسود
bārūd aswad
blade نصل . ريشة
naṣl, rīša
blade grader مدرجة ممهدة
mudarraja mumahhida
blading back ظهر الريشة
ẓahr ar-rīša
blank carburizing كربنة بدون كربون
karbana bidūn karbūn
blank flange شفة أنبوب مغلقة
šafat unbūb muġlaqa
blank nitriding تصليد الفولاذ بالحرارة
taṣlīd al-fūlāḏ bil-ḥarāra
blast furnace فرن الصهر
furn aṣ-ṣahr
blast-furnace cement إسمنت الفرن العالي
ismant al-furn al-ʿālī
blast-furnace slag خبث الفرن العالي
ḵubṯ al-furn al-ʿālī
blasting سفع
safʿ
blasting fuse فتيل النسف
fatīl an-nasf
blasting machine مكنة تفجير
makinat tafjīr
blasting record سجل التفجير
sijill at-tafjīr
bleaching تبييض . تقصير
tabyīḍ, taqṣīr
bleeding نزف . استنزاف
nazf, istinzāf
blind مستتر
mustatir
blind arch قوس حجري
qaus ḥajarī
blind drain مصرف حجري
muṣarrif ḥajarī
blinding حجب . رش الحصباء على القار
ḥajb, rašš al-ḥaṣbāʾ ʿalal-qār

blinding concrete بقة خرسانة تحتية
ṭabaqat ḵarasāna taḥtīya
blinding layer بقة تغطية
ṭabaqat taġṭīya
bloated clay كام خفيف الوزن
rukām ḵafīf al-wazn
block كرة لرفع الأثقال . كتلة
bakara li-rafʿ al-aṯqāl, kuṯla
block clutch قابض الاحتكاكي
al-qābiḍ al-iḥtikākī
block-in-course تشييد بالحجارة المنحوتة
at-tašyīd bil-ḥijāra al-manḥūta
block of dwellings اية سكنية
bināya sakanīya
block of flats اية شقق
bināyat šuqaq
block pavement تبليط بكتل مستطيلة
at-tablīṭ bi-kutal mustaṭīla
blockwork ناء من الكتل
bināʾ min al-kutal
blockyard احة صب الكتل
saḥat ṣabb al-kutal
Blondin ريق حبلي
ṭarīq ḥablī
bloom وير . كتلة حديد
tanwīr, kuṯla ḥadīd
blow خة . طرقة
nafḵa, ṭarqa
blow down صريف سفلي
taṣrīf suflī
blow off صريف (البخار)
taṣrīf al-buḵār
blow out طفاء . إنطلاق
iṭfāʾ, inṭilāq
blue bricks وب أزرق صلب
ṭaub azraq ṣalib
blue edge lights ضواء جانبية زرقاء
aḍwāʾ janibīya zarqāʾ
blueing سخين الفولاذ حتى الزرقة
taskīn al-fūlāḏ hatta az-zurqa
blueprint سخة زرقاء
nusḵa zarqāʾ
board ح
lauḥ
body سم . هيكل
jism, haikal
bodying-up صقل التحضيري
aṣ-ṣaql at-taḥḍīrī

B

body injury liability مسؤولية عن الحوادث
mas'ūlīya 'an al-hawādiṯ
body track أثر السير
aṯar as-sair
boil غلي . غليان
ǧalā, ǧalayān
boiled oil زيت بزر الكتان المغلي
zait bizr al-kuttān al-maǧlī
boiler غلاية . مرجل
ǧallāya, mirjal
boiler group مجموعة غلايات
majmū'at ǧallāyāt
boiler house مبنى الغلاية
mabnā al-ǧallāya
boiler house foundations أساسات مبنى الغلايات
asāsāt mabnā al-ǧallāyāt
boiler inspection التفتيش على المراجل
at-taftīš 'ala al-marājil
boiler rating قدرة الغلاية
qudrat al-ǧallāya
boiler support columns أعمدة دعم الغلاية
a'midat da'm al-gallāya
boiling water reactor مفاعل ذري يبرد بالماء
mufā'il ḏarrī yulaṭṭaf bil-mā'
bollard عمود ربط الحبال
'amūd rabṭ al-ḥibāl
bolster مسند . سناد
masnad sinād
bolt مسمار ملولب
mismār mulaulab
bolt connections توصيلات بالمسامير الملولبة
tauṣīlāt bil-masāmīr al-mulaulaba
bolt sleeve مجرى المسمار الملولب
majrā al-mismār al-mulaulab
bond ترابط
tarābuṭ
bond breaker مفكك الترابط
mufakkik at-tarābuṭ
bonder حجر الربط
hajar ar-rabṭ
bond form عامل الترابط
'āmil at-tarābuṭ
bond length فترة الترابط
fatrat at-tarābuṭ
bond stress إجهاد الترابط
ijhād at-tarābuṭ
boning دق الأوتاد ومحاذاتها
daqq al-autād wa-muḥāḏātiha
boning rod وتد محاذاة
watad muḥāḏāt
bonus scheme نظام المنح
niẓām al-minaḥ
boogie box صندوق حقن الاسمنت
ṣandūq ḥaqn al-ismant
boojee pump مضخة حقن الاسمنت
miḍakkat ḥaqn al-ismant
booking تدوين
tadwīn
Boolean logic instruction تعليم منطق الحساب البوليني
ta'līm manṭiq al-ḥisāb al-būlīnī
boom ذراع تطويل
ḏirā' taṭwīl
boom joint وصلة ذراع التطويل
waṣlat ḏirā' at-taṭwīl
boom-lifting cylinder إسطوانة رفع ذراع التطويل
isṭiwānat raf' ḏirā' at-taṭwīl
booster جهاز تقوية
jihāz taqwiya
boot نعل الدلو
na'l ad-dalū
boot man عامل خرسانة
'āmil ḵarasāna
bootstrapping برنامج كمبيوتر
barnāmij kumbjūtar
Bordeaux connection وصلة «بوردو»
waṣlat būrdū
border stone حجر حافة
ḥajar ḥāffa
bore ثقب . تجويف
ṯuqb, tajwīf
bored pile ركيزة مصبوبة في الموقع
rakīza maṣbūba fil-mauqi'
borehole ثقب الحفر
ṯuqb al-ḥafr
borehole equipment معدات الحفر
mu'iddāt al-ḥafr
borehole log سجل الحفر
sijill al-ḥafr
borehole pump مضخة الحفر
miḍakkat al-ḥafr
borehole samples عينات الحفر
' aināt al-ḥafr
borehole surveying مسح الحفر
masḥ al-ḥafr

B

boring — حفر
ḥafr

borrow — مواد الردم المستعارة
mawādd ar-radm al-mustaʿāra

borrow excavation — حفر مواد الردم
ḥafr mawādd ar-radm

borrow pit — حفرة الإمداد
ḥufrat al-imdād

bort — بورت
būrt

Boston caisson — قيسون « بوستون »
qaisūn būstūn

bottom bar — القضيب السفلي
al-qaḍīb as-suflī

bottom cut — الحفرة السفلية
al-ḥufra as-suflīya

bottom heading — نفق سفلي
nafaq suflī

bottoming — الطبقة السفلى
aṭ-ṭabaqa as-suflā

bottom-opening skip — قادوس بفتحة سفلية
qādūs bi-fatḥa suflīya

bottom outlet — مخرج سفلي
makraj suflī

bottom sampler — أداة استخراج العينات
adāt istikrāj al-ʿayyināt

boulder clay — طين جلمودي
ṭīn jalmūdī

boulders — حجر جلمودي
ḥajar jalmūdī

boulevard — شارع عريض
šāriʿ ʿarīḍ

boundary fence — سياج حدود
sīyāj ḥudūd

boundary mark — علامة حدود
ʿalāmat ḥudūd

boundary survey — مسح الحدود
masḥ al-ḥudūd

bound water — ماء محجوز
māʾ maḥjūz

Bourdon pressure gauge — مقياس «بوردون» لضغط السوائل
miqyās būrdūn li-ḍaġṭ as-sawāʾil

Boussinesq equation — معادلة «بوسينزك»
muʿādalat būsīnisk

Bowditch's rule — قاعدة «بوديتش»
qāʿidat būdītš

bowk — قادوس
qādūs

bowl scraper — كاشطة على عجلات
kāšita ʿala ʿajalāt

bowstring girder — عارضة مقوسة
ʿāriḍa muqawwasa

bowstring truss — كتيفة مسنّمة
katīfa musannama

bow window — نافذة ناتئة مدورة
nāfiḏa nātiʾa mudawwara

box — صندوق
ṣundūq

box beam — عارضة صندوقية
ʿāriḍa ṣundūqīya

box caisson — قيسون صندوقي
qaisūn ṣundūqī

box culvert — مجرى سفلي ذو منفذ مستطيل
majrā suflī ḏū manfaḏ mustaṭīl

box dam — سد صندوقي
sadd ṣundūqī

box drain — مصرف صندوقي
maṣrif ṣundūqī

box frame construction — بناء ذو إطار صندوقي
bināʾ ḏū iṭār ṣundūqī

box girder — عارضة صندوقية
ʿāriḍa ṣundūqīya

box girder bridge — جسر بجائز صندوقي
jisr bi-jāʾiz ṣundūqī

box heading — نفق أفقي صندوقي
nafaq ufuqī ṣundūqī

boxing — تعشيق اللسان في النقر
taʿšīq al-lisān fī-naqr

boxing up — وضع الحصى تحت السكة
waḍʿ al-ḥaṣā taḥt as-sikka

box pile — ركيزة صندوقية
rakīza ṣundūqīya

box sextant — سدسية صندوقية
sudsīya ṣundūqīya

box shear test — اختبار القص الصندوقي
iktibār al-qaṣṣ aṣ-ṣundūqī

brace — شُكال
šakkāl

braced framework — هيكل مقوى بالشكالات
haikal muqawwā bi-šakkālāt

braced-timber framing — هيكل خشب مقوى بالشكالات
haikal kašab muqawwā bi-šakkālāt

brace of a scaffold — شكال سقالة
šakkāl siqāla

braces — أقواس مزدوجة
aqwāṣ muzdawwaja

B

bracing تقوية بالشكل
taqwiya biš-šukal
bracket كتيفة
katīfa
bracket baluster عمود درابزين الدرج
'amūd drābzīn ad-durj
bracketed cornice إفريز ذو كتيفات
ifrīz ḏū katīfāt
bracketed stairs درج بحافة مزخرفة
durj bi-ḥāffa muzakrafa
bracketing محمل ذو كتيفات
maḥmal ḏū katīfāt
brackets on columns كتيفات مدعومة على أعمدة
katīfāt mad'ūma 'ala a'mida
Braithwaite piles ركائز ملولبة
rakā'iz mulaulaba
branch instruction تعليمات جزئية
ta'līmāt juz'īya
branch line خط فرعي
katt far'ī
branch pipe أنبوب متفرع
unbūb mutafarri'
branch sewer مجرور متفرع
majrūr mutafarri'
brander شرائح خشبية معترضة
šarā'iḥ kašabīya mu'tariḍa
brandering خُشبان (تمليط)
kašbān, tamlīṭ
branding الدمغ . الوسم
ad-damġ, al-wasm
branding iron ميسم . مرصن
maisam, marṣan
brass النحاس الأصفر
an-nuḥās al-aṣfar
braze welding اللحام بالنحاس
al-liḥām bin-nuḥās
brazing spelter زنك لحام
zink liḥām
breaker قاطع التيار
qāṭi' at-tayyār
breaking ground تفتيت الأرض في المنجم
taftīt al-arḍ fil-manjam
breaking load حمل الكسر
ḥiml al-kasr
breaking piece قطعة الكسر
qiṭ'at al-kasr
breaking point نقطة الانكسار
nuqṭat al-inkisār
breaking strength مقاومة الانكسار
muqāwamat al-inkisār
breaking stress إجهاد الانكسار
ijhād al-inkisār
break-pressure tank خزان تخفيف الضغط
kazzān takfīf aḍ-ḍaġṭ
breakwater حاجز أمواج
ḥājiz amwāj
breakwater pier ركيزة حاجز الموج
rakīzat ḥājiz al-mauj
breast قلابة الجرافة
qallābat al-jarrāfa
breast boards قلابة المحراث
qallābat al-miḥrāṯ
breast drill مثقاب صدر
miṯqāb ṣadr
breasting dolphin جدار مرسى
jidār marsa
breasting jack عمود الارساء
'amūd al-irsā'
breathing apparatus جهاز تنفس
jihāz tanaffus
breeze concrete خرسانة الكوك
karasānat al-kūk
brick قرميدة . طوبة
qarmīda, ṭūba
bricklayer بنّاء
bannā'
bricklayer's tools أدوات البناء
adawāt al-binā'
brick moulding machine ماكينة صب القوالب
mākinat ṣabb al-qawālib
brick pier ركيزة من الطوب
rakīza min aṭ-ṭaub
brickwork بناء الطوب
binā' aṭ-ṭaub
bridge جسر
jisr
bridge bearing محمل الجسر
maḥmal al-jisr
bridge cap رأس الجسر
ra's al-jisr
bridge centre line خط المركز للجسر
katt al-markaz lil-jisr
bridge deck سطح الجسر
saṭḥ al-jisr
bridge excavation حفر أساس الجسر
ḥafr asās al-jisr
bridge pier ركيزة جسر
rakīzat jisr

B

bridge pier cap رأس ركيزة الجسر
ra'sʾ rakīzat al-jisr
bridges جسور
jusūr
bridge thrust قوة دفع الجسر
qūwwat dafʿ al-jisr
bridge truss جُملون الجسر
jamlūn al-jisr
bridging وصل قنطري
waṣl qanṭarī
bright bolt مسمار ملولب مخروط
mismār mulaulab maḵrūṭ
Brinell hardness test إختبار الصلادة البرينيلية
iḵtibār aṣ-ṣalāda al-brīnīlīya
briquette قالب من السقاط
qālab min as-suqaṭ
British Thermal Unit وحدة حرارية بريطانية
wiḥda ḥarārīya brīṭānīya
brittle fracture صدع إجهادي
ṣadaʿ ijhādī
broach channelling حفر بالثقوب المتقاربة
ḥafr bit-ṯuqūb al-mutaqāriba
broaching ضبط الثقوب
ḍabṭ aṯ-ṯuqūb
broad-base tower برج عريض القاعدة
burj ʿarīḍ al-qāʿida
broad-crested weir سد عريض القمة
sadd ʿarīḍ al-qumma
broad gauge سكة حديدية عريضة
sikka ḥadīdīya ʿarīḍa
broad irrigation الري الرحيب
ar-rayy ar-raḥīb
broken rail سكة مكسورة
sikka maksūra
broken stone حجر مكسّر
ḥajar mukassar
bronze البرونز
al-brūnz
bronze welding اللحام بالبرونز
al-liḥam bil-brūnz
brook جدول . غدير
jadwal, ġadīr
brooming الكنس
al-kans
Brown and Sharp wire gauge مقياس الأسلاك الأمريكي
miqyās al-aslāk al-amrīkī
brushwood أغصان مقطوعة
aġṣān maqṭūʿa
brushwood fascine mattress فرشة من أغصان مقطوعة
farša min aġṣān maqṭūʿa
bubble فقاعة . نفاخة
fuqāʿa, nufāḵa
bubble trier فاحص الفقاعات
fāḥiṣ al-fuqāʿāt
bubble tube ميزان تسوية بفقاعة
mīzān taswiya bi-fuqāʿa
bucket دلو
dalū
bucket dredging الكرّاءة بالقواديس
al-karrāʾa bil-qawādīs
bucket elevator رافعة بالقواديس
rāfiʿa bil-qawādīs
bucket energy dissipator مبدد طاقة قادوسي
mubaddid ṭāqa qādūsī
bucket-ladder dredger جرافة بقواديس دوارة
jarrāfa bi-qawādīs dawwāra
bucket-ladder excavator حفار بالقواديس الدوارة
ḥaffār bil-qawādīs ad-dawwāra
bucket-wheel excavator حفار بدولاب ذو قواديس
ḥaffār bi-dūlāb ḏū qawādīs
buckle تحديب
taḥdīb
buckling load حمل التحديب
ḥiml at-taḥdīb
buckling strain توتر التحديب
tawattur at-taḥdīb
buckling stress إجهاد التحديب
ijhād at-taḥdīb
budgetary control مراقبة الميزانية
murāqabat al-mīzānīya
buffer stop مصد عربات القطار
miṣadd ʿarabāt al-qiṭār
buffing wheel رحى صقل
ruḥā ṣaql
buggy عربة نقل الخرسانة
ʿarabat naql al-ḵarasāna
building مبنى
mabnā
building association جمعية البناء
jamʿīyat al-bināʾ
building blocks كتل بناء
kutal al-bināʾ
building code قوانين البناء
qawānīn al-bināʾ
building construction إنشاءات المباني
inšāʾāt al-mabānī

B

building contractor مقاول بناء
muqāwil binā'
building equipment معدات البناء
mu'iddāt al-binā'
building for electronic equipment مبنى الأجهزة الالكترونية
mabnā al-ajhiza al-iliktrūnīya
building line حدود البناء
ḥudūd al-binā'
building materials مواد البناء
mawādd al-binā'
building owner صاحب البناء
ṣāḥib al-binā'
building paper ورق عزل الصوت
waraq 'azl aṣ-ṣauṭ
building regulations لوائح وقوانين البناء
lawā'iḥ wa-qawānīn al-binā'
buildings مباني
mabānī
building sewer مجرور المبنى
majrūr al-mabnā
building site موقع البناء
mauqi' al-binā'
buildings layout مخطط البناء
mukaṭṭaṭ al-binā'
building system نظام البناء
niẓām al-binā'
built up مركب . مجمع
murakkab, mujamma'
bulb angle مقطع فولاذي ثخين
maqṭa' fūlāḏī ṯakīn
bulb of pressure التربة المضغوطة تحت الأساس
at-turba al-maḍgūṭa taḥt al-asās
bulk cargo حمولة سائبة
ḥumūla sā'iba
bulk cargo terminal محطة البضائع السائبة
muḥaṭṭat al-baḍā'i' as-sā'iba
bulk concrete خرسانة سائبة
karasāna sā'iba
bulk density الكثافة الظاهرية
al-kaṯāfa aẓ-ẓāhirīya
bulkhead حاجز انشائي
ḥājiz inšā'ī
bulkhead wharf رصيف حاجز
raṣīf ḥājiz
bulking ناتج الحفر
nātij al-ḥafr
bulk modulus معامل المرونة
mu'āmil al-murūna
bulk spreader آلة فرش المواد السائبة
ālat farš al-mawādd as-sā'iba
bulk storage تخزين سائب
takzīn sā'ib
bulldog grip قامطة ركابية
qāmiṭa rukābīya
bulldozer بولدوزر
buldūzar
bullet-proof glass زجاج لا يخترقه الرصاص
zujāj lā-yakruquh ar-raṣāṣ
bullhead rail قضيب دائري الطرفين
qaḍīb dā'irī aṭ-ṭarafair
bull wheel دولاب إدارة
dūlāb idārā
bump رطم
raṭm
bump cutter قاطعة نتوءات
qāṭi'at nutū'āt
bunker مخزن وقود
makzan wuqūd
bunkering facilities تسهيلات التخزين
tashīlāt at-takzīn
buoyancy قابلية الطفو
qābilīyat aṭ-ṭafū
buoyant foundation أساس عائم
asās 'ā'im
buoyant raft طوف عائم
ṭauf 'ā'im
burden حمل . عبء
ḥiml, 'ibb'
buried drain مصرف مطمور
maṣrif maṭmūr
burn حرقة . حرق
ḥarqa, ḥarq
burner interlocks توصيلات الموقد
tauṣīlāt al-mauqid
burnt shale طين صفحي كربوني
ṭīn ṣafḥī karbūnī
busbar قضيب التوصيل
qaḍīb at-tauṣīl
bush hammer منحات مسنن
minḥāt musannan
bush hammering نقش الحجارة بالمنحات المسنن
nagš al-ḥijāra bil-minḥāt al-musannan
butane غاز البيوتان
gāz al-bīyūtān

butterfly valve صمام مروحي
sammām mirwaḥī
buttress دعامة. كتف
da'āma, katif
buttress dam سد مدعم
sadd mud'am
buttress drain مصرف لولبي مزدوج
maṣrif laulabī muzdawaj
buttressed wall جدار مدعم
jidar mud'am
buttress screw thread سن لولبي كتفي
sinn laulabī katifī
butt strap شريحة ربط تناكبي
šarīḥat rabṭ tanākubī
butt weld لحام تناكبي
liḥām tanākubī
butt-welded rail قضيب ملحوم تناكبياً
qaḍīb malḥūm tanākubīyan
butt-welded tube أنبوب ملحوم تناكبياً
unbūb malḥūm tanākubīyan
buzzer أزاز. رنان
azzāz, rannān
byatt دعامة أفقية
da'āma ufuqīya
bye channel قناة تحويل
qanāt taḥwīl
by-pass طريق تجاوز
ṭarīq tajāwuz
cabinet projection إسقاط مائل
isqāṭ mā'il
cable كبل
kabl
cable-braced girder system نظام عوارض مقوى بالكبلات
niẓam 'awāriḍ muqawwa bil-kablāt
cable drill حفارة كبلية
ḥaffāra kablīya
cable duct ماسورة الأسلاك
māsūrat al-aslāk
cable-laid rope حبل ذو جدائل عادية
ḥabl ḏū jadā'il 'ādīya
cable railway سكة حديد معلقة
sikkat ḥadīd mu'allaqa
cable-stayed bridge جسر مدعم بالحبال
jisr mud'am bil-ḥibāl
cableway مصعد كبلي
mas'ad kablī
cableway excavator حفارة كبلية
ḥaffāra kablīya
cableway transporter ناقلة كبلية
nāqila kablīya
cache memory كمبيوتر سريع
kumbyītar sarī'
cage قفص
gafaṣ
caisson قيسون
qaisūn
caisson disease شلل الغواص
šalal al-ġawwāṣ
caisson pile ركيزة قيسون
rakīzat qaisūn
calcine كلس بالتحميص
kallasa bil-taḥmīṣ
calcium chloride كلوريد الكالسيوم
klūrīd al-kalsyūm
calibrate عاير. درّج
'āyara, darraja
calibre عيار
'ijār
calliper log سجل قطر الحفر
sijill quṭr al-ḥafr
callipers فرجار قياس القطر
firjār qiyās al-quṭr
calorific value القيمة الحرارية
al-qīma al-ḥarārīya
camber احديداب. تقوس
iḥdīdāb, taqawwus
camp sheathing جدار ساند
jidār sānid
canal قناة
qanāt
canalization شق القنوات
šaqq al-qanawāt
canal lift رافعة هويسية
rafi'a hawīsīya
canal lock chamber غرفة الهويس
ġurfat al-hawīs
candela كنديلا
kandīla
cant أمال. مائل. إنعطاف
amāl, mā'il, in'iṭāf
cant deficiency قصور الانعطاف
quṣūr al-in'iṭāf
canteen كانتين
kantīn
cantilever كابول. كتيفة معلقة
kābūl, katīfa mu'allaqa
cantilever arm ذراع كابولي
ḏirā' kābūlī

cantilever bridge قنطرة كابولية
qanṭara kabūlīya
cantilever crane مرفاع كابولي
mirfāʿ kābūlī
cantilever footing قاعدة كابولية
qāʿida kabūlīya
cantilever formwork هيكل كابولي
haikal kābūlī
cantilever foundation أساس كابولي
asās kābūlī
cantilever wall جدار كابولي
jidār kābūlī
cap قلنسوة
qulansuwa
capacitor مكثف
mukaṯṯif
capacity curve منحنى السعة
munhana as-siʿa
cap block كتلة رأس الدعامة
kutlat raʾs ad-duʿāma
capel كمية الكبل
kammīyat al-kabl
capillarity الجاذبية الشعرية
al-jaḏibīya aš-šaʿrīya
capillary fringe أرض شعرية
arḍ šaʿrīya
capillary pressure ضغط شعري
ḍaġṭ šaʿrī
capillary rise إرتفاع شعري
irtifāʿ šaʿrī
capillary water ماء شعري
māʾ šaʿrī
capital تاج العمود
tāj al-ʿāmūd
capital construction cost التكاليف الرأسمالية للانشاءات
at-takālīf ar-raʾsmālīya lil-inšāʾāt
capstans رحوية
raḥawīya
carbon كربون . فحم
karbūn, faḥm
carbon-arc welding لحام بقوس الكربون
liḥām bi-qaus al-karbūn
carbon dioxide extinguisher مطفأة بثاني أكسيد الكربون
miṭfaʾa bi-ṯānī aksīd al-karbūn
carbon dioxide recorder مسجل ثاني أكسيد الكربون
musajjil ṯānī aksīd al-karbūn
carbon dioxide treatment plant محطة معالجة بثاني أكسيد الكربون
muḥaṭṭat muʿālajat bi-ṯānī aksīd al-karbūn
carbon dioxide welding لحام بثاني أكسيد الكربون
liḥām bi-ṯānī aksīd al-karbūn
carboniferous limestone حجر جيري كربوني
hajar jīrī karbūnī
carbon monoxide أول أكسيد الكربون
awwal aksīd al-karbūn
carbon monoxide analyser جهاز تحليل أول أكسيد الكربون
jihāz taḥlīl awwal aksīd al-karbūn
carbon steel فولاذ كربوني
fūlāḏ karbūnī
Carborundum (TM) كاربورندم
kārburundum
car dumpers عربة قلابة
ʿaraba qallāba
car ferry apron رصيف معبرة السيارات
raṣīf maʿbarat as-sayyārāt
cargo terminal محطة البضائع
muḥaṭṭat al-baḍāʾiʿ
carousel حامل دوار
ḥamil dawwār
car parking area منطقة وقوف السيارات
minṭaqat wuqūf as-sayyārat
carpenter نجّار
najjār
car-repair track سكة إصلاح العربات
sikkat iṣlaḥ al-ʿarabāt
carriage of a stair حامل السلم
ḥamil as-sullam
carriageway طريق العربات
ṭarīq al-ʿarabāt
carriers ناقلات . حاملات
nāqilāt, ḥāmilāt
cartographer رسام خرائط
rassām ḵarāʾiṭ
cartographic cover تغطية بالخرائط
taġṭīya bil-ḵarāʾiṭ
cartography علم رسم الخرائط
ʿilm rasm al-ḵarāʾiṭ
cartridge لفيفة فيلم
lafīfat film
cartridge brass نحاس الطلقات
nuḥās aṭ-ṭalaqāt

C

cartridge paper ورق لف
waraq laff
cascade شاغور . شلال صغير
šāġūr, šallāl ṣaġīr
cased pile ركيزة خرسانية مغلفة
rakīza k̲arasānīya muġallafa
case-hardening التصليد المغلف
at-taṣlīd al-muġallaf
casement نافذة بابية
nāfiḏa
casement window شباك ذو مفصلات رأسية
šubbāk ḏū mafṣilāt ra'sīya
cash flow control مراقبة السيولة النقدية
murāqabat as-suyūla an-naqdīya
cash flow forecast تنبؤ السيولة النقدية
tanabbu' as-suyūla an-naqdīya
cash payment مدفوعات نقدية
madfū'āt naqdīya
cash register آلة تسجيل النقد
ālat tasjīl an-naqd
casing غلاف
ġilāf
cast صب . سبك
ṣabb, sabk
castellated beam عارضة برجية
'āriḍa burjīya
casting سبك
sabk
cast-in-place صب في الموقع
ṣabb fil-mauqi'
cast-in-place concrete خرسانة الصب في الموقع
k̲arasānat aṣ-ṣabb fil-mauqi
cast-in-situ صب في الموقع
ṣabba fil-mauqi'
cast iron حديد الزهر
ḥadīd az-zahr
cast iron pipe أنبوب من حديد الزهر
unbūb min ḥadīd az-zahr
cast steel فولاذ الصب
fūlāḏ aṣ-ṣabb
cast-welded rail joint وصلة قضيب ملحومة بالصب
waṣlat qaḍīb malḥūma biṣ-ṣabb
catch basin حوض تجميع
ḥauḍ tajmī'
catchment area مستجمع
mustajma'
catchment boundary حدود المستجمع
ḥudūd al-mustajma'
catch pit حفرة تجميع
ḥufrat tajmī'
catch points نقط تحويل
nuqaṭ taḥwīl
catchwater مياه الأمطار
miyāh al-amṭār
catenary suspension تعليق سلسلي
ta'līq silsilī
caterpillar gate بوابة تحكم ضخمة
bawwābat taḥakkum ḍak̲ma
caterpillars زنجير
zanjīr
cat-head sheave بكارة بأعلى هيكل الركيزة
bakāra bi-a'la haikal ar-rakīza
cathode كاثود . مهبط
kaṯūd – mihbaṭ
cathodic protection الوقاية الكاثودية
al-wiqāya al-kaṯūdīya
cation كاتيون . شاردة موجبة
katyūn, šārida mūjaba
catwalk ممر ضيق (على جسر)
mamarr ḍayyiq 'ala jisr
caulking جلفطة . تغليف
jalfaṭa, taġlīf
caulking tool عدة جلفطة
'uddat jalfaṭa
causeway جسر بحري
jisr baḥrī
cavitation تجويف
tajwīf
cavity bricks طوب مجوف
ṭaub mujawwaf
cavity tanking تصميد للماء بالفراغات الهوائية
taṣmīd al-mā' bil-farāġāt al- hawā'īya
cavity wall جدار فجوي
jidār fajawī
ceiling سقف
saqf
ceiling construction إنشاء السقف
inšā' as-saqf
ceiling joists روافد السقف
rawāfid as-saqf
ceiling laths ألواح تغطية السقف
alwāḥ taġṭiyat as-saqf
ceiling membrane غشاء السقف
ġišā' as-saqf

ceiling price السعر الأقصى
as-si'r al-aqṣa
cellular cofferdam سد إنضاب خلوي
sadd inḍāb kalawī
cellular concrete خرسانة خلوية
karasāna kalawīya
cellular construction إنشاء خلوي
inšā' kalawī
cellular lava حمم بركانية خلوية
ḥumam burqānīya kalawīya
cellular steel panel لوح فولاذ خلوي
lauḥ fūlāḍ kalawī
cellulose acetate أسيتات السليولوز
asītāt as-salyūlūs
cellulose nitrate نترات السليولوز
nītrat as-salyūlūs
cement إسمنت
ismant
cementation سمتنة . تمليط
samtana, tamlīṭ
cement blocks كتلة إسمنت
kutlat ismant
cement concrete خرسانة إسمنت
karasānat ismant
cemented carbides كربيد ملبد
karbīd mulabbad
cement grout إسمنت سائل
ismant sā'il
cement gun مسدس إسمنت
musaddas ismant
cementitious material مادة إسمنتية
mādda ismantīya
cement mortar ملاط الاسمنت
milāṭ al-ismant
cement rendering طلاء إسمنتي
ṭilā' ismantī
cement roof tiles بلاط سطح إسمنتي
bilāṭ saṭḥ ismantī
cemetery مقبرة
maqbara
centimetre سنتيمتر
sintimītar
central core قلب مركزي
qalb markazī
central heating تدفئة مركزية
tadfīa markazīya tadfi'a
centralised مركزي
markazī
centralised handling مناولة مركزية
munāwala markazīya

centralised traffic control مراقبة مركزية للمرور
murāqaba markazīya lil-murūr
central processing unit وحدة معالجة مركزية
waḥdat mu'ālaja markazīya
central processor بروسسور مركزي
brūsisūr markazī
central reservation أرض فاصلة بين طريقين
arḍ fāṣila baina ṭarīqain
central reserve الاحتياطي المركزي
al-iḥtiyāṭi al-markazī
centre of gravity مركز الثقل
markaz aṯ-ṯiqal
centre of pressure مركز الضغط
markaz aḍ-ḍağṭ
centrifugal brake مكبح نابذ
mikbaḥ nābiḏ
centrifugal clutch قابض نابذ
qābiṣ nābiḏ
centrifugal compressor ضاغط بالقوة النابذة
ḍāğiṭ bil-qūwwa an-nābiḏa
centrifugal dredge pump مضخة حفر بالقوة النابذة
miḍakkat ḥafr bil-qūwwa an-nābiḏa
centrifugal force قوة نابذة
qūwwa nābiḏa
centrifugal pump مضخة نابذة
miḍakka nābiḏa
centrifuge نابذ . طرد مركزي
nābiḏ, ṭard markazī
centroid مركز متوسط
markaz mutawassiṭ
ceramic tiles بلاط خزفي
bilāṭ kazafī
ceramic veneer قشور خزفية
qušūr kazafīya
cesspit بالوعة المجاري
bālū'at al-majārī
cesspool حفرة أقذار المجاري
ḥufrat aqḏār al-majārī
chain سلسلة
silsila
chain block بكارة بسلسلة
bakāra bi-silsila
chain book سجل المسّاح
sijill al-massāḥ
chain-bucket dredger كراءة سلسلية
karrā'a silsilīya
chaining قياس بالسلسلة
qiyas bis-silsila

C

chainman حامل السلسلة
ḥāmil al-silsila
chain of locks سلسلة هويسات
silsilat huwaisāt
chain pump مضخة سلسلية
miḍakka silsilīya
chain saw منشار آلي سلسلي
minšār ālī silsilī
chain sling حبل رفع سلسلي
ḥabl raf' silsilī
chain survey مسح بالسلسلة
masḥ bis-silsila
chain suspension تعليق بالسلاسل
ta'līq bis-salāsil
chair مثبت خط حديدي
muṯabbit ḵaṭṭ ḥadīdī
chair bolt مسمار لمثبت الخط الحديدي
mismār muṯabbit al-ḵaṭṭ al-ḥadīdī
chalking تجيُّر السطح المدهون
tajayyur as-saṭḥ al-madhūn
chalk line حبل تسوية
ḥabl taswiya
chamber حجرة . غرفة
ḥujra, ġurfa
chamber drain مصرف الحجرة
maṣrif al-ḥujra
chambered-level tube مسواة بفتحة لزيادة الهواء
musawwāt bi-fatḥa li-ziyādat al-hawā'
chambering فرقعة . تفجير
farqa'a, tafjīr
change face عكس الواجهة
'aks al-wājiha
change of gradient تغيير معدل الانحدار
taġyīr mu'addal al-inḥidār
change order تغيير التسلسل
taġyīr at-tasalsul
change point النقطة الحرجة
an-nuqṭa al-ḥarija
channel قناة . مجرى
qanā, majrā
channel alignment محاذاة القناة
muḥāḏāt al-qanā
channel depth عمق القناة
'amq al-qanā
channel excavation حفر القناة
ḥafr al-qanā
channeller آلة شق القناة
ālat šaqq al-qanā
channel spillway قناة تصريف الفائض
qanāt taṣrīf al-fā'iḍ
channel width عرض القناة
'arḍ al-qanā
characteristic strength المتانة المميزة
al-matāna al-mumayyiza
charge شحن . عبأ
šaḥn, 'ibb'
chargehand مراقب
murāqib
charging hopper قادوس التعبئة
qādūs at-ta'bi'a
Charpy test إختبار الصدم
iḵtibār aṣ-ṣadm
chartered civil engineer مهندس مدني
muhandis madanī
chartered municipal engineer مهندس بلدي
muhandis baladī
chartered structural engineer مهندس إنشاءات
muhandis inšā'āt
chase سن اللولب
sinn al-laulab
chassis هيكل . شاسيه
haikal, šāsī
check كبح . تدقيق
kabḥ, tadqīq
checker مراقب . فاحص
murāqib, fāḥiṣ
checking of plans تدقيق الخطط
tadqīq al-ḵuṭaṭ
check of door مصد الباب
miṣadd al-bāb
check rail سكة المراقبة
sikkat al-murāqaba
check valve صمام غير مرجع
ṣimām ġair murajja'
chemical composition تركيب كيماوي
tarkīb kīmāwī
chemical gauging القياس الكيماوي
al-qiyās al-kīmāwī
chemical grouting حقن كيماوي
ḥaqn kīmāwī
chemical incident unit وحدة الحوادث الكيماوية
wiḥdat al-ḥawādiṯ al-kīmāwīya
chemical oxygen demand طلب الكيماويات للأكسجين
ṭalab al-kīmāwīyāt lil-aksijīn

chemical precipitation ترسيب كيماوي
tarsīb kīmāwī
chemical-resistant concrete خرسانة مقاومة للكيماويات
karasāna muqāwima lil-kīmāwīyāt
chemical toilet مرحاض كيماوي
mirḥāḍ kīmāwī
chemi-hydrometry قياس التدفق الكيماوي
qiyās at-tadaffuq al-kīmāwī
chemise جدار داعم لصد التراب
jidar dā'im li-ṣadd at-turāb
chequer plate لوح فولاذي مثقوب
lauḥ fūlāḏī maṯqūb
chevron drain مصرف لولبي مزدوج
maṣrif laulabī muzdawaj
Chicago caisson قيسون شيكاغو
qaisūn šīkāğū
chief draughtsman رئيس الرسامين
ra'īs ar-rassāmīn
chilled cast iron حديد زهر مصلد
ḥadīd zahr muṣallad
chimney مدخنة
midkana
chimney drain مصرف المدخنة
maṣrif al-midkana
Chinaman chute مسقط قنطري
misqaṭ qanṭarī
chipping goggles نظارات نحاتة
naẓārāt niḥāta
chipping hammer مطرقة نحاته
miṭraqa niḥāta
chippings نحاتة . جذاذة
niḥāta, jiḏāḏa
chip set طقم نحاتة
ṭaqm niḥāta
chisel إزميل . أجنة
izmīl, ajana
chlorination معالجة بالكلور
mu'ālaja bil-klūr
chlorination plant محطة معالجة بالكلور
muḥaṭṭat mu'ālaja bil-klūr
chord وتر
watar
chrome-tanned leather gloves قفاز جلد مدبوغ بالكروم
qaffāz jild madbūğ bil-krūm
chuck قابض لقم المثقب
qābiḍ laqm al-miṯqab
church كنيسة
kanīsa
churn drill حفر كبلي بالدق
ḥafr kabli bid-daqq
chute مجرى مائل . مسقط
majrā mā'il, misqaṭ
chute channel قناة المجرى المائل
qanāt al-majrā al-mā'il
chute spillway قناة تصريف مائلة
qanāt taṣrīf mā'ila
Cipolletti weir سد قياس (سيبوليتي)
sadd qiyās sībūlītī
circle دائرة
dā'ira
circuit دائرة . دارة كهربائية
dā'ira, dāra kahrabā'īya
circuit breaker قاطع الدائرة
qāṭi' ad-dā'ira
circular-arc method طريقة القوس الدائري
ṭarīqat al-qaus ad-dā'irī
circular curve منحنى دائري
munḥana dā'irī
circular level مسواة المسّاح الدائرية
musawwāt al-massāḥ ad-dā'irīya
circular road طريق دائري
ṭarīq dā'irī
circular tunnel section قطاع نفقي دائري
qiṭā' nafaqī dā'irī
circulating water ماء جارٍ في دائرة محصورة
mā' jārī fi-dā'ira maḥṣūra
cistern صهريج
ṣahrīj
civil airport مطار مدني
maṭār madanī
civil engineer مهندس مدني
muhandis madanī
civil engineering هندسة مدنية
handasa madanīya
civil engineering assistant مساعد مهندس مدني
musā'id muhandis madanī
civil engineering draughtsman رسام الهندسة المدنية
rassām al-handasa al-madanīya
civil engineering technician فني الهندسة المدنية
fannī al-handasa al-madanīya

C

cladding *taǧlīf, taṣfīḥ*	تغليف . تصفيح
clad steel *fūlāḏ karbūnī muǧallaf*	فولاذ كربوني مغلف
clamp *qāmiṭa, mišbak*	قامطة . مشبك
clamp handle *miqbaḍ al-mišbak*	مقبض المشبك
clamping bolt *mismār qamṭ mulaulab*	مسمار قمط ملولب
clamping screw *burǧī qamṭ*	برغي قمط
clamshell *dalū majārī*	دلو محاري
clarification *tanqīya, tarwīq*	تنقية . ترويق
classification of soils *taṣnīf at-turba*	تصنيف التربة
classification yard *sāḥat taṣnīf*	ساحة تصنيف
classifier *muṣannif, munassiq*	مُصنف . منَسق
clay *ṣalṣāl, ṭafāl*	صلصال . طفال
clay cutter *luqmaṭ ḥafr aṭ-ṭīn*	لقمة حفر الطين
clay dam *sadd ṭīnī*	سد طيني
clay pockets *jiyūb ṭīnīya*	جيوب طينية
clay-puddle cofferdam *sadd ṭīnī*	سد طيني
clay roofing tile *bilāṭat saṭḥ ṣalṣālīya*	بلاطة سطح صلصالية
cleaning *tanẓīf*	تنظيف
cleansing hydrant *ṣanbūr tanẓīf*	صنبور تنظيف
clearance *ḥayyiz'majal*	حيز. مجال
clearance hole *ṯuqb ḵulūṣī*	ثقب خلوصي
clear glass *zujāj šaffāf*	زجاج شفاف
clearing *izāla*	إزالة
clear span *bā' ṣāfī*	باع صافي
clear-water reservoir *ḵazzān mā' naqī*	خزان ماء نقي
clearway *mamnū' al-intiẓār*	ممنوع الانتظار
clear window glass *zujāj šaffāf lin-nawāfiḏ*	زجاج شفاف للنوافذ
cleat *mirbaṭ, kallāb*	مربط . كلاّب
cleavage fracture *ṣada' tašaqquqī*	صدع تشققي
clerk of works *kātib ā'māl inšā'īya*	كاتب أعمال إنشائية
clevis *šauka mifṣalīya*	شوكة مفصلية
client *zabūn, 'amīl*	زبون . عميل
climbing crane *mirfā' mutasalliq*	مرفاع متسلق
clinker *ḵabṯ al-faḥm au al-ma'ādin*	خبث الفحم أو المعادن
clinker concrete *ismant al-ḵabṯ*	إسمنت الخبث
clinograph *mirsamat al-mail*	مرسمة الميل
clinometer *klīnūmitar, miqyās al-mail*	كلينومتر. مقياس الميل
closed-box caisson *qaisūn sandūqī muǧlaq*	قيسون صندوقي مغلق
closed-circuit television camera *muṣawwira li-talafizyūnāt ad-dawā'ir al-muǧlaqa*	مصورة لتلفزيونات الدوائر المغلقة
closed-conduit spillway *qanāt taṣrīf bi-majrā muǧlaq*	قناة تصريف بمجرى مغلق
closed loop system *niẓām ad-dāra al-muǧlaqa*	نظام الدارة المغلقة
closer *ṭauba nisfīya*	طوبة نصفية
close timbering *taǧṭiya bil-alwāḥ al-ḵašabīya*	تغطية بالألواح الخشبية
clough *wādī ḍayyiq*	واد ضيق
clutch *qābiḍ, klatš*	قابض (كلتش)
coach screw direct fastening *taṯbīt mubāšir bi-burǧī kabīr*	تثبيت مباشر ببرغي كبير
coagulation *takaṯṯur, tarwīb*	تخثر . ترويب

C

coal فحم
faḥm
coarse aggregate حصباء خشنة
ḥaṣbā' kašina
coarse-grained soil تربة خشنة الحبيبات
turba kašinat al-hubaibāt
coated chippings حصى مدهون بالقار
ḥaṣā madhūn bil-qār
cobbles حجر الرصف
ḥajar ar-raṣf
coded control مراقبة بالشيفرة
murāqaba biš-šifra
code of practice قوانين ممارسة العمل
qawanīn mumārasat al-'aml
coefficient of compressibility مُعامل الانضغاط
mu'āmil al-inḍigāṭ
coefficient of contraction مُعامل الانكماش
mu'āmil al-inkimāš
coefficient of expansion مُعامل التمدد
mu'āmil at-tamaddud
coefficient of friction مُعامل الاحتكاك
mu'āmil al-iḥtikāk
cofferdam سد مؤقت
sadd mu'aqqat
cogging تشكيل بالطرق
taškīl biṭ-ṭarq
cohesive soil تربة متماسكة
turba mutamāsika
Colcrete (TM) كولكريت
kūlkrīt
cold chisel إزميل قطع على البارد
izmīl qaṭ' 'ala al-bārid
cold drawing سحب على البارد
saḥb 'ala al-bārid
cold-formed steel فولاذ مسحوب على البارد
fūlāḏ masḥūb 'ala al-bārid
cold rolling دلفنة على البارد
dalfana 'ala al-bārid
cold storage building مخزن مبرَّد
makzan mubarrad
cold working معالجة المعادن على البارد
mu'ālajat al-ma'ādin 'ala al-bārid
Colgrout (TM) ملاط «كولجروت»
milāṭ kūlgrūt
collapsible cradle حمالة قابلة للطي
hammāla qābila liṭ-ṭayy
collar beam شداد علوي
šiddād 'alawī
collecting system شبكة تجميع مياه المجاري
mišbakat tajmī' al-majārī
collecting tank خزان تجميع
kazzān tajmī'
collection of data جمع بيانات
jam' bayānāt
colloidal concrete خرسانة حقن
karasānat ḥaqn
coloured glass زجاج ملون
zujāj mulawwan
column عمود
'amūd
column analogy المقارنة بالعمود
al-muqārana bil-'amūd
column head رأس العمود
ra's al-'amūd
combined base قاعدة مشتركة
qā'ida muštaraka
combined footing أساس مشترك
asās muštarak
combined sewers مجارير مشتركة
majārīr muštaraka
combined stresses إجهاد موحد
ijhād muwaḥḥad
combined system نظام تصريف مشترك
niẓām taṣrīf muštarak
combustible material مادة قابلة للاحتراق
mādda qābila lil-iḥtirāq
combustion control مقاومة الاحتراق
muqāwamat al-iḥtirāq
commencement of work الشروع في العمل
aš-šurū' fil-'amal
comminutor ماكينة تفتيت
mākinat taftīt
commissioning فحص المنشآت والمعدات وإعدادها للعمل
faḥṣ al-munša'āt wal-mu'iddāt wa i'dādaha lil-'amal
commissioning schedule جدول مواعيد الاختبار
jadwal mawā'īd al-iktibār
common crossing معبر مشترك
ma'bar muštarak
common data area مكان المعطيات المشتركة
makān al-mu'ṭayāt al-muštaraka
common sewer مجرور مشترك
majrūr muštarak
common wall جدار مشترك
jidār muštarak

C

communication facilities وسائل الاتصال
wasā'il al-ittiṣāl
commuter coach حافلة ركاب
ḥafilat rukkāb
commuter lines خطوط ركاب الضواحي
ḵuṭūṭ rukkāb aḍ-ḍawāḥī
commuter systems شبكة خطوط الضواحي
šabakat ḵuṭuṭ aḍ-ḍawāḥī
compacted hardcore حجر أساس صلد مضغوط
ḥajar asās ṣalid maḍḡuṭ
compacted sand رمل متضام
raml mutaḍām
compacting factor test إختبار معامل الدموج
iḵtibār mu'āmil ad-dumūj
compaction دموج
dumūj
compact material مادة مدمجة
mādda mudmaja
compactor ماكينة ضغط
mākinat ḍaḡṭ
comparator مقارن
muqārin
comparator base قاعدة المقارنة
qā'idat al-muqārana
compass بوصلة . فرجار
būṣala, firjār
compensating diaphragm رق تعويض
raqq ta'wīḍ
compensating error خطأ متكافئ
ḵaṭa' mutakāfi'
compensation water مياه التعويض
miyāh at-ta'wīḍ
competent examiner فاحص ذوكفاءة
fāḥis ḏu kafū'
completion date تاريخ الإتمام
tārīḵ al-ıtmām
composite breakwater حاجز أمواج مركب
ḥajiz amwaj murakkab
composite construction إنشاء مركب
inšā' murakkab
composite dam سد مركب
sadd murakkab
composite-girder bridge جسر بعوارض مركبة
jisr bi-'awāriḍ murakkaba
composite member عضو مركب (لهيكل إنشائي)
'uḍu murakkab li-haikal inšā'ī
composite pavement رصيف مركب
raṣīf murakkab
composite pile ركيزة مركبة
rakīza murakkaba
composite pipe أنبوب مركب
unbūb murakkab
composites مواد إنشائية مركبة
mawād inšā'īya murakkaba
composting تحول النباتات الى سماد عضوي
taḥawwul an-nabātāt ila samād 'uḍwi
compound air lift مضخة هوائية مركبة
miḍakka hawā'īya murakkaba
compound curve منحنى مركب
munhana murakkab
compound dredger كراءة بالقواديس الدوارة
karrā'a bil-qawādīs ad-dawwāra
compound engine محرك مركب
muḥarrik murakkab
compound girder عارضة مركبة
'ārida murakkaba
compound pipe أنبوب مركب
unbūb murakkab
comprehensive general liability مسؤولية قانونية عامة
mas'ūlīya qānūnīya 'āmma
compressed air هواء مضغوط
hawā' maḍḡūṭ
compressed air caisson قيسون يعمل بالهواء المضغوط
qaisūn ya'mal bil-hawā' al-maḍḡuṭ
compressed air tunnelling شق الأنفاق بالهواء المضغوط
šaqq al-anfāq bil-hawā' al-maḍḡuṭ
compressed asphalt أسفلت مضغوط
isfalt maḍḡūṭ
compression إنضغاط
inḍiḡāṭ
compression flange شفة الانضغاط
šaffat al-inḍiḡāṭ
compressive force قوة ضاغطة
quwwa ḍāḡiṭa
compressive strength مقاومة الانضغاط
muqāwamat al-inḍiḡāṭ
compressor ضاغط
ḍāḡiṭ
compromise joint وصلة تعويض
waṣlat ta'wīḍ
computer كمبيوتر
kumbyūtar

computer control room غرفة المراقبة بالكمبيوتر
ġurfat al-murāqaba bil-kumbyūtar
concentrated load حمل مركز
ḥiml murakkaz
concentrating تركيز
tarkīz
concentrating plant وحدة تركيز
waḥdat tarkīz
concentrically loaded تحميل متحد المركز
taḥmīl muttaḥid al-markaz
concept design تصميم المفاهيم
taṣmīm al-mafāhīm
concertina shutter doors أبواب بمصاريع كونسرتينية
abwāb bi-maṣārīʿ kunsirtīnīya
conchoidal محاري الشكل
muḥārī aš-šakl
concrete خرسانة
k̲arasāna
concrete base قاعدة خرسانية
qāʿida k̲arasānīya
concrete batching/mixing plant وحدة خلط الخرسانة على دفعات
waḥdat k̲alṭ al-k̲arasāna ʿala dufuʿāt
concrete bedding فرشة خرسانة
faršat k̲arasāna
concrete benching مصطبة خرسانية
maṣṭaba k̲arasānīya
concrete block قالب خرساني
qālab k̲arasānī
concrete breaker كسّارة خرسانة
kassārat k̲arasāna
concrete bridge جسر خرساني
jisr k̲arasāna
concrete caisson قيسون خرساني
qaisūn k̲arasānī
concrete cancer سرطان الخرسانة
saraṭān al-k̲arasāna
concrete column عمود خرساني
ʿamūd k̲arasānī
concrete construction إنشاء خرساني
inšāʾ k̲arasānī
concrete cover غطاء خرساني
ġiṭāʾ k̲arasānī
concrete culvert مجرى خرساني
majra k̲arasānī
concrete cutting قطع الخرسانة
qaṭʿ al-k̲arasāna

concrete dam سد خرساني
sadd k̲arasānī
concrete deck سطح خرساني
saṭḥ k̲arasānī
concrete delivery hose خرطوم تصريف الخرسانة
k̲arṭūm taṣrīf al-k̲arasāna
concrete delivery pipe أنبوب تصريف الخرسانة
unbūb taṣrīf al-k̲arasāna
concrete deterioration تلف الخرسانة
talaf al-k̲arasāna
concrete displacement pile ركيزة إزاحة الخرسانة
rakīzat izāḥat al-k̲arasāna
concrete-encased steelwork تركيبات فولاذية مغلفة بالخرسانة
tarkībāt fūlāḏīya muġallafa bil-k̲arasāna
concrete fill حشوة خرسانية
ḥašwa k̲arasānīya
concrete-finishing machine ماكينة تشطيب الخرسانة
mākinat tašṭīb al-k̲arasāna
concrete floor أرضية خرسانية
arḍīya k̲arasānīya
concrete foundation أساس خرساني
asās k̲arasānī
concrete gravity dam سد ثقالي خرساني
sadd ṯuqālī k̲arasānī
concrete infill تعبئة بالخرسانة
taʿbiʾa bil-k̲arasāna
concrete layer طبقة خرسانية
ṭabaqa k̲arasānīya
concrete lining تبطين بالخرسانة
tabṭīn bil-k̲arasāna
concrete mixer خلاطة خرسانة
k̲allāṭat k̲arasāna
concrete outlet مخرج الخرسانة
mak̲raj al-k̲arasāna
concrete pavement رصيف خرساني
raṣīf k̲arasānī
concrete paver راصفة الخرسانة
rāṣifat al-k̲arasāna
concrete pier دعامة خرسانية
duʿāma k̲arasānīya
concrete pile ركيزة خرسانية
rakīza k̲arasānīya
concrete pipe ماسورة خرسانية
māsūra k̲arasānīya
concrete placer وحدة صب الخرسانة
waḥdat ṣabb al-k̲arasāna

C

concrete plant مصنع خرسانة
maṣnaʿ karasāna
concrete protection وقاية خرسانية
wiqāya karasānīya
concrete pump مضخة خرسانة
midakkat karasāna
concrete retaining wall جدار ساند من الخرسانة
jidār sānid min al-karasāna
concrete roof سطح خرساني
saṭḥ karasānī
concrete slab بلاطة خرسانية
bilāṭa karasānīya
concrete sleeper عارضة خرسانية
ʿāriḍa karasānīya
concrete spreader فارشة خرسانة
fārišat karasāna
concrete-vibrating machine هزاز لدمج الخرسانة
hazzāz li-damj al-karasāna
concrete wall جدار خرساني
jidār karasānī
concreting boom ذراع مرفاع الخرسانة
ḏirāʿ mirfāʿ al-karasāna
concreting equipment معدات الخرسانة
muʿiddāt al-karasāna
condensate ناتج التكثيف
nātij at-takṯīf
condensation تكثيف . تكاثف
takṯīf, takaṯṯuf
condenser مُكثف
mukaṯṯif
conditional branch instruction تعليمات جزئية مشروطة
taʿlīmāt juz ʾīya mašrūṭa
condition code مكشاف الرموز
mikšāf ar-ramūz
conditions of contract شروط العقد
šurūṭ al-ʿaqd
conductor موصل . ناقل
muwaṣṣil, nāqil
conduit ماسورة الأسلاك
māsūrat al-aslak
cone clutch قابض مخروطي
qābiḍ makrūṭī
cone penetration resistance مقاومة الاختراق المخروطي
muqāwamat al-iktirāq al-makrūṭī
cone penetration test إختبار الاختراق المخروطي
iktibār al-iktirāq al-makrūṭī
cone valve صمام مخروطي
ṣimām makrūṭī
connected frogs تقاطعات متصلة
taqāṭuʿāt muttaṣila
connecting pipe أنبوب الوصل
unbūb al-waṣl
connecting rod ذراع توصيل
ḏirāʿ tauṣīl
connecting transitions تحويلات توصيل
taḥwīlāt tauṣīl
connection وصل
waṣl
connection flange شفة وصل
šaffat waṣl
consistency تماسك
tamāsuk
consistency index دليل التماسك
dalīl at-tamāsuk
construction إنشاء
inšāʾ
constructional engineer مهندس إنشاءات
muhandis inšāʾāt
constructional erector عامل تركيب الانشاءات
ʿāmil tarkīb al-inšāʾāt
constructional fitter أخصائي تجميع الانشاءات
akiṣṣāʾi tajmiʿ al-inšāʾāt
construction contract عقد الانشاءات
ʿaqd al-inšāʾāt
construction detail تفاصيل الانشاءات
tafāṣīl al-inšāʾāt
construction joint وصلة إنشائية
waṣla inšāʾīya
construction-management agreement عقد إدارة الانشاءات
ʿaqd idārat al-inšāʾāt
construction period فترة الانشاءات
fatrat al-inšāʾāt
construction requirements متطلبات الانشاءات
mutaṭallabāt al-inšāʾāt
construction spanner مفتاح ربط الهياكل الانشائية
miftaḥ rabṭ al-hayākil al-inšāʾīya
construction survey مسح الانشاءات
masḥ al-inšāʾāt
construction way مسار موقت
masār muʾaqqat
consultant مستشار
mustašār
consulting engineer مهندس إستشاري
muhandis istišārī

consumer's voltage فلطية المستهلك
fulṭīyat al-mustahlik
contact aerator خزان تهوية مضغوطة
ḵazzān tahwiya maḍġūṭa
contact bed طبقة ملامسة
ṭabaqa mulāmisa
contactor قاطع تلقائي
qāṭiʿ tilqāʾī
contact pressure under foundations ضغط التلامس تحت الأساسات
ḍaġṭ at-talāmus taḥt al-asāsāt
container وعاء . حاوية
wiʿaʾ, ḥāwīya
container-handling terminal محطة مناولة الحاويات
muḥaṭṭat munāwalat al-ḥāwīyāt
contaminated drinking water ماء شرب ملوث
māʾ šurb mulawwaṯ
contiguous bored pile wall جدار على ركائز مصبوبة في الموقع
jidār ʿala rakāʾiz maṣbūba fil-mauqiʿ
continuity إستمرار
istimrār
continuous beam عتبة متواصلة
ʿataba mutawāṣila
continuous cab signals إشارات مستمرة بحجرة القيادة
išārāt mustamirra bi-ḥujrat al-qiyādā
continuous filter مرشح مستمر
muraššiḥ mustamirr
continuous footing قاعدة متواصلة
qāʿida mutawāṣila
continuous girder عارضة متواصلة
ʿāriḍa mutawāṣila
continuously welded track سكة ملحومة بصورة متواصلة
sikka malḥūma bi-ṣūra mutawāṣila
continuous mixer خلاطة متواصلة العمل
ḵallāṭa mutawāṣilat al-ʿamal
continuous ropeway طريق حبلي متواصل
ṭarīq ḥablī mutawāṣil
continuous welded rail قضيب ملحوم متواصل
qaḍīb malḥūm mutawāṣil
contour كنتور
kuntūr
contour interval مسافة كنتورية
masāfa kuntūrīya
contour line خط مناسيب
ḵaṭṭ manāsīb
contract عقد
ʿaqd
contract-agreement form نموذج اتفاقية العقد
namūḏaj itifaqiat al-ʿaqd
contract bond تعهد بتنفيذ العقد
taʿahhud bi-tanfīḏ al-ʿaqd
contract documents مستندات العقد
mustanadāt al-ʿaqd
contract drawings رسوم العقد
rusūm al-ʿaqd
contracting procedure إجراءات التعاقد
ijrāʾāt at-taʿāqud
contraction in area تقلص بالمساحة
taqalluṣ bil-masāḥa
contraction joint وصلة تقلص
waṣlat taqalluṣ
contract manager مدير العقود
mudīr al-ʿuqūd
contractor مقاول . متعهد
muqāwil, mutaʿahhid
contractor's agent وكيل المقاول
wakīl al-muqāwil
contractor's comprehensive general liability مسؤولية المقاول العامة الشاملة
masʾūlīyat al-muqāwil al-ʿāmma aš-šāmila
contractor's protective liability insurance تأمين ضد المسؤولية القانونية لحماية المقاول
taʾmīn ḍid al-masʾūlīya al-qānūnīya li-himāyat al-muqāwil
contract revision تعديل العقد
taʿdīl al-ʿaqd
contractual liability المسؤولية التعاقدية
al-masʾūlīya at-taʿāqudīya
contraflexure إنحناء معاكس
inḥināʾ muʿākis
control مراقبة . تحكم
murāqaba, taḥakkum
control block مبنى المراقبة والتحكم
mabna al-murāqaba wat-taḥakkum
control cabin مقصورة المراقبة
maqṣūrat al-murāqaba

C

control console وحدة التحكم
waḥdat at-taḥakkum
controlled filtration ترشيح موجّه
taršiḥ muwajjah
controlled junction ملتقى خاضع للمراقبة
multaqa ḵāḍiʿ lil-murāqaba
controlled tipping تفريغ خاضع للتحكم
tafrīḡ ḵāḍiʿ lit-taḥakkum
control lever ذراع توجيه
ḏirāʿ taujīh
controlling valve صمام تحكم
ṣimām taḥakkum
control panel لوحة تحكم
lauḥat taḥakkum
control platform منصة تحكم
minaṣṣat taḥakkum
control point نقطة مراقبة
nuqṭat murāqaba
control room غرفة مراقبة
ḡurfat murāqaba
control system نظام تحكم
niẓām taḥakkum
control test إختبار المراقبة
iḵtibār al-murāqaba
control tower برج مراقبة
burj murāqaba
control valve صمام تحكم
ṣimām taḥakkum
conversion factor مُعامل تحويل
muʿāmil taḥwīl
conveyor ناقل
nāqil
conveyor belt سير ناقل
sair nāqil
cooling chamber غرفة تبريد
ḡurfat tabrīd
cooling pond بركة تبريد
burkat tabrīd
cooling tower برج تبريد
burj tabrīd
cooling-water culvert مجرى ماء التبريد
majrā māʾ at-tabrīd
cooling-water pumphouse مبنى مضخة ماء التبريد
mabnā miḍakkat māʾ at-tabrīd
cooling-water system نظام ماء التبريد
niẓām māʾ at-tabrīd
coordinates إحداثيات
iḥdāṯīyāt
coping إفريز مائل
ifrīz māʾil
copper welding لحام بالنحاس
liḥām bin-nuḥās
coprecipitation ترسب مشترك
tarassub muštarak
corbel ركيزة
rakīza
corduroy road طريق أخشاب مرصوفة بالعرض
ṭarīq akšāb marṣūfa bil-ʿarḍ
core قلب
qalb
core barrel ماسورة حفظ العينة
māsurat ḥifẓ al-ʿayyina
cored hole ثقب مشكل بقالب
ṯuqb mušakkal bi-qālab
core drill مثقاب لاستخراج العينات
miṯqāb li-istikrāj al-ʿayyināt
cores عينات جوفية
ʿayyinat jaufīya
core wall جدار مانع لتسرب الماء
jidār māniʿ lit-tasarrub al-māʾ
coring tools أدوات تجويف
adawāt tajwīf
cork فلين
fillīn
cork tile بلاطة فلين
bilāṭat fillīn
corner ركن . زاوية
rukn, zāwīya
corporate member عضو مشارك
ʿuḍu mušārik
correcting unit وحدة تصحيح
waḥdat taṣḥīḥ
corridor دهليز . ممر
dahlīz, mamarr
corrosion تأكل
taʾākul
corrosion control مكافحة التأكل
mukāfaḥat at-taʾākul
corrosion fatigue كلال التأكل
kilāl at-taʾākul
corrosion prevention منع التأكل
manʿ at-taʾākul
corrosive أكال . حات
akkāl, ḥāt
corrugated sheet لوح مموج
lauḥ mumawwaj

corrugated steel pipe
unbūb min al-fūlāḏ al-mumawwaj
أنبوب من الفولاذ المموج

corrugated steel sheet
lauḥ min al-fūlāḏ al-mumawwaj
لوح من الفولاذ المموج

corundum
kurundum, yaqūt
كورندم . ياقوت

cost-benefit analysis
taḥlīl manāfiʿ at-takālīf
تحليل منافع التكاليف

cost clerk
kātib takālīf
كاتب تكاليف

cost control
murāqabat at-takālīf
مراقبة التكاليف

cost control system
niẓam murāqabat at-takālīf
نظام مراقبة التكاليف

cost estimate
taqdīr at-takālīf
تقدير التكاليف

cost-reimbursable agreement
ittifāq tasdīd at-takālīf
إتفاق تسديد التكاليف

costs
takālīf
تكاليف

coulomb
kulūm
كولوم

counter bore
taqwīr
تقوير

counter bracing
taqwiya ʿaksīya biš-šakkālāt
تقوية عكسية بالشكالات

counterfort
katf jānibīya
كتف جانبية

counterfort wall
jidār iḥtijāz
جدار إحتجاز

counterweight
wazn muʿādil
وزن معادل

coupling
waṣl, taqārun
وصل . تقارن

course
ittijāh ḵaṭṭ al-masḥ
إتجاه خط المسح

coursed rubble
bināʾ bi-hijāra muḵtalifa
بناء بحجارة مختلفة

cover
sumk al-ḵarasāna, ḡiṭāʾ
سمك الخرسانة . غطاء

cover seating
qāʿidat al-ḡiṭāʾ
قاعدة الغطاء

crack
ṣadʿ
صدع

cracked rail
qaḍīb mutaṣaddiʿ
قضيب متصدع

crack inducers
fawāṣil li-manʿ takassur al-ḵarasāna
فواصل لمنع تكسر الخرسانة

cracking
taksīr, takassur
تكسير . تكسر

cracking in concrete
taṣadduʿ al-ḵarasāna
تصدع الخرسانة

cradle
ḥammāla
حمالة

craftsman
ḥirafi
حرفي

crane
mirfāʿ, winš
مرفاع (ونش)

crane-and-ball method
ṭarīqat al-mirfāʿ wal-kura
طريقة المرفاع والكرة

crane gantry
qanṭarat al-mirfāʿ
قنطرة المرفاع

crane post
ʿamūd al-mirfāʿ
عمود المرفاع

crane slinger
ḥalaqat taʿlīq al-mirfāʿ
حلقة تعليق المرفاع

crane tower
burj al-mirfāʿ
برج المرفاع

crash barrier
ḥājiz irtiṭām
حاجز إرتطام

crash services
ḵadamāt inqāḏ
خدمات الانقاذ

crash tenders
ʿarabāt inqāḏ
عربات الانقاذ

crawler mounted excavator
ḥaffāra muzanjara
حفارة مزنجرة

crazing
tajazzuʿ
تجزع

creek
jadwal, ḵaur
جدول . خور

creep
zaḥf, taharruk baṭīʾ
زحف . تحرك بطيء

creeper cranes
marāfiʿ mutasalliqa
مرافع متسلقة

creosote
krīyūsūt
كريوزوت

crest gate
bawwābat qanāt taṣrīf
بوابة قناة تصريف

crest surfacing
saṭh muzaḵraf
سطح مزخرف

crib
akšāb daʿm
أخشاب دعم

crib dam
sadd min akšāb dāʿima
سد من أخشاب داعمة

C

33

C

cribwork أعمدة دعم خشبية
ʿāmidat daʿm ḵašabīya
crimper مدحلة تحزيز
midhalat taḥzīz
crippling load الحمل الحرج
al-ḥiml al-ḥarj
critical height الارتفاع الحرج
al-irtifāʿ al-ḥarij
critical hydraulic gradient الميل الهيدرولي الحرج
al-mail al-hīdrūlī al-ḥarij
critical path scheduling تنظيم المسار الحرج
tanẓīm al-masār al-ḥarij
critical point نقطة حرجة
nuqṭa ḥarija
critical voids ratio of sands نسبة الثبات الحرج للرمل
nisbat aṯ-ṯabāt al-ḥarij lil-raml
cropper آلة قطع أسياخ الفولاذ
ālat qaṭʿ as-siyyaḵ al-fūlāḏ
cross-assembler جامعة معترضة
jāmiʿa muʿtarriḍa
crossbeam عتبة معترضة
ʿataba muʿtarriḍa
crosshead طربوش الوصل
ṭarbūš al-waṣl
crossing تقاطع . عبور
taqāṭuʿ, ʿubūr
crossing frog تقاطع خطوط حديدية
taqāṭuʿ ḵuṭūṭ ḥadīdīya
crossing gates بوابات عبور
bawwābāt ʿubūr
crossing tie رافدة معترضة
rāfida muʿtarrida
crossover وصلة تحويل
waṣlat taḥwīl
cross poling ألواح دعم
alwāḥ daʿm
crossroads تقاطع طرق
taqāṭuʿ ṭuruq
cross-section مقطع عرضي
maqṭaʿ ʿarḍī
cross-sectional area مساحة المقطع العرضي
masāḥat al-maqṭaʿ al-ʿarḍī
cross-section features خصائص القطاع العرضي
ḵaṣāʾiṣ al-qiṭāʿ al-ʿarḍī
cross-section levelling تسوية القطاع العرضي
taswiyat al-qiṭāʿ al-ʿarḍī
cross-under يمر تحت طريق
yamurr taḥt ṭariq

cross-wind component أداة الرياح المعترضة
adāt ar-riyyaḥ al-muʿtarrida
crowbar عتلة . مخل
ʿatala, mikall
crown تاج . قمة
tāj, qumma
crude oil النفط
an-nafṭ
crump weir سد ذو ذروة
sadd ḏū ḏurwa
crushed slag خبث مفتت
ḵabṯ mufattat
crushing strength مقاومة التفتت
muqāwamat at-tafattut
crushing test إختبار التفتت
iḵtibār at-tafattut
cube strength مقاومة الخرسانة
muqāwamat al-ḵarasāna
cube test إختبار مقاومة الخرسانة
iḵtibār muqāwamat al-karasāna
cubic metre متر مكعب
mitr mukaʿʿab
cubit ذراع (وحدة قياس)
ḏirāʿ, waḥdat qiyas
culvert بربخ . مجرى
barbaḵ, majrā
cumulative errors أخطاء متراكمه
akṭāʾ mutarākima
cupola سطح مقبب
saṭḥ muqabbab
curb حافة الرصيف
ḥafat ar-raṣīf
curing إنضاج الاسمنت بالترطيب
inḍāj al-ismant bit-tarṭīb
curing compound مركّب إنضاج
murakkab inḍāj
curing membrane غشاء الانضاج
ġišāʾ al-indāj
curing period فترة الانضاج
fatrat al-indāj
current forces قوى التيار
qiwa at-tayyār
current meter مقياس التيار
miqyās at-tayyār
cursor دليل الشاشة
dalīl aš-šāša
curtain walling تغليف الجدران
taġlīf al-judrān

curvature إنحناء . تقوس
inhinā', taqawwus
curve منحنى . منعطف
munhanā, mun'aṭaf
curved track سكة منحنية
sikka munhanīya
curve ranging تحديد نقاط المنحنى
taḥdīd nuqāṭ al-munḥanā
customs جمارك
jamārik
cut-and-cover طريقة القطع والردم
ṭarīqat al-qaṭ' war-radm
cut-and-cover tunnel نفق محفور بالقطع والردم
nafaq maḥfūr bil-qaṭ' war-radm
cut and fill قطع وردم
qaṭ' war-radm
cut holes حفر ثقوب التفجير
ḥafr ṯuqūb at-tafjīr
cut-off depth عمق القطع
'umq al-qaṭ'
cut-off wall جدار مانع لتسرب الماء
jidār māni' lit-tasarrub al-mā'
cutout قاطع الدائرة الكهربائية
qāṭi' ad-dā'ira al-kahrabā'īya
cut stones حجر طبيعي منحوت
ḥajar ṭabī'ī manḥūt
cutter suction dredger كراءة حفر ماصة
karrā'at ḥafr māṣṣa
cutting قطع
qaṭ'
cutwater جانب دعامة الجسر
jānib du'āmat al-jisr
cyclone إعصار
i'ṣār
cyclonic spray scrubber كاشطة رش حلزونية
kāšiṭat rašš ḥalazūnīya
cyclopean concrete خرسانة سيكلوبية
ḵarasāna sīklūbīya
cylinder أسطوانة
isṭiwāna
cylinder gate بوابة أسطوانية
bawwāba isṭiwānīya
cylindrical rubber fender مصد مطاطي أسطواني
miṣadd maṭāṭī isṭiwānī
dam سد
sadd
damage claim مطالبة بالتعويض عن الأضرار
muṭālaba bit-ta'wīḍ 'an al-aḍrār
dam axis محور السد
miḥwar as-sadd
damper مخمد
muḵammid
damp proof عازل للرطوبة
'āzil lir-ruṭūba
damp-proof course طبقة عازلة للرطوبة
ṭabaqa 'āzila lir-ruṭūba
damp-proofing معالجة لمنع الرطوبة
mu'ālaja li-man' ar-ruṭūba
damp-proof membrane غشاء مانع للرطوبة
ġišā' māni' lir-ruṭūba
danger sign إشارة خطر
išārat ḵaṭr
data معطيات
muc'ṭayāt
data storage تخزين البيانات
taḵzīn al-bayānāt
data synchronisation مزامنة البيانات
muzāmanat al-bayānāt
date of recharging تاريخ إعادة الشحن
tārīḵ i'ādat aš-šaḥn
datum مرجع إسناد
marja' isnād
daylight ضوء النهار
ḍau' an-nahār
dead bolt مسمار ملولب مربع المقطع
mismār mulaulab murabba' al-maqṭa'
dead load حمل ساكن
ḥiml sākin
dead lock إخفاق المسعى
iḵfāq al-mas'ā
deadman مرسى
marsa
deadman control التحكم في المرسى
at-taḥakkum fil-marsa
deadweight tonnage وزن الحمل الساكن بالطن
wazn al-ḥiml as-sākin biṭ-ṭann
death watch beetle خنفساء الخشب
ḵunfuṣā' al-ḵašab
debris dam سد أنقاض
sadd anqāḍ
decanting صب . تصفية
ṣabb, taṣfiya

D

deceleration lane — ممر إبطاء السرعة
mamarr ibṭā' as-sur'a
decentralised — لا مركزي
lā-markazī
decentralised handling — مناولة لا مركزية
munāwala lā-markazīya
deck — أرضية صب الخرسانة
arḍīyat ṣabb al-karsāna
deck bridge — جسر سطحي
jisr saṭḥī
deck framing — هيكل السطح
haikal as-saṭḥ
deck-plate girder — عارضة ألواح السطح
'āriḍat alwāḥ as-saṭḥ
decommissioning — ايقاف العمل
īqāf al-'amal
decompression — تخفيف الضغط
takfīf aḍ-ḍağṭ
decontamination unit — وحدة إزالة التلوث
waḥdat izālat at-talāwuṯ
decoration — زخرفة
zakrafa
deep basement — طابق سفلي عميق
ṭābiq suflī 'amīq
deep beam — ركيزة عميقة
rakīza 'amīqa
deep blasting — تفجير عميق
tafjīr 'amīq
deep compaction — تدميج عميق
tadmīj 'amīq
deep foundations — أساسات عميقة
asāsāt 'amīqa
deep manhole — فتحة تفتيش عميقة
fatḥat taftīš 'amīqa
deep-penetration electrodes — إلكترود لحام عميق الصهر
iliktrūd liḥām 'amiq aṣ-ṣahr
deep vibration — إهتزاز عميق
ihtizāz 'amīq
deep well — بئر عميقة
bi'r 'amīqa
deep well pump — مضخة البئر العميقة
miḍakkat al-bi'r al-'amīqa
default of contract — إخلال بالعقد
iklāl bil-'aqd
defective tools — أدوات عاطلة
adawāt 'āṭila
defects — خلل
kalal
deflection — إنحراف
inḥirāf
deflection of supports — إنحراف الدعائم
inḥirāf ad-da'ā'im
deflectometer — مقياس الانحراف
miqyās al-inḥirāf
deformation — تشوّه
tašawwuh
deformed bars — قضبان مشوّهة
quḍbān mušawwaha
degree — درجة
daraja
degree of saturation — درجة التشبع
darajat at-tašabbu'
delay-action detonator — مفجر متأخر الفعل
mufajjir muta'akkir al-fi'l
deliberate collapse — تقوض متعمد
taqawwuḍ muta'ammad
delivery — تصريف
taṣrīf
delivery hose — خرطوم التصريف
kurṭūm at-taṣrīf
delivery pipe — أنبوب التصريف
unbūb at-taṣrīf
de-mineralized water — ماء مزال المعدن
mā' muzāl al-ma'adan
demolition — هدم
hadm
dense concrete — خرسانة غليظة
karasāna ğalīẓa
dense tar surfacing — رصف بالقار الغليظ
raṣf bil-qār al-ğalīẓ
density — كثافة
kaṯāfa
dentated sill — عتبة محززة
'ataba muḥazzaza
department store — متجر كبير
matjar kabīr
departure yard — ساحة المغادرة
sāḥat al-muğādara
depth of foundation — عمق الأساس
'amq al-asās
derailment — خروج عن الخط
kurūj 'an al-kaṭṭ
derrick — برج الحفر
burj al-ḥafr
derrick crane — مرفاع برج الحفر
mirfā' burj al-ḥafr
derrick, guy — برج مدعم بالحبال
burj mud'am bil-ḥibāl

derrick, oil-well برج حفر آبار الزيت
burj ḥafr ābār az-zait
derrick, standing برج منصوب
burj manṣūb
derrick stone حجر البرج
ḥajar al-burj
derrick, three legged برج ثلاثي القوائم
burj ṯulāṯī al-qawā'im
derrick tower gantry حامل قنطري لبرج الحفر
ḥāmil qanṭarī li-burj al-ḥafr
desalination إزالة الملوحة
izālat al-mulūḥa
desalination plant محطة إزالة الملوحة
muḥaṭṭat izālat al-mulūḥa
descriptive specification مواصفات تصويرية
muwāṣafāt taṣwīrīya
design تصميم
taṣmīm
design agency مكتب تصميم
maktab taṣmīm
design capacity إمكانية التصميم
imkānīyat at-taṣmīm
design criteria معيار التصميم
mi'yār at-taṣmīm
designed mix خلط مصمم
ḵalṭ muṣammam
designer مصمم
muṣammim
design lanes تصميم الطرقات
taṣmīm aṭ-ṭuruqāt
design load الحمل المصمم
al-ḥiml al-musammam
design team فريق التصميم
farīq at-taṣmīm
desired value القيمة المفضلة
al-qīma al-mufaḍḍala
detachable bit لقمة حفر يمكن فصلها
luqmat ḥafr yumkin faṣluha
detail drawing رسم تفصيلي
rasm tafṣīlī
detailed design تصميم مفصل
taṣmīm mufaṣṣal
detailed requirements متطلبات مفصلة
mutaṭallabāt mufaṣṣala
detecting element عنصر كشف
'unṣur kašf
detection device وسيلة كشف
wasīlat kašf
detonating fuse صمامة التفجير
ṣimmāmat at-tafjīr
detonation تفجير
tafjīr
detonator مُفجر
mufajjir
detritus settlement ترسب الحصى
tarassub al-ḥaṣā
detritus slide إنزلاق بطيء
inzilāq baṭī'
development pattern نمط التنمية
namaṭ at-tanmiya
development plan خطة التنمية
ḵuṭṭat at-tanmiya
deviation إنحراف
inḥirāf
dewatering نزح الماء
nazḥ al-mā'
dewpoint نقطة الندى
nuqṭat an-nadā
diagonal brace شكال مائل
šakkal mā'il
diagonally braced frame tower برج هيكلي مقوى بشكالات مائلة
burj haikalī muqawwa bi-šakkālāt mā'ila
diagrid floor أرضية متشابكة
arḍīya mutašābika
dial gauge مقياس مدرج
miqyās mudarraj
diamond ألماسة
al-māsa
diamond crossing تقاطع معين الشكل
taqāṭu' ma'īn aš-šakl
diamond drilling حفر بلقم ألماسية
ḥafr bi-luqam māsīya
diamond interchange تقاطع بشكل معين
taqāṭu' bi-šakl ma'īn
diamond saw منشار دائري لقطع الأحجار
minšār dā'irī li-qaṭ' al-aḥjār
diaphragm pump مضخة ذات رق
miḍaḵḵa ḏāt riqq
diaphragm wall جدار رقي
jidār riqqī
die لقمة لولبة
luqmat laulaba
die-formed strand سلك مجدول بلقمة لولبة
silk majdūl bi-luqmat laulaba
dielectric heating تسخين العازل
tasḵīn al-'āzil

diesel-electric — ديزل كهربائي
dīzl kahrabā'ī
diesel-electric locomotive — قاطرة ديزل كهربائية
qāṭira dīzl kahrabā'īya
diesel engine — محرك ديزل
muḥarrik dīzl
diesel hammer — مطرقة بمحرك ديزل
miṭraqat muḥarrik dīzl
diesel oil — زيت الديزل
zait ad-dīzl
differential cylinder — أسطوانة تفاضلية
isṭiwāna tafāḍulīya
differential levelling — تسوية تفاضلية
taswiya tafāḍulīya
differential pulley — بكارة تفاضلية
bakara tafāḍulīya
differential pulley block — مجموعة بكرات تفاضلية
majmū'at bakarāt tafāḍulīya
differential settlement — هبوط متفاوت
hubūṭ mutafāwit
diffused-air system — نظام هواء منتشر
niẓām hawā' muntaššir
diffuser — ناشرة . رذاذة
nāšira, raḏāḏa
diffusion — إنتشار
intišār
digestion — هضم
haḍm
digging face — سطح الحفر
saṭḥ al-ḥafr
digital computer — كمبيوتر رقمي
kumbyūtar raqmī
digital sequential data — بيانات تتابعية رقمية
bayānāt tatābu'īya raqmīya
digital-to-analogue conversion — تحويل الأرقام إلى نسب
taḥwīl al-arqām ila nisab
dimension — بُعد
bu'd
dimensional analysis — تحليل الأبعاد
taḥlīl al-ab'ād
dining room — غرفة الطعام
ġurfat aṭ-ṭa'ām
diode transistor logic — دائرة منطقية
dā'ira manṭiqīya
dip needle — بوصلة الميل المغنطيسي
būṣalat al-mail al-muġnaṭīsī
dipper — مغرفة
miġrafa
direct arc furnace — فرن بالقوس المباشر
furn bil-qaus al-mubāšir
direct cooling — تبريد مباشر
tabrīd mubāšir
direct incineration — حرق مباشر
ḥarq mubāšir
directional drilling — حفر إتجاهي
ḥafr ittijāhī
direct labour — عمل مباشر
'amal mubāšir
direct lighting — إضاءة مباشرة
iḍā'a mubāšir
direct memory access interface — وحدة دخول مباشر على الذاكرة
waḥdat duḵūl mubāšir 'ala ad-ḏākira
directors suite — جناح المدراء
janāḥ al-mudarā'
directory — دليل
dalīl
direct rail fastening — ربط مباشر للقضبان
rabṭ mubāšir lil-quḍbān
direct stress — إجهاد تضاغطي مستقيم
ijhād taḍāġuṭī mustaqīm
disassembly — تفكيك . تفكك
tafkīk, tafakkuk
disc — قرص
qurṣ
disc cutter — قطاعة قرصية
qaṭṭā'a qurṣīya
discharge — تصريف
taṣrīf
discharge nozzle — فوهة تصريف
fauhat taṣrīf
discharge regulator — منظم تصريف
munaẓẓim taṣrīf
discharge valve — صمام تصريف
ṣimām taṣrīf
disclaimer — متخلي
mutaḵallī
discounted cash flow — سيولة نقدية مخصومة
suyūla naqdīya maḵṣūma
disinfection — تطهير
taṭhīr
dispersing agent — عامل تشتيت
'āmil taštīt
dispersion — تشتيت
taštīt
displacement light — ضوء الازاحة
ḍau' al-izāḥa

displacement pile ركيزة إزاحية
rakīza izāḥīya
displacement pump مضخة إزاحية
miḍakka izāḥīya
display عرض
ʿarḍ
disposal well بئر التصريف
biʾr at-taṣrīf
distomat آلة قياس أطوال الكترونية
ālat qiyās aṭwāl iliktrūnīya
distributed load حمل موزع
ḥiml muwazzaʿ
distribution توزيع
tauzīʿ
distribution bar قضيب التوزيع
qaḍīb at-tauzīʿ
distribution curve منحنى التوزيع
munḥana at-tauzīʿ
distribution network شبكة توزيع
šabakat tauzīʿ
distribution of loading توزيع الحمل
tauzīʿ al-ḥiml
distribution steel فولاذ توزيع الحمل
fūlāḏ tauzīʿ al-ḥiml
ditch خندق . حفرة
ḵandaq, ḥufra
ditcher حفارة خنادق
ḥaffārat ḵanādiq
diurnal tides مد أو جزر نهاري
madd au jazr nahārī
diver غواص
ḡawwāṣ
diverging section قطاع إنحراف
qiṭāʿ inḥirāf
dividend حصة
ḥuṣṣa
diving bell ناقوس غوص
nāqūs ḡauṣ
diving cap لباس الغوص
libās al-ḡauṣ
dock مرفأ
marfaʾ
dock design تصميم المرفأ
taṣmīm al-marfaʾ
dock fenders مصدات السفن بالمرفأ
miṣaddāt as-sufun bil-marfaʾ
docking blocks كتل الارساء
kutal al-irsāʾ
docking of ships إرساء السفينة بالمرفأ
irsāʾ as-safīna bil-marfaʾ
dog كلّاب
kallāb
dolly منصة صغيرة نقالة
minaṣṣa ṣaḡīra naqqāla
dolomite دولوميت
dulūmīt
dolphin دلفين
dalfīn
dome قبة
qubba
domestic consumption الاستهلاك المحلي
al-istihlāk al-maḥallī
domestic sewage مياه المجاري المنزلية
mīyah al-majārī al-manzilīya
domestic wastewater مياه بواليع منزلية
mīyah bawālīʿ manzilīya
door باب
bāb
doorhandle مقبض الباب
miqbaḍ al-bāb
dosing siphon سيفون الجرعات الكيماوية
sīfūn al-jurʿāt al-kīmāwīya
dosing tank خزان الجرعات الكيماوية
ḵazzān al-jurʿāt al-kīmāwīya
double-acting مزدوج الفعل
muzdawij al-fiʿl
double-acting hammer مطرقة مزدوجة الفعل
miṭraqa muzdawijat al-fiʿl
double air valve صمام هواء مزدوج
ṣimām hawāʾ muzdawij
double door باب مزدوج
bāb muzdawij
double duct مجرى مزدوج
majrā muzdawij
double duct all-air system نظام هوائي كامل مزدوج المجرى
niẓām hawāʾī kāmil muzdawij al-majrā
double filtration ترشيح مزدوج
taršiḥ muzdawij
double-hung window نافذة مزدوجة المفصلة
nāfiḏa muzdawijat al-mifṣala
double-layer grid شبكة ثنائية الاتجاه
šabaka ṯunāʾīyat al-ittijāh
double lock هويس مزدوج
huwais muzdawij

D

D

double-seal manhole cover غطاء فتحة تفتيش مزدوج
ġiṭā' fatḥat taftīš muzdawij
double-track مسار مزدوج
masār muzdawij
double-track railway خط حديد مزدوج السكة
ḵaṭṭ ḥadīd muzdawij as-sikka
double-wall cofferdam سد إنضاب بجدار مزدوج
sadd inḍāb bi-jidār muzdawij
dowel bar قضيب دسر
qaḍīb dasir
down-line loading تحويل المعلومات
taḥwīl al-ma'lūmāt
down-the-hole drill حفارة الحفر العميقة
ḥaffārat al-ḥufar al-'amīqa
dozer بولدوزر
būldūzar
draft of a ship غاطس السفينة
ġāṭis as-safīna
draft tube أنبوب تصريف
unbūb taṣrīf
drag جر
jarr
drag forces قوى المقاومة
qiwa al-muqāwama
draghead كراءة مقطورة
karrā'a maqṭūra
dragline كبل السحب
kabl as-saḥb
dragline bucket قادوس كبل السحب
qādūs kabl as-saḥb
dragline scraper مكشطة ذات كبل
mikšaṭa ḏāt kabl
drain صرف . نزح
ṣarf, nazḥ
drainage تجفيف . تصريف
tajfīf, taṣrīf
drainage area مستجمع الصرف
mustajma' aṣ-ṣarf
drainage basin حوض الصرف
ḥauḍ aṣ-ṣarf
drainage channel قناة صرف
qanāt ṣarf
drainage excavation حفر قناة الصرف
ḥafr qanāt aṣ-ṣarf
drainage gallery مجرى قناة الصرف
majrā qanāt aṣ-ṣarf
drainage inlet منفذ التصريف
manfaḏ at-taṣrīf
drainage opening فتحة التصريف
fatḥat at-taṣrīf
drainage plant وحدة التصريف
waḥdat at-taṣrīf
drainage pump مضخة التصريف
miḍakkat at-taṣrīf
drainage system نظام التصريف
niẓām at-taṣrīf
drainage tunnel نفق التصريف
nafaq at-taṣrīf
drainpipe مواسير تصريف
mawāsīr taṣrīf
drain plug سدادة تفريغ
sidādat tafrīġ
drain rods قضبان التسليك للمجاري
quḍbān at-taslīk lil-majārī
drain tile صرف زراعي
ṣarf zirā'ī
drain trench خندق الصرف
ḵandaq aṣ-ṣarf
drain valve صمام تصريف
ṣimām taṣrīf
drain well بئر تصريف
bi'r taṣrīf
draped rubber fender مصد مطاطي متدلي
miṣadd maṭāṭī mutadallī
draughtsman رسام
rassām
draught tube ماسورة سحب
māsūrat saḥb
drawbar قضيب جر
qaḍīb jarr
drawbar pull قوة الجر
quwwat al-jarr
drawbridge جسر متحرك
jisr mutaḥarrik
draw cut قطع سفلي
qaṭ' suflī
draw-door weir سد ببوابات تفتح رأسياً
sadd bi-bawwābāt tuftaḥ ra'sīyan
drawdown خفض منسوب الماء
ḵafḍ mansūb al-mā'
draw-file مبرد مستعرض
mibrad musta'raḍ
drawing board لوحة رسم
lauḥat rasm

draw-off culvert مجرى سحب سفلي
majrā saḥb suflī
dredged bottom قاع مجروف
qāʿ majrūf
dredge excavation حفر بالكراءة
ḥafr bil-karrāʾa
dredge pipeline خط أنابيب الحفر
ḵaṭṭ anābīb al-ḥafr
dredger جرافة
jarrāfa
dredging الحفر بالجرف
al-ḥafr bil-jarf
dredging plant وحدة الكراءة
waḥdat al-karrāʾa
dressed stone حجر مسوّى
hajar musawwā
drifting sand terrain تضاريس رملية متنقلة
taḍāris ramlīya mutanaqqila
drill ثقب . مثقاب
ṯuqb, miṯqāb
drill barge صندل حفر
ṣandal ḥafr
drilled shaft بئر محفور
biʾr maḥfūr
drilling fluid سائل الحفر
sāʾil al-ḥafr
drilling rate معدل الحفر
muʿaddal al-ḥafr
drinking trough حوض ماء الشرب
ḥauḍ māʾ aš-šurb
drinking water ماء الشرب
māʾ aš-šurb
drip irrigation ريّ بالتقطر
rayy bit-taqaṭṭur
drive head رأس حفر
raʾs ḥafr
driven cast-in-place pile ركيزة خرسانية مصبوبة بالموقع
rakīza ḵarsānīya maṣbūba bil-mauqiʿ
driven pile ركيزة مدقوقة
rakīza madqūqa
drive pipe ماسورة حفر
māsūrat ḥafr
driver سائق
sāʾiq
driver's cabin مقصورة السائق
maqṣūrat as-sāʾiq
driving band طوق الدفع
ṭauq ad-dafʿ
drop-bottom bucket قادوس بفتحة سفلية
qādūs bi-fatḥa suflīya
drop hammer مطرقة ساقطة آلية
miṭraqa sāqiṭa ālīya
drum برميل . أسطوانة
barmīl, isṭiwāna
drum gate بوابة قناة دائرية
bawwābat qanāt dāʾirīya
drum screen مصفاة أسطوانية
miṣfāt isṭiwānīya
drums method طريقة الرفع بالأسطوانة
ṭarīqat ar-rafʿ bil-isṭiwāna
dry brook جدول جاف
jadwal jāff
dry chemical extinguisher مطفأة بالكيماويات الجافة
miṭfāʾa bil-kīmāwīyāt al-jāffa
dry dock حوض جاف للسفن
ḥauḍ jāff lis-sufun
drying room غرفة تجفيف
ġurfat tajfīf
dry joint وصلة جافة
waṣla jāffa
dry lean concrete خرسانة رقيقة جافة
ḵarasāna raqīqa jāffa
dry pack إسمنت حشو جاف
ismant ḥašū jāff
dry-packed concrete خرسانة ردم جافة
ḵarasānat radm jāffa
dual carriageway طريق مزدوج للسيارات
ṭarīq muzdawij lis-sayyārāt
dual-conduit ماسورة مزدوجة
māsūra muzdawija
duct مسلك . مجرى
maslak, majrā
ductile cast iron حديد زهر طروق
ḥadīd zahr ṭarūq
ductile iron حديد مطروق مرن
ḥadīd maṭrūq marin
ductility قابلية الليونة
qābilīyat al-liyūna
ductube أنبوب منفوخ
unbūb manfūḵ
dumb barge نقالة مقطورة
naqqāla maqṭūra
dumbwaiter مصعد صغير بين طابقين
maṣʿad ṣaġīr baina ṭabiqain
dummy argument خلاف كاذب
ḵilāf kāḏib

D

dummy joint وصلة تقلص
waṣlat taqalluṣ
dump تفريغ البيانات
tafrīg al-bayānāt
dump barge نقالة قلاّبة
naqqāla qallāba
dumper قلاّبة
qallāba
duo-rail system نظام مزدوج السكة
niẓām muzdawij as-sikka
duration curve منحنى الدوام
munḥana ad-dawām
dusting إزالة الغبار
izālat al-ġubār
dust respirators جهاز تنفس اصطناعي
jihāz tanaffus iṣtinā'i
dynamic load حمل دينامي
ḥiml dīnāmī
dynamic loading تحميل دينامي
taḥmīl dīnāmī
dynamic pressure ضغط دينامي
ḍaġṭ dīnāmī
dynamite ديناميت
dīnāmīt
dynamite cap قلنسوة الديناميت
qulunsuwat ad-dīnāmīt
earth bank جسر ترابي
jisr turābī
earth borer جهاز حفر
jihāz ḥafr
earthen dam سد ترابي
sadd turābī
earthen embankment سد دعم ترابي
sadd da'm turābī
earthenware pipe أنبوب فخار
unbūb fukkār
earth excavation حفر التربة
ḥafr at-turba
earthing equipment معدات تأريض
mu'iddāt ta'rīḍ
earthing of equipment تأريض المعدات
ta'rīd al-mu'iddāt
earth-leakage protection وقاية من التسرب الأرضي
wiqāya min at-tasarrub al-arḍī
earth-moving plant معدات حفر ونقل التربة
mu'iddat ḥafr wa-naql at-turba
earth pressure ضغط التربة
ḍaġṭ at-turba
earth pressure at rest ضغط التربة الثابت
ḍaġṭ at-turba aṯ-ṯābit
earthquake زلزال
zilzāl
earthquake force قوة الزلزال
quwwat az-zilzāl
earth road طريق ترابي
ṭarīq turābī
earthwire spacing مباعدة سلك التأريض
muba'adat silk at-ta'rīḍ
earthwork الأعمال الترابية
al- a'māl at-turābīya
earthworks design standards مواصفات التصميم للأعمال الترابية
muwāṣafāt at-taṣmīm lil-a'māl at-turābīya
easting إتجاه نحو الشرق
ittijāh naḥwa aš-šarq
eaves إفريز السطح المائل
ifrīz, as-saṭḥ al-mā'il
ebb channel قناة الجزر
qanāt al-jazr
eccentric لا تمركزي
lā-tamarkuzī
eccentric load حمل لا تمركزي
ḥiml lā-tamarkuzī
eccentric loading تحميل لا تمركزي
taḥmīl lā-tamarkuzī
echo sounder مسبار بالصدى
misbār biṣ-ṣada
economic aspects النواحي الاقتصادية
an-nawāḥī al-iqtiṣādīya
economic ratio النسبة الاقتصادية
an-nisba al-iqtiṣādīya
economics علم الاقتصاد
'ilm al-iqtiṣād
economic viability الجدوى الاقتصادية
al-jadwa al-iqtiṣādīya
economist عالم في الاقتصاد
'ālim fil-iqtiṣāḍ
economizer مقتصد
muqtaṣid
eddy loss فقد دوامي
fuqd dawwāmī
edge marking تعليم الحافة
ta'līm al-ḥāffa
edge preparation إعداد الحافة
i'dād al-ḥāffa
effective depth العمق الفعال
al-'umq al-fa''āl

Effective Perceived Noise Decibels — الضوضاء الملحوظة الفعالة بالديسبل
aḍ-ḍauḍā' al-malḥūẓa al-fa''āla bid-disibl
effective span — الباع الفعال
al-bā' al-fa''āl
effective stress — الإجهاد الفعال
al-ijhād al-fa''āl
effects of radiation — تأثيرات الاشعاع
ta'ṯīrāt al-iš'ā'
effluent — نفاية
nifāya
effluent stream — جدول فرعي للنفايات
jadwal far'ī lin-nifayāt
egg-shaped sewer — مجرور بيضاوي الشكل
majrūr baiḍāwī aš-šakl
eight — ثمانية
ṯamānya
eighteen — ثمانية عشر
ṯamānya 'ašar
eighty — ثمانون
ṯamānūn
ejector — طارد
ṭārid
elastic — مرن . مطاط
marin
elastic curve — منحنى المرونة
munḥana al-murūna
elastic design — تصميم المرونة
taṣmīm al-murūna
elastic limit — حد المرونة
ḥadd al-murūna
elastic modulus — مُعامل المرونة
mu'āmil al-murūna
elastic strain — تشوّه مرن
tašawwuh marin
elastomeric bearing — تحميل مرن
taḥmīl marin
elbow — مرفق . كوع
mirfaq, kū'
electric — كهربائي
kahrabā'ī
electrical engineering work — أعمال الهندسة الكهربائية
a'māl al-handasa al-kahrabā'īya
electrical fire — مدفأة كهربائية
midfa'a kahrabā'īya
electrical installation — تركيبات كهربائية
tarkībat kahrabā'iya
electrically insulated gloves — قفازات معزولة كهربائياً
qaffāzāt ma'zūla kahrabā'īyan
electrical power transmission — نقل الطاقة الكهربائية
naql aṭ-ṭāqa al-kahrabā'īya
electrical property — الخاصة الكهربائية
al-ḵāṣṣa al-kahrabā'īya
electrical switchgear — مجموعة المفاتيح الكهربائية
majmū'at al-mafātiḥ al-kahrabā'īya
electric-arc welding — لحام بالقوس الكهربائي
liḥām bil-qaus al-kahrabā'ī
electric cable — كبل كهربائي
kabl kahrabā'ī
electric drill — مثقاب كهربائي
miṯqāb kahrabā'ī
electric eye — عين كهربائية
'ain kahrabā'ī
electric furnace — فرن كهربائي
firn kahrabā'ī
electricity — كهرباء
kahrabā'
electricity supply regulations — أنظمة إمداد الكهرباء
unẓimat imdād al-kahrabā'
electric lift — مصعد كهربائي
mis'ad kahrabā'ī
electric lighting — إضاءة كهربائية
iḍā'a kahrabā'īya
electric locomotive — قاطرة كهربائية
qāṭira kahrabā'īya
electric motor — محرك كهربائي
muḥarrik kahrabā'ī
electric power — طاقة كهربائية
ṭāqa kahrabā'īya
electric shock — صدمة كهربائية
ṣadma kahrabā'īya
electric tools — أدوات كهربائية
adawāt kahrabā'īya
electric traction — الجر بالقوة الكهربائية
al-jarr bil-quwwa al-kahrabā'īya
electric welding — لحام بالكهرباء
liḥām bil-kahrabā'
electrode — إلكترود . قطب
iliktrūd, quṭb
electrode stub — قاعدة الالكترود
qā'idat al-iliktrūd

E

E

electro-dialysis	*farz intišārī kahrabā'ī*	فرز إنتشاري كهربائي
electro-hydraulic lift	*mis'ad hīdrūlī kahrabā'ī*	مصعد هيدرولي كهربائي
electrolysis	*taḥlīl kahrabā'ī*	تحليل كهربائي
electrolyte	*iliktrūlīt*	إلكتروليت
electrolytic corrosion	*ta'ākul iliktrūlītī*	تأكل إلكتروليتي
electromagnet	*muġnaṭīsī kahrabā'ī*	مغنطيسي كهربائي
electronic amplifier	*muḍakkim iliktrūnī*	مضخم إلكتروني
electronic analyser	*muḥallila iliktrūnīya*	محللة إلكترونية
electronic components	*ajzā' iliktrūnīya*	أجزاء إلكترونية
electronic distance measurement	*qiyās al-masāfa al-iliktrūnīya*	قياس المسافة الالكترونية
electro-osmosis	*at-tanāḍuḥ al-kahrabā'ī*	التناضح الكهربائي
electroplating	*taġlīf kahrabā'ī*	تغليف كهربائي
electroslag welding	*liḥām kahrabā'ī mustamirr*	لحام كهربائي مستمر
elevated railway	*sikkat ḥadīd murtafi'a*	سكة حديدية مرتفعة
elevated track	*masār murtafi'*	مسار مرتفع
elevating grader	*mumahhida rāfi'a*	ممهدة رافعة
elevating plant	*mu'iddāt raf'*	معدات رفع
elevation	*irtifā', masqaṭ ra'sī*	إرتفاع . مسقط رأسي
elevation of curve	*irtifā' al-munhana*	إرتفاع المنحنى
elevator	*maṣ'd*	مصعد
eleven	*aḥada'ašar*	أحد عشر
elongation	*istiṭāla*	إستطالة
embanked road	*ṭarīq bi-ḥājiz turābī*	طريق بحاجز ترابي
embankment	*ḥājiz turābī*	حاجز ترابي
embankment design	*taṣmīm al-ḥājiz at-turābī*	تصميم الحاجز الترابي
embankment wall	*jidār al-ḥājiz at-turābī*	جدار الحاجز الترابي
emergency call	*nidā' ṭāri'*	نداء طارئ
emergency door	*bāb liṭ-ṭawāri'*	باب للطوارئ
emergency generator	*muwallid liṭ-ṭawāri'*	مولد للطوارئ
emergency tender	*'aṭā' ṭāri'*	عطاء طارئ
emergency water strainer	*miṣfāt miyāh liṭ-ṭawāri'*	مصفاة مياه للطوارئ
emery	*sanfara*	سنفرة
emitter coupled logic	*dā'ira manṭiqīya maqrūnat al-bā'iṯ*	دائرة منطقية مقرونة الباعث
empirical formula	*ṣīġa tajrībīya*	صيغة تجريبية
employee training	*tadrīb al-muwaẓẓafīn*	تدريب الموظفين
emulsifier	*'āmil istiḥlāb*	عامل إستحلاب
emulsion	*mustaḥlab*	مستحلب
emulsion injection	*ḥaqn al-mustaḥlab*	حقن المستحلب
encroachment	*ta'addī*	تعدي
end-bearing pile	*rakīzat taḥmīl ṭarafī*	ركيزة تحميل طرفي
end bearings	*maḥāmil ṭarafīya*	محامل طرفية
end block	*kutla karasānīya ṭarafīya*	كتلة خرسانية طرفية
end contraction	*taqalluṣ ṭarafī*	تقلص طرفي
end-fixed	*muṯabbatat aṭ-ṭarf*	مثبتة الطرف
end frogs	*'atabāt ṭarafīya*	عتبات طرفية
end gable	*jamlūn ṭarafī*	جُملون طرفي
end span	*bā' ṭarafī*	باع طرفي
end thrust	*daf' ṭarafī*	دفع طرفي

E

endurance limit حد التحمل
ḥadd at-taḥammul
energy طاقة
ṭāqa
energy dissipation تبديد الطاقة
tabdīd aṭ-ṭāqa
engine محرك
muharrik
engineer مهندس
muhandis
engineering geology الجيولوجيا التطبيقية
al-jiyūlūjīya at-taṭbīqīya
engineering staff موظفي الهندسة
muwaẓẓafī al-handasa
engineer's chain سلسلة المهندس
silsilat al-muhandis
engineer's level مسواة المهندس
musawwāt al-muhandis
engine house مبنى المحركات
mabnā al-muḥarrikāt
enlargement تكبير
takbīr
enrockment فرش بالصخور
farš biṣ-ṣuk̲ūr
entrance مدخل
madk̲al
entrance gates بوابات الدخول
bawwābāt ad-duk̲ūl
entrance hall بهو الدخول
bahū ad-duk̲ūl
entrance lock هويس دخول
huwais duk̲ūl
environmental aspects النواحي البيئية
an-nawāḥī al-bī'īya
environmental engineering هندسة البيئة
handasat al-bī'a
environmental impact تأثير البيئة
ta'ṯīr al-bī'a
environmental study دراسة البيئة
dirāsat al-bī'a
epoxy resin راتينج الإيبوكسي
ratīnaj al-ībūksī
equalization of boundaries معادلة الحدود
mu'ādalat al-ḥudūd
equalizing bed فرشة تسوية
faršat taswiya
equilibrium توازن
tawāzun
equipment-storage building مخزن المعدات
mak̲zan al-mu'iddāt
equipotential lines خطوط متساوية الجهد
k̲uṭūṭ mutasāwīyat al-juhd
erecting shop ورشة تركيب
wiršat tarkīb
erection procedure إجراءات التركيب
ijrā'āt at-tarkīb
erector arm ذراع التركيب
ḏirā' at-tarkīb
erosion تحات
taḥāt
error خطأ
k̲aṭṭa
escalator سلم متحرك
sullam mutaḥarrik
escalator capacity إستطاعة السلم المتحرك
istiṭā'at as-sullam al-mutaḥarrik
escalator speed سرعة السلم المتحرك
sur'at as-sullam al-mutaḥarrik
escalator width عرض السلم المتحرك
'arḍ as-sullam al-mutaḥarrik
escape هرب
harab
Escherichia coli بكتيريا عضوية الشكل
baktīrīya 'uḍwiyat aš-šakl
estimating draughtsman رسام مثمّن
rassām muṯammin
estimation تقدير. تخمين
taqdīr, tak̲mīn
estimation of traffic تقدير حركة المرور
taqdīr ḥarakat al-murūr
estimator مثمّن. مقيم
muṯammin, muqayyim
ethane إيثان
īṯān
Euler crippling stress الحمل الحرج
al-ḥiml al-ḥarij
evaluation of bids تقييم العطاءات
taqyīm al-'aṭa'āt
evaporative condenser مكثف تبخري
mukaṯṯif tabak̲urī
excavating cableway مصعد حفر كبلي
maṣ'ad ḥafr kablī
excavating equipment معدات الحفر
mu'iddāt al-ḥafr

excavation حفر. تنقيب
ḥafr, tanqīb
excavation limit حد الحفر
ḥadd al-ḥafr
excavator حفار
ḥaffār
exciter محرض. مثير
muḥarriḍ, muṯīr
execution تنفيذ
tanfīḏ
execution of contract تنفيذ العقد
tanfīḏ al-ʿaqd
execution phase مرحلة التنفيذ
marḥalat at-tanfīḏ
executive office المكتب التنفيذي
al-maktab at-tanfīḏī
exfiltration تسرب للخارج
tasarrub lil-ḵārij
exhaust-air system نظام هواء العادم
niẓām hawāʾ al-ʿādim
exhaust gases غازات العادم
ġāzāt al-ʿādim
exhaust valve صمام العادم
ṣimām al-ʿādim
exhibition hall صالة العرض
ṣālat al-ʿarḍ
exit مخرج
maḵraj
expanded clay صلصال ممدد
ṣalṣāl mumaddad
expanded metal شبك معدني ممدد
šabk maʿdanī mumaddad
expanding cement إسمنت تمددي
ismant tamaddudī
expanding-cement concrete خرسانة إسمنت تمددية
ḵarasānat ismant tamaddudīya
expanse of water رقعة ماء
ruqʿat māʾ
expansion bearing محمل تمدد
maḥmal tamaddud
expansion bend مرفق تمدد
mirfaq tamaddud
expansion bolt مسمار تمددي ملولب
mismār tamaddudī mulaulab
expansion joint وصلة تمددية
waṣla tamaddudīya
expansion rollers دحاريج التمدد
daḥārīj at-tamaddud

expansive soil تربة قابلة للتمدد
turba qābila lit-tamaddud
experiment تجربة
tajruba
expert خبير. أخصائي
ḵabīr, aḵiṣṣāʾī
exploration إستكشاف
istikšāf
exploratory boring حفر إستكشافي
ḥafr istikšāfī
explosive compaction دمج بالتفجير العميق
damj bit-tafjīr al-ʿamīq
explosives متفجرات
mutafajjirāt
extended addressing أمر عنونة موسع
amr ʿanwana muwassaʿ
extension of time تطويل المدة
taṭwīl al-mudda
exterior panel لوح خارجي
lauḥ ḵārijī
external fabric نسيج خارجي
nasīj ḵārijī
external scaffolding سقالة خارجية
saqāla ḵārijīya
external vibration إهتزاز خارجي
ihtizāz ḵarijī
external vibrator هزازة خارجية
hazzāza ḵarijīya
external wall جدار خارجي
jidār ḵarijī
extrapolate إستكمل بالاستقراء
istakmala bil-isṯiqrāʾ
extruded sections قطاعات مشكلة بالبثق
qiṭaʿāt mušakkala bil-baṯq
extrusion بروز
burūz
eye-and-face protection وقاية العينين والوجه
wiqāyat al-ʿainain wal-wajh
eye bolt مسمار ذو عروة ملولب
mismār ḏū ʿurwa mulaulab
fabric نسيج. بنية
nasīj, bunya
fabricated mesh شبكة إصطناعية
šabaka iṣṭināʿīya
fabrication تصنيع
taṣnīʿ
fabric reinforcement تسليح بنية الخرسانة
taslīḥ bunyat al-ḵarasāna
fabridam سد ليفي
sadd līfī

facade واجهة
wājiha
face وجه . سطح
wajh, saṭḥ
faced wall جدار مكسو
jidār maksū
face shovel جاروف أمامي
jārūf amāmī
facilities for mooring تسهيلات إرساء
tashīlāt irsā'
facing تلبيس
talbīs
facing bricks طوب التلبيس
ṭaub at-talbīs
facing wall جدار طرف الخندق
jidār ṭaraf al-ḵardaq
factor of safety عامل الأمان
'āmil al-amān
factory precast مسبق الصب بالمصنع
misbaq aṣ-ṣabb bil-maṣna'
factory-sealed double glazing زجاج مزدوج محكم بالمصنع
zujāj muzdawaj muḥakkam bil-maṣna'
fairlead دليل إمرار الحبل
dalīl imrār al-ḥabl
fall سقوط
suqūṭ
fall block بكرة متحركة
bakara mutaḥarrika
falling weir sluice بوابة سد ساقطة
bawwābat sadd sāqiṭa
false leaders صاري دليلي من الفولاذ
sārī dalīlī min al-fūlāḏ
fan مروحة
mirwaḥa
fan control التحكم في المروحة
at-taḥakkum fil-mirwaḥa
fanlight نافذة السطح
nāfiḏat as-saṭḥ
farad فاراد
farād
farm مزرعة
mazra'a
farm road طريق زراعي
ṭarīq zarā'i
fastenings أدوات تثبيت
adawāt taṯbīt
fast road طريق سريع
ṭarīq sarī'
fatal accident حادث مميت
ḥadiṯ mumīt
fathom قامة
qāma
fathometer مقياس أعماق
miqyās a'māq
fatigue كلال
kilāl
fatigue test إختبار الكلال
iḵtibār al-kilāl
fault عطل . خلل
'aṭal, ḵalal
feasibility study دراسة الجدوى
dirāsat al-jadwa
feed تغذية
taḡḏiya
feeder مُغذ
muḡaḏḏi
feedpump مضخة التغذية
miḏakḵat at-taḡḏiya
feedwater ماء التغذية
mā' at-taḡḏiya
felt لُباد
lubbād
fence سياج
siyāj
fencing لوازم التسييج
lawāzim at-tasyīj
fender وقاء . مصد
wiqā' miṣadd
fender pile ركيزة صد
rakīzat ṣadd
ferry معبرة
mi'bara
festoon lighting إضاءة الزينة
iḍā'at az-zīna
fetch phase طور الجلب
ṭaur al-jalb
fibre board لوح ليفي
lauḥ līfī
fibreglass-wrapped pipe أنبوب مغلف بزجاج ليفي
unbūb muḡallaf bi-zujāj līfī
fibre-reinforced concrete خرسانة مسلحة بالألياف
ḵarasāna musallaḥa bil-alyāf
field book سجل قياسات المسّاح
sijill qiyāsat al-massāḥ
field ditch خندق حقلي
ḵandaq ḥaqlī
field drain مصرف حقلي
maṣrif ḥaqlī

F

F

field drainage تجفيف الحقل
tajfīf al-ḥaql
field survey دراسة التضاريس
dirāsat at-taḍārīs
fifteen خمسة عشر
ḵamsata'ašar
fifty خمسون
ḵamsūn
file مبرد . مصنف
mibrad, muṣannif
file structure أسلوب التصنيف
uslūb at-taṣnīf
fill دكة ترابية
dakka turābīya
filled bitumen قار يحتوي على حشوة
qār yaḥtawī 'ala ḥašwa
filler حشوة
ḥašwa
filler joist floor بلاطة خرسانة الحشوة
bilāṭat ḵarasānat al-ḥašwa
fillet weld لحام زاوي
liḥām zāwī
filling تعبئة . حشو
ta'bi'a, hašī
filling-up تعبئة الحفر
ta'bi'at al-ḥufar
filter مرشح
muraššiḥ
filter bed طبقة ترشيح
ṭabaqat taršīḥ
filter blocks كتل ترشيح
kutal ṭaršīḥ
filter layer طبقة ترشيح
ṭabaqat taršīḥ
filter material مادة ترشيح
māddat taršīḥ
filter separator فاصل المرشح
fāṣil al-muraššiḥ
filter zone منطقة ترشيح
minṭaqat taršīḥ
filtrate راشح
rāšiḥ
filtration ترشيح
taršīḥ
filtration plant وحدة ترشيح
waḥdat taršīḥ
final design تصميم نهائي
taṣmīm nihā'ī
final set تصلد الخرسانة النهائي
taṣallud al-ḵarasāna an- nihā'ī
final setting time زمن التصلد النهائي
zaman at-taṣallud an-nihā'ī
financial control مراقبة مالية
murāqaba mālīya
financing تمويل
tamwīl
fine-adjustment screw برغي الضبط الدقيق
burġī aḍ-ḍabṭ ad-daqīq
fine aggregate ركام دقيق
rukām daqīq
fine cold asphalt أسفلت رصف دقيق
isfalt raṣf daqīq
fine-grained soil تربة دقيقة الحبيبات
turba daqīqat al-ḥubaibāt
fines دقائق الخام
daqā'iq al-ḵām
fine screen غربال دقيق الشبيكية
ġirbāl daqīq aš-šubaikīya
finger system نظام الدفق
niẓām ad-dafq
finishing إنهاء . إتمام
inhā', itmām
finishing coat طلية نهائية
ṭaliya nihā'īya
fire alarm إنذار بالحريق
inḏār bil-ḥarīq
fire alarm box صندوق الانذار بالحريق
sandūq al-inḏār bil-ḥarīq
fire-break خندق حائل للحريق
ḵandaq ḥā'il lil-ḥarīq
fire-crash-rescue facilities وسائل إخماد الحريق والانقاذ
wasā'il iḵmād al-ḥarīq
wal inqḍ
fire department دائرة المطافئ
dā'irat al-maṭāfi'
fire engine سيارة إطفاء
sayyārat iṭfā'
fire escape سلم النجاة من الحريق
sullam an-najāt min al-ḥarīq
fire extinguisher مطفأة حريق
miṭfa'at ḥarīq
fire fighting مكافحة الحريق
mukāfaḥat al-ḥarīq
fire hose خرطوم الاطفاء
ḵarṭūm al-iṭfā'
fire-hose coupling قارنة خرطوم الاطفاء
qārinat ḵarṭūm al-iṭfā'

fire hydrant — حنفية مطافئ
ḥanafīyat maṭāfi'
fire mains — المأخذ الرئيسي لمياه الاطفاء
al-ma'kaḏ ar-ra'īsī li-miyah al-iṭfā'
fireman's lift — مصعد رجل المطافئ
maṣ'ad rajul al-muṭāfi'
fireproof cement — إسمنت صامد للنار
ismant ṣāmid lin-nār
fireproof door — باب صامد للنار
bāb ṣāmid lin-nār
fireproof material — مادة صامدة للنار
mādda ṣāmida lin-nār
fire protection — الوقاية من الحريق
al-wiqāya min al-ḥarīq
fire-resistant concrete — خرسانة صامدة للنار
ḵarasāna ṣāmida lin-nār
fire setting — تفتيت التربة بالنار
taftīt at-turba bin-nār
fire stops — مصدات الحريق
miṣaddat al-ḥarīq
firing — إشعال
iš'āl
first aid supplies — لوازم الاسعاف الأولي
lawāzim al-is'āf al-awwalī
first floor (level 2) — الطابق الأول
aṭ-ṭābiq al-awwal
first order levelling — التسوية بالتثليث الأولي
at-taswiya bit-taṯlīṯ al-awwalī
first underlayer — الطبقة السفلية الأولى
aṭ-ṭabaqa as-sufliya al-ūlā
fish passes — معابر السمك
ma'ābir as-samak
fishplate — عارضة وصل
'āriḍat waṣl
fishplated joints — وصلات تراكبية
waṣlāt tarākubīya
fishplate section — قطاع عارضة الوصل
qiṭā' 'ariḍat al-waṣl
fishtail bit — لقمة بشكل ذيل السمكة
luqma bi-šakl ḏail as-samaka
fishtail bolt — برغي بشكل ذيل السمكة
burğī bi-šakl ḏail as-samaka
fist fastening — تثبيت باليد
taṯbīt bilyad
fitness of design — ملاءمة التصميم
mulā'amat at-taṣmīm
fitter — ميكانيكي تجميع
mikānīkī tajmi'
fitting — تركيب
tarkīb
fittings — تجهيزات . تركيبات
tajhīzat, tarkībāt
five — خمسة
ḵamsa
fixed beam — عارضة مثبتة
'āriḍa muṯabbata
fixed bearing — محمل مثبت
maḥmal muṯabbat
fixed bridge — جسر ثابت
jisr ṯābit
fixed cone-sleeve valve — صمام بغلاف مخروطي ثابت
ṣimmām bi-ğilāf maḵrūṭī ṯābit
fixed end — طرف مثبت
ṭaraf muṯabbat
fixed-end beam — عارضة ثابتة الطرف
'āriḍa ṯābitat aṭ-ṭaraf
fixed head disc — قرص مثبت الرأس
qurṣ muṯabbat ar-ra's
fixed mooring berth — مرسى ثابت للسفن
marsa ṯābit lis-sufun
fixed pick — إلتقاط ثابت
iltiqāṭ ṯābit
fixed-price contract — عقد بسعر محدد
'aqd bi-si'r muḥaddad
fixed retaining walls — جدران إحتجاز ثابتة
judrān iḥtijāz ṯābita
fixed-temperature detector — مكشاف ثابت الحرارة
mikšāf ṯābit al-harāra
fixing moment — عزم التثبيت
'azm at-taṯbīt
flame cutting — قطع باللهب
qaṭ' bil-lahab
flame detector — مكشاف لهب
mikšāf lahab
flammable gases — غازات سريعة الالتهاب
ğāzāt sarī'at al-iltihāb
flammable liquids — سوائل سريعة الالتهاب
sawā'il sarī'at al-iltihāb
flange — شفة
šaffa
flanged — مشفّه
mušaffa
flap gate — بوابة قلاّبة
bawwāba qallāba

flap trap مصيدة قلاّبية
miṣyada qallābīya
flap valve صمام لا رجعي قلاّب
ṣimmām lā-rajʿī qallāb
flared column head رأس عمود بوقي
raʾs ʿamūd būqī
flashboard عارضة في جدار السد
ʿāriḍa fi-jidār as-sadd
flashing حشوة معدنية لمنع التسرب
ḥašwa maʿdanīya li-manʿ at-tasarrub
flash welding لحام ومضي
liḥām wamḍī
flat شقة سكنية . مستوي
šuqqa sakanīya, mustawī
flat-bottomed rail قضيب مسطح القعر
qaḍīb musaṭṭaḥ al-qaʿr
flat escalator سلم متحرك مستو
sullam mutaḥarrik mustawī
flat-plate construction إنشاء ذو لوحة مستوية
inšāʾ ḏū lauḥa mustawīya
flat roof سطح مستو
saṭḥ mustawī
flat slab بلاطة مسطحة
bilāṭa musaṭṭaḥa
flat-slab construction إنشاء ذو بلاطة مسطحة
inšāʾ ḏū bilāṭa musaṭṭaḥa
flexible مرن
marin
flexible carpet سجادة مرنة
sijjāda marina
flexible coupling قارنة مرنة
qārina marina
flexible hose خرطوم مرن
karṭūm marin
flexible membrane غشاء مرن
ġišāʾ marin
flexible pavement رصيف مرن
raṣīf marin
flexible pipe أنبوب مرن
unbūb marin
flexible rubber disc قرص مطاط مرن
qurṣ maṭṭāṭ marin
flexible shaft عمود إدارة مرن
ʿamūd idāra marin
flexible surfacing تسطيح مرن
tasṭīḥ marin
flexible wall جدار مرن
jidār marin
flexural strength مقاومة الثني
muqāwamat aṯ-ṯanī
flexure formula صيغة الالتواء
ṣīġat al-iltiwā
flight طيران
ṭayarān
flight of stairs درج السلم
durj as-sullam
flip bucket قادوس قلاّب
qādūs qallāb
float طفا . عام
ṭafāʾ ʿāma
float glass زجاج عائم
zujāj ʿāʾim
floating boom حاجز عائم
ḥājiz ʿāʾim
floating breakwater حاجز أمواج عائم
ḥājiz amwāj ʿāʾim
floating caisson قيسون عائم
qaisūn ʿāʾim
floating control ضبط الطفو
ḍabṭ aṭ-ṭaffū
floating crane مرفاع عائم
mirfaʿ ʿāʾim
floating dock حوض عائم
ḥauḍ ʿāʾim
floating foundation أساسات عائمة
asāsāt ʿāʾima
floating pipeline خط أنابيب عائم
kaṭṭ anābīb ʿāʾim
flocculation تدمج . إندماج
tadammuj, indimāj
flood فيضان
fayaḍān
flood channel قناة الفيضان
qanāt al-fayaḍān
flood irrigation ري بالغمر
rayy bil-ġamr
floor أرضية . طابق
arḍīya, ṭābiq
floor beam عتبة الأرضية
ʿatabat al-arḍīya
floor boards ألواح الأرضية
alwaḥ al-arḍīya
floor plan مخطط الأرضية
mukaṭṭaṭ al-arḍīya
floor slab بلاطة الأرضية
bilāṭat al-arḍīya
floor tiles بلاط الأرضيات
bilāṭ al-arḍīyāt

floppy disk قرص مرن
qurṣ marin
flotation طفو . تعويم
ṭafū, ta'wīm
flotation structure إنشاء عائم
inšā' 'ā'im
flotation thickener مُغلظ طفو
muġliẓ ṭafū
flow control valve صمام التحكم في الدفق
ṣimām at-taḥakkum fid-dafq
flow curve منحنى التدفق
munhana at-tadaffuq
flow index دليل التدفق
dalīl at-tadaffuq
flowing ground أرض التدفق
arḍ at-tadaffuq
flow limiting device وسيلة حدية لمعدل الدفق
wasīla ḥaddīya li-mu'addal ad-dafq
flow lines خطوط التدفق
kuṭūṭ at-tadaffuq
flow meter مقياس التدفق
miqyās at-tadaffuq
flow metering قياس التدفق
miqyās at-tadaffuq
flow net صورة ثنائية الأبعاد لتدفق الماء الجوفي
ṣūra ṯunā'īyat al-ab'ād li-tadaffuq al-mā' al-jaufī
flow pipe ماسورة الدفق
māsūrat ad-dafq
flow regulation تنظيم الدفق
tanẓīm ad-dafq
flue design تصميم مجرى الغازات
taṣmīm majra al-ġazāt
flue gas غاز المداخن
ġāz al-madākin
flue lining تبطين المداخن
tabṭīn al-madākin
fluorescent light إضاءة فلورية
iḍā'a flūrīya
flushing رحض . شطف
raḥḍ, šaṭf
flushing outlet مخرج الرحض
makraj ar-raḥḍ
flush irrigation ري بالرحض
rayy bir-raḥḍ
fluxes مواد صهورة
mawādd ṣahūra
flying buttress كتف زافرة
katif zāfira
flying shores دعائم زافرة
dawā'im zāfira
foamed-slag concrete خرسانة خبث مهواة
karasānat kabṯ mihwāt
foamite extinguisher مطفأة رغوية
miṭfa'a raġawīya
foam tender عربة إطفاء رغوي
'arabat iṭfā' raġawī
folded-plate construction إنشاء ذو سطح مضلع
inšā' ḏū saṭḥ muḍalla'
folding doors أبواب تنطوي
abwāb tanṭawī
follower تابع
tābi'
foot قدم . قاعدة
qadam, qā'ida
foot guard واقي القدم
wāqi al-qadam
footing أساس . قاعدة
asās, qā'ida
footing excavation حفر الأساسات
ḥafr al-asāsāt
footings قواعد . أساسات
qawā'id, asāsāt
footpath ممر المشاة
mamarr al-mušāt
footway طريق للمشاة
ṭarīq lil-mušāt
force قوة
quwwa
forced ventilation تهوية قسرية
tahwiya qasrīya
force mains موصلات القوة الرئيسية
muwaṣṣilāt al-quwwa ar-ra'īsīya
force pump مضخة دافعة
miḍakka dafi'a
ford معبر ضحل
mi'bar ḍaḥil
forecast تنبؤ
tanabbu'
foreman ملاحظ عمال
mulāhiẓ 'ummāl
foreman training تدريب ملاحظ العمال
tadrīb mulāhiẓ al-'ummāl
forepoling تدعيم النفق الأولي
tad'īm an-nafaq al-awwalī
forest غابة
ġāba

F

forge ورشة حدادة
wiršat ḥidāda
forge welding لحام بالحرارة والتطريق
liḥām bil-ḥarāra wat-taṭrīq
forging تشكيل بالحرارة والتطريق
taškīl bil-ḥarāra wat-taṭrīq
fork-lift truck عربة بمرفاع شوكي
'araba bi-mirfā' šaukī
formwork قالب صب الخرسانة
qālab ṣabb al-ḵarasāna
forty أربعون
arba'ūn
foul sewer مجرور الأوساخ
majrūr al-ausāḵ
foul water ماء قذر
mā' qaḏir
foul water sewerage شبكة مجارير المياه القذرة
šabakat majārīr al-miyah al-qaḏira
foundation أساس
asās
foundation block كتلة الأساس
kutlat al-asās
foundation failure تداعي الأساس
tadā'ī al-asās
foundation footing قاعدة الأساس
qā'idat al-asās
foundry مسبك . مصهر
misbak, mishar
fountain نافورة
nāfūra
four أربعة
arba'a
four-leg sling حمالة رباعية الأرجل
hammāla ruba'īyat al-arjul
fourteen أربعة عشر
arba'ata'ašar
fourth floor (level 5) الطابق الرابع
aṭ-ṭābiq ar-rābi'
foyer ردهة
radha
fractional sampling أخذ العينات جزئياً
akḏ al-'ayyināt juz'īyan
frame هيكل . إطار
haikal, iṭār
frame construction إنشاء الهيكل
inšā' al-haikal
framed partition حاجز باطار
ḥājiz bi-iṭār
framed structure إنشاء ذو هيكل
inšā' ḏū haikal
frame saw منشار إطاري
minšār iṭāri
framework هيكل
haikal
framework stiffness صلابة الهيكل
ṣallābat al-haikal
Francis turbine توربين «فرانسز»
tūrbīn fransis
freeboard جزء السفينة الظاهر من الماء
juz' as-safīna aẓ-ẓāhir min al-mā'
free-falling arm ذراع حر السقوط
ḏirā' ḥurr as-suqūṭ
free-nappe profile weir سد خالي من الصخور المغتربة
sadd ḵālī min aṣ-ṣuḵūr al-muġtariba
free-piston compressor مضغط بمكبس حر
miḍġaṭ bi-miqbas ḥurr
free port ميناء حر
mīnā' hurr
free water ماء يسري بثقل الجاذبية
mā' yasrī bi-ṯuql al-jāḏibīya
freeway طريق حرة
ṭarīq hurra
freezing تجمد
tajammud
freight elevator مصعد نقل البضائع
miṣ'ad naql al-baḍā'i'
freight terminal محطة شحن البضائع
muḥaṭṭaṭ šaḥn al-baḍā'i'
freight tonnage الحمولة بالطن
al-ḥumūla biṭ-ṭann
freight wagon عربة الشحن
'arabat aš-šaḥn
frequency curve منحنى التردد
munḥana at-taraddud
frequency diagram مخطط التردد
mukaṭṭat at-taraddud
frequency distribution توزيع التردد
tauzi' at-taraddud
frequency of inspection تواتر التفتيش
tawātur at-taftīš
frequency rate معدل التردد
mu'addal at-taraddud
friction إحتكاك
iḥtikāk
frictional loading تحميل إحتكاكي
taḥmīl iḥtikākī

friction clutch قابض إحتكاكي
qābiḍ iḥtikākī
friction head علو الاحتكاك
'ulū al-iḥtikāk
friction pile ركيزة يدعمها الاحتكاك
rakīza yad'amuha al-iḥtikāk
frog مفرق خطوط حديدية
mafraq kuṭūṭ ḥadīdīya
frog plate لوح المفرق
lauḥ al-mafraq
frontal layout مخطط أمامي
mukaṭṭaṭ amāmī
frost صقيع
ṣaqī'
frost-proof layer طبقة صامدة للصقيع
ṭabaqa ṣāmida liṣ-ṣaqī'
fuel consumption إستهلاك الوقود
istihlāk al-wuqūd
fuel flow line خط تدفق الوقود
kaṭṭ tadaffuq al-wuqūd
fuel installation تركيبات للوقود
tarkībāt lil-wuqūd
fuelling machine مضخة الوقود
miḍakkat al-wuqūd
fuel oil وقود بترولي
waqūd batrulī
fuel storage تخزين الوقود
takzīn al-wuqūd
fuel tank خزان الوقود
kazzān al-wuqūd
full-face tunnelling حفر النفق بالكامل
ḥafr an-nafaq bil-kāmil
full-size drawing رسم بالحجم الكامل
rasm bil-ḥajm al-kāmil
fully-fixed ثابت تماماً
tābit tamāman
functional block diagram رسم تخطيطي للمراحل الوظيفية
rasm takṭīṭī lil-marāḥil al-wazīfīya
fungicidal paint دهان مبيد للفطريات
duhān mubīd lil-fiṭrīyat
fungus فطر
fiṭr
funicular railway سكة حديد شديدة الانحدار
sikkat ḥadīd šadīdat al-inḥidār
furlong فيرلونج
firlūng
furnace فرن . أتون
furn, atūn
furnace insulation عزل الفرن
'azl al-furn
furrow irrigation ري فلاحي أخدودي
rayy fallaḥī ukdūdī
fuse مصهر. فاصمة
miṣhar, fāṣima
fusible plug قابس بمصهر
qābis bi-miṣhar
fusion welding اللحام بالصهر
al-liḥām biṣ-ṣahr
gabion سد صغير مؤقت
sadd ṣaǧīr mu'aqqat
gable جُملون
jumlūn
gale ريح عاصفة
rīḥ 'āṣifa
gallery رواق
riwāq
gallon غالون
ǧalūn
galvanize غلفن
ǧalfana
galvanized iron حديد مغلفن
ḥadīd muǧalfan
galvanized pipe ماسورة مغلفنة
māsūra muǧalfana
galvanized steel ribbon شريط فولاذ مغلفن
šarīṭ fūlāḏ muǧalfan
galvanized woven wire سلك منسوج مغلفن
silk mansūj muǧalfan
gang زمرة
zumra
ganger ناظر
nāẓir
gantry جسر الرافعة المتنقلة
jisr ar-rāfi'a al-mutanaqqila
gantry crane مرفاع قنطري متحرك
mirfā' qanṭarī mutaḥarrik
garage جراج . (كراج)
ǧarāj
garage door باب جراج
bāb ǧarāj
garden حديقة
ḥadīqa
garland drain حوض جمع الماء
hawḍ jam' al-mā'
gas analysis تحليل غازي
taḥlīl ǧāzī
gas circulator مسخن غازي
musakkin ǧāzī

G

G

gas concrete خرسانة خلوية غازية
karasāna kalawīya gāzīya
gas-cooled reactor مفاعل يبرد بالغاز
mufā'il yubarrad bil-gāz
gasket حشية
ḥašya
gas oil زيت الغاز
zait al-gāz
gasometer مستودع غاز
mustauda' gāz
gas-turbine electric تربين غازي كهربائي
turbīn gāzī kahrabā'ī
gas-turbine hydraulic تربين غازي هيدرولي
turbīn gāzī hīdrūlī
gas welding لحام بالغاز
liḥām bil-gāz
gate بوابة
bawwāba
gate chamber غرفة الهويس
gurfat al-hawīs
gate guide دليل بوابة
dalīl bawwāba
gate post قائم بوابة
qā'im bawwāba
gate seal سد بوابة
sadd bawwāba
gate valve صمام بوابي
ṣimām bawwābī
gauge مقياس
miqyās
gauged arch قنطرة معيارية
qanṭara mi'yārīya
gauged mortar ملاط معياري
milāṭ mi'yārī
gauge length طول معياري
ṭūl mi'yārī
gauging قياس . معايرة
qiyās mu'āyara
geared coupling قارن معشق
qārin mu'aššaq
geared machine آلة تدار بمسننات
āla tudār bi-musannanāt
gearless machine آلة بدون مسننات
āla bidūn musannanāt
gear pump مضخة ذات مسننات
miḍakka ḏāt musannanāt
gelatine جيلاتين
jilātīn
gelatine explosives متفجرات جيلاتينية
mutafajjirāt jilātīnīya

gelignite جليغنايت
jiliğnīt
general cargo بضائع عامة
baḍḍā'i' 'āmma
general cargo handling مناولة البضائع العامة
munāwalat al-baḍḍā'i' al-'āmma
general conditions شروط عامة
šurūṭ 'āmma
general contractor مقاول عام
muqāwil 'āmm
general foreman ملاحظ عمال عام
mulāḥiẓ 'ummāl 'āmm
general office مكتب عمومي
maktab 'umūmī
general provision إمدادات عامة
imdādāt 'āmma
general-purpose cement إسمنت للاغراض العامة
ismant lil-ağrāḍ al-'āmma
general-purpose register سجل أغراض عامة
sijill ağrāḍ 'āmma
generating plant محطة توليد
muḥaṭṭat taulīd
generator مولد
muwallid
geodetic mark علامة جيوديسية
'alāma jiyūdisīya
geodetic surveying مسح جيوديسي
masḥ jiyūdīsī
geodimeter آلة قياس الأطوال الالكترونية
ālat qīyās al-aṭwāl al-iliktrūniya
geological map خريطة جيولوجية
karīṭa jiyūlūjīya
geologist عالم جيولوجي
'ālim jiyulūjī
geology جيولوجيا
jiyulūjīya
geophysical investigation بحث جيوفيزيائي
baḥt jiyufiziyā'ī
geophysical surveying مسح جيوفيزيائي
masḥ jiyufiziyā'ī
gin pole قائم المرفاع
qā'im al-mirfā'
girder عارضة . جائز
'āriḍa, jā'iz
girder bridge قنطرة ذات عوارض
qanṭara ḏāt 'awāriḍ
girder design تصميم العارضة
taṣmīm al-'āriḍa

gland جلبة حشو
jalabat ḥašū
gland bolt برغي جلبة الحشو
burġī jalabat al-ḥašū
gland joint وصلة بجلبة
waṣla bi-jalaba
glass زجاج
zujāj
glass block طوب زجاجي
ṭaub zujājī
glass-concrete construction إنشاء خرساني مقوى بالزجاج
inšā' ḵarasānī muqawwa biz-zujāj
glass cutter مقطع زجاج
muqṭa' zujāj
glass fibre ألياف زجاجية
alyāf zujājīya
glassfibre-reinforced concrete خرسانة مقواة بالألياف الزجاجية
ḵarasāna muqawwāt bil-alyāf az-zujājīya
glass tiles بلاط زجاجي
bilāṭ zujājī
glass wool صوف زجاجي
ṣūf zujājī
glazed bricks طوب مزجج
ṭaub muzajjaj
glazed earthenware pipe أنبوب فخار مزجج
unbūb faḵār muzajjaj
glazed stoneware مصنوعات خزفية مزججة
maṣnū'āt ḵazafīya muzajjaja
glazed tiles بلاط مزجج
bilāṭ muzajjaj
glazier مركب الزجاج
murakkab az-zujāj
glazier's putty معجونة الزجاج
ma'jūnat az-zujāj
glazing تركيب الزجاج
tarkīb az-zujāj
glazing bar قضيب تثبيت الزجاج
qaḍīb taṯbīt az-zujāj
globe valve صمام كروي
ṣimām kurawī
gloss paint دهان لماع
dihān lammā'
gloves قفازات
quffāzāt
glued joint وصلة مغرّاة
waṣla muġarrāt
glued-laminated timber خشب رقائقي مغرّى
ḵašab raqā'iqī muġarra
Goliath crane مرفاع نقالي ضخم
mirfā' niqālī ḍaḵm
good site housekeeping تنظيف جيد للموقع
tanẓīf jayyid lil-mauqi'
goods lift مصعد البضائع
miṣ'ad al-baḍā'i'
government agency وكالة حكومية
wakāla ḥukūmīya
Gow caisson قيسون « بوستون »
qaisūn būstūn
grab خطاف
ḵaṭṭāf
grab-dredger كراءة ذات كباش
karrā'a ḏāt kabbāš
grab dredging تطهير بالكباش
taṭhīr bil-kabbāš
grab sampling أخذ العينات العشوائي
akḏ al-'ayyināt al-'ašwā'ī
graded aggregate ركام مصنف
rukām muṣannaf
graded filter مرشح متدرج الطبقات
muraššiḥ mutadarrij aṭ-ṭabaqāt
graded sand رمل مصنف
raml muṣannaf
grade of steel مرتبة الفولاذ
martabat al-fūlāḏ
grader مدرجة
mudarrija
gradient درجة الميل
darajat al-mail
grading تصنيف
taṣnīf
grain إتجاه الألياف في الخشب
ittijah al-alyāf fil-ḵašab
granite غرانيت
ġranīt
granolithic concrete خرسانة غرانوليتية
ḵarasāna ġrānūlītīya
granular filling ردم حبيبي
radm ḥubaibī
granular filter مرشح محبب
muraššiḥ muḥabbab
graph خط بياني
ḵaṭṭ bayānī

G

G

graphics الفنون التخطيطية
al-funūn at-takṭīṭīya
graphite غرافيت
ğrāfīt
grass cutter قاطعة العشب
qāṭi'at al-'ušb
grassed slope منحدر عشبي
munhadar 'ušbī
grating حاجز شبكي
ḥājiz šabakī
grave قبر
qabr
gravel حصى
ḥaṣā
gravel pump مضخة حصى
miḍakka ḥaṣā
graving dock حوض سفن جاف
ḥauḍ sufun jāf
gravitational water ماء يجري بالجاذبية
mā' yajrī bil-jāḏibīya
gravitation drainage تصريف بالجاذبية
taṣrīf bil-jāḏibīya
gravity الجاذبية
al-jāḏibīya
gravity circulation دوران بالجاذبية
dawarān bil-jāḏibīya
gravity dam سد ثقالي
sadd ṯuqālī
gravity-quay wall جدار تحميل بالثقل
jidār taḥmīl biṯ-ṯuql
gravity retaining wall جدار إحتجاز ثقالي
jidār iḥtijāz ṯuqālī
gravity supply إمداد بالجاذبية
imdād bil-jāḏibīya
gravity thickener مُغلظ ثقلي
muğliẓ ṯuqlī
gravity wall جدار ثقالي
jidār ṯuqālī
green centreline lights أضواء خط المركز الخضراء
aḍwā' kaṭṭ al-markaz al-kaḍrā'
grid شبكة
šabaka
grid bearing الزاوية الاتجاهية التسامتية
az-zāwīya al-ittijāhīya at-tasāmutīya
grillage شبيكة
šubaika
grinding طحن . تجليخ
ṭaḥn tajlīk
grip مقبض
miqbaḍ
grit blasting سفع بالحصباء
saf' bil-ḥaṣbā'
grit chamber خزان ترسيب
kazzān tarsīb
gritter آلة فرش الحصباء
ālat farš al-ḥaṣbā'
gritting material مادة فرش الطبقة الحبيبية
māddat farš aṭ-ṭabaqa al-hubaibīya
grommet حلقة . عروة
ḥalaqa, 'urwa
gross aircraft weight وزن الطائرة الاجمالي
wazn aṭ-ṭā'ira al-ijmālī
gross error خطأ كبير
kaṭa' kabīr
gross tonnage الحمولة الاجمالية بالطن
al-ḥumūla al-ijmālīya biṭ-ṭann
ground anchor مثبت أرضي
muṯabbit arḍī
ground beam عتبة أرضية
'ataba arḍīya
ground floor (level 1) الطابق الأرضي
aṭ-ṭābiq al-arḍī
ground movement signs علامات التحركات الأرضية
'alāmāt at-taḥarrukāt al-arḍīya
ground plan مخطط الأساس
mukaṭṭaṭ al-asās
ground survey مسح أرضي
masḥ arḍī
groundwater مياه جوفية
miyāh jaufīya
grout ملاط سائل
milāṭ sā'il
grouted masonry مبنى محقون
mabnā maḥqūn
grouting حقن بالاسمنت
ḥaqn bil-asmant
growth ring حلقة نمو
ḥalaqat numūw
groyne مرطم أمواج
mirṭam amwāj
guarantee period فترة الضمان
fatrat aḍ-ḍamān
guard يحمي
yaḥmī

guardhouse مبنى الحرس
mabnā al-ḥaras
guard lock هويس واق
hawīs wāqī
guard railing خط التحرز
kaṭṭ at-taḥarruz
guard screen حاجب واق
ḥājib wāqī
gudgeon pin مسمار المفصلة
mismār al-mifṣala
guide pile ركيزة دليلية
rakīza dalīlīya
guide rail قضيب دليلي
qaḍīb dalīlī
guide runner عمود دليلي
ʿamūd dalīlī
guide track سكة دليلية
sikka dalīlīya
guideway alignment محاذاة الشق
muḥāḏāt aš-šaqq
gullet أخدود
ukdūd
gully قناة البالوعة
qanāt al-bālūʿa
gully cleansing تنظيف الأخدود
tanẓīf al-ukdūd
gully trap محبس المجرور
miḥbas al-majrūr
gully vacuum tanker عربة تفريغ المجرور
ʿarabat tafrīḡ al-majrūr
guncotton قطن البارود
quṭn al-bārūd
gunite غونيت
ḡūnīt
gunmetal برونز المدافع
brūnz al-madāfiʿ
Gunter's chain سلسلة «غنتر»
silsilat ḡantar
gusset plate لوح تقوية
lauḥ taqwiya
gutter ميزاب . مزراب
mīzāb, mizrāb
gutter bearer حامل المزراب
ḥāmil al-mizrāb
gutter cleansing تنظيف المزراب
tanẓīf al-mizrāb
guy شدادة
šiddāda
guy derrick برج مدعم بالحبال
burj mudʿam bil-ḥibāl
guyed-mast صاري مدعم بالحبال
ṣārī mudʿam bil-ḥibāl
gypsum جبس
jibs
gypsum cement إسمنت جبسي
ismant jibsī
gypsum plaster تجصيص بالجبس
tajṣīṣ bil-jibs
gyratory breaker قاطع دوار
qāṭiʿ dawwār
gyratory crusher كسارة لفافة
kassāra laffāfa
hacksaw منشار معادن
minšār maʿādin
hair crack صدع شعري
ṣadaʿ šaʿrī
half-brick wall جدار بسمك نصف طوبة
jidār bi-sumk niṣf ṭauba
half-joist رافدة نصفية
rāfida niṣfīya
half landing مصطبة درج السلم
masṭabat durj as-sullam
half-lattice girder جملون مثلثي
jumlūn muṯallaṯī
half-sized aggregate ركام متوسط الحجم
rukām mutawassiṭ al-ḥajm
half-socket pipe أنبوب بنصف جلبة
unbūb bi-niṣf jalaba
half-tide cofferdam سد للمد النصفي
sadd lil-madd an-niṣfī
half timber نصف خشبي
niṣf kašabī
half-track tractor جرار نصف مزنجر
jarrār niṣf muzanjar
Hallinger shield حجاب مجري
ḥijāb majra
halogenated hydrocarbon هيدروكربون مهجن
hīdrūkarbūn muhaljan
halved joint وصلة تنصيفية
waṣla tanṣīfīya
hammer مطرقة
miṭraqa
hammer base قاعدة المطرقة
qāʿidat al-miṭraqa
hammer cushion وسادة المطرقة
wisādat al-miṭraqa
hammer drill مثقاب مطرقي
miṯqāb miṭraqī
hammer grab كباش المطرقة
kabbaš al-miṭraqa

H

H

hammer-mills method   طريقة الكسارة المطرقية
ṭarīqat al-kassāra al-miṭraqīya

hammer ram   مدك مطرقة
midakk miṭraqa

hand boring   حفر يدوي
ḥafr yadawī

hand demolition   هدم يدوي
hadm yadawī

hand distributor   موزع يدوي
muwazziʻ yadawī

hand-dressed stone   حجر ملبس باليد
ḥajar mulabbas bil-yad

hand finisher   آلة إنهاء يدوي
ālat inhāʼ yadawī

hand lead   مسبار يدوي
misbā yadawī

hand level   ميزان تسوية يدوي
mīzān taswiya yadawī

handling cargo   مناولة البضائع
munāwalat al-baḍāʼiʻ

handling goods   مناولة السلع
munāwalat as-silaʻ

handover date   تاريخ التسليم
tārīḵ at-taslīm

handrail   درابزين
darābzīn

handsaw   منشار يدوي
minšār yadawī

hand sprayer   مرش يدوي
mirašš yadawī

hangar   حظيرة طائرات
ḥazīrat ṭāʼirāt

hanging leaders   هيكل فولاذي معلق
haikal fūlāḏī muʻallaq

hanging steps   درجات معلقة
darajāt muʻallaqa

harbour   ميناء
mīnāʼ

harbour engineering   هندسة الموانئ
handasat al-mawānīʼ

harbour entrance   مدخل الميناء
madḵal al-mīnāʼ

harbour layout   مخطط الميناء
muḵaṭṭaṭ al-mīnāʼ

harbour models   نماذج الموانئ
namāḏij al-mawānī

harbour of refuge   ميناء اللجوء
mīnāʼ al-lujūʼ

harbour planning   تخطيط المرفأ
taḵṭīṭ al-marfāʼ

harbour protection   وقاية المرفأ
wiqāyat al-marfāʼ

hardboard   لوح صلد
lauḥ ṣalid

hard-burnt bricks   طوب مصلّد بالاحماء
ṭaub muṣallad bil-iḥmāʼ

hard core   فرشة صلبة
farša ṣaliba

hard-drawn copper   نحاس مسحوب بصلابة
nuḥās masḥūb bi-ṣalāba

hardenability   قابلية التصليد
qābilīyat at-taṣlīd

hardened cement   إسمنت مصلد
ismant muṣallad

hardened concrete   خرسانة مصلدة
ḵarasāna muṣallada

hardening   تصلد
taṣallud

hard facing   تصليد السطح
taṣlīd as-saṭḥ

hard finish   طلية ملساء
ṭalya malsāʼ

hard hat   قبعة صلبة
qubbaʻa ṣaliba

hardness   صلادة
ṣalāda

hardpan   كتيم
katīm

hard standing   سطح صلب
saṭḥ ṣalib

hard water   ماء عسر
māʼ ʻasir

hardwood   خشب صلد
ḵašab ṣalid

Hardy Cross method   طريقة «هاردي كروس»
ṭarīqat hārdī krūs

hatchet   فأس صغير
faʼs ṣaġīr

hatching   تهشير الرسم الهندسي
tahšīr ar-rasm al-handasī

haulage rope   حبل الجر
ḥabl al-jarr

hauling plant   معدات جر
muʻiddāt jarr

haunch   كتف العقد
katif al-ʻaqd

hazardous waste   نفايات خطرة
nufāyāt ḵaṭira

haze control التحكم في الرهج
at-taḥakkum fir-rahj
Hazen's law قانون «هازين»
qānūn hāzīn
head رأس
ra's
head bay حوز أمامي
hauz amāmī
head board لوح أفقي
lauḥ ufuqī
header أنبوب رئيسي للمياه
unbūb ra'īsī lil-miyāh
head gate بوابة السد الرئيسية
bawwābat as-sadd ar-ra'īsīya
heading نفق أفقي
nafaq ufuqī
headline عنوان رئيسي
'unwān ra'īsī
head of water علو الماء
'ulū al-mā'
headphones سماعات رأس
sammā'āt ra's
head protection وقاية الرأس
wiqāyat ar-ra's
head race مجرى الماء الرأسي
majra al-mā' ar-ra'sī
headroom إرتفاع السقف
irtifā' as-saqf
head tree لوح دعم جانبي
lauḥ da'm jānibī
head wall جدار دعم
jidār da'm
headwater منبع النهر
manba' an-nahr
headway تقدم
taqaddum
headworks أعمال فكرية
a'māl fikrīya
heat-absorbing glass زجاج ماص للحرارة
zujāj māṣṣ lil-ḥarāra
heater مسخن . مدفأة
musak̲k̲in, midfa'a
heating coil ملف المسخن
milaff al-musak̲k̲in
heating element عنصر تسخين
'unsur task̲īn
heating oil زيت تسخين
zait task̲īn
heating plant محطة تسخين
muḥaṭṭaṭ task̲īn
heating unit وحدة تسخين
waḥdat task̲īn
heating value القيمة الحرارية
al-qīma al-ḥarārīya
heat-insulating concrete خرسانة عازلة للحرارة
k̲arasāna 'āzila lil-ḥarāra
heat-proof صامد للحرارة
ṣāmid lil-ḥarāra
heat-resistant concrete خرسانة صامدة للحرارة
k̲arasāna ṣamida lil-ḥarāra
heat-resistant paint دهان صامد للحرارة
dihān ṣamid lil-ḥarāra
heat-transfer إنتقال الحرارة
intiqāl al-ḥarāra
heat-treated steel فولاذ معالج بالحرارة
fūlād̲ mu'ālaja bil-ḥarāra
heat treatment معالجة حرارية
mu'ālaja ḥarārīya
heave رفع بجهد
raf' bi-juhd
heaving shale طين صفحي صدعي
ṭīn ṣafḥī ṣada'ī
heavy aggregate ركام ثقيل
rukām t̲aqīl
heavy lift رفع ثقيل
raf' t̲aqīl
heavy soil تربة ثقيلة
turba t̲aqīla
heavy-weight concrete خرسانة ثقيلة الوزن
k̲arasāna t̲aqīlat al-wazn
hectare هكتار
hiktār
hecto- هكتو
hiktū
hedge سياج طبيعي
sīyāj ṭabī'ī
hedge trimmers مهذبات الوشيع
muhad̲d̲ibāt al-wašī'
heel كعب
ka'b
heel blocks كتل حجر الزاوية
kutal ḥajar az-zāwīya
heel post عمود الركن
'amūd ar-rukn
height of instrument method طريقة علو آلة القياس
ṭarīqat 'ulū ālat al-qiyās

H

height of wall إرتفاع الجدار
irtifāʿ al-jidār
held water مياه محجوزة
miyāh maḥjūza
helical conveyor ناقل حلزوني
nāqil ḥalazūnī
helical reinforcement تعزيز حلزوني
taʿzīz ḥalazūnī
helical stairs درج حلزوني
durj ḥalazūnī
helicopter هليكوبتر
hilikubtar
heliport محط طائرات الهليكوبتر
maḥaṭṭ ṭāʾirāt al-hilikubtar
helium هيليوم
hīlyūm
helium diving bell ناقوس غوص بالهيليوم
nāqūs ġauṣ bil-hīlyūm
hemp قنب
qinnab
herringbone drain مصرف لولبي
maṣrif laulabī
hertz هرتز
hirts
hexadecimal عشري سداسي
ʿušri sudāsī
hexapods سداسية الأرجل
sudāsīyat al-arjul
HF communications الاتصالات بالتردد العالي
al-ittiṣālāt bit-tarradud al-ʿālī
high accident rate معدل حوادث مرتفع
muʿaddal ḥawādiṯ murtafiʿ
high-alumina cement إسمنت بنسبة ألومينا مرتفعة
ismant bi-nisbat ālūmīna murtafiʿa
high-carbon steel فولاذ بنسبة كربون مرتفعة
fūlāḏ bi-nisbat karbūn murtafiʿa
high density airport مطار ذو كثافة عالية
maṭār ḏū kaṯāfa ʿālīya
high density baling رزم كثيف
razm kaṯīf
high-early-strength cement إسمنت سريع التصلب
ismant sarīʿ at-taṣallub
high explosive شديد الانفجار
šadīd al-infijār
high gear مسنن عال
musannan ʿāl
high-intensity runway lights أضواء مدرج قوية الاضاءة
aḍwāʾ madraj qawīyat al-iḍāʾa
high level مستوى عال
mustawa ʿāli
high level decks سطوح ذات مستوى مرتفع
suṭūḥ ḏāt mustawa murtafiʿ
high mast lighting إضاءة الصاري العالي
iḍāʾat aṣ-ṣārī al-ʿālī
high-output boiler غلاية ذات إنتاجية عالية
ġallāya ḏāt intājīya ʿālīya
high-pressure sodium vapour lamp مصباح بخار صوديوم عالي الضغط
misbāh bukār sūdyūm ʿālī aḍ-ḍaġṭ
high-pressure steam-curing تصليد ببخار ذو ضغط عال
taṣlīd bi-bukār ḏū ḍaġṭ ʿālī
high productivity إنتاجية عالية
intājīya ʿālīya
high-rate filter مرشح ذو سعة عالية
muraššiḥ ḏū siʿa ʿālīya
high-strength friction-grip bolts براغي زنق إحتكاكي عالية المتانة
barāġī zanq iḥtikākī ʿālīyat al-matāna
high-strength steel فولاذ قوي المتانة
fūlāḏ qawī al-matāna
high-tensile steel فولاذ قوي الشد
fūlāḏ qawī aš-šadd
high tolerance تفاوت كبير
tafāwut kabīr
high-torque motor محرك ذو عزم مرتفع
muḥarrik ḏu ʿazm murtafiʿ
high-velocity jet نفاثة عالية السرعة
naffāṯa ʿālīyat as-surʿa
high-voltage line خط جهد عال
ḵaṭṭ juhd ʿālī
high-voltage power cable كبل طاقة للجهد العالي
kabl ṭāqa lil-juhd al-ʿālī
highway طريق عام
ṭarīq ʿāmm
highway authority هيئة المرور
haiʾat al-murūr
highway bridge جسر طريق عام
jisr ṭarīq ʿāmm
highway bridge load حمولة جسر الطريق العام
ḥumūlat jisr aṭ-ṭarīq al-ʿāmm

highway legislation قوانين المرور
qawānīn al-murūr
highway maintenance صيانة الطرق
ṣīyānat aṭ-ṭuruq
highway tunnel نفق طريق عام
nafaq ṭarīq ʿāmm
highway under construction طريق عام قيد الانشاء
ṭarīq ʿāmm qaid al-inšāʾ
high-yield bar قضيب عالي المطاوعة
qaḍīb ʿālī al-mutāwāʿa
Hiley's formula معادلة «هيلي»
muʿādalat hīlī
hindered settling ترسب معاق
tarassub muʿāq
hinge مفصلة
mifṣala
hinged cylinder إسطوانة مفصلية
istiwāna mifṣalīya
hinged door باب مفصلي
bāb mifṣalī
hinged leaf gate باب بمصراع مفصلي
bāb bi-miṣrāʿ mifṣalī
hip rafter رافدة وركية
rāfida warkīya
hip roof سقف مسنم
saqf musannam
hip tiles بلاط سنامي
bilāṭ sanāmī
hire charges رسوم التأجير
rusūm at-taʾjīr
hire company شركة تأجير
šarikat taʾjīr
hoardings أسوار خشبية مؤقتة
aswār ḵašabīya muʾaqqata
hoe scraper مكشطة معزقية
mikšaṭa maʿzaqīya
hoggin حصباء رملية
ḥasbāʾ ramlīya
hogging girder عارضة مقوسة
ʿāriḍa muqawwasa
hogging moment عزم التقوس
ʿazm at-taqawwus
hoist مرفاع
mirfāʿ
hoist controller ضابط سرعة المرفاع
ḍābiṭ surʿat al-mirfāʿ
hoisting engine محرك رفع
muḥarrik rafʿ
hoisting rope حبل الرفع
ḥabl ar-rafʿ
hoistway construction إنشاء هيكل الرفع
inšāʾ haikal ar-rafʿ
hoistway enclosure مبنى هيكل الرفع
mabna haikal ar-rafʿ
holdfast مربط
mirbaṭ
holding-down bolt مسمار ربط
mismār rabṭ
hollow-block floor أرضية من القوالب المفرغة
arḍīya min al-qawālib al-mufaraġa
hollow clay blocks آجر مفرغ
ājurr mufarraġ
hollow dam سد أجوف
sadd ajwaf
hollow jet valve صمام نافوري
ṣimām nāfūrī
hollow masonry unit وحدة حجرية جوفاء
waḥda ḥajarīya jaufāʾ
hollow partition فاصل من الطوب المفرغ
fāṣil min aṭ-ṭaub al-mufarraġ
hollow quoin حجر زاوية مفرغ
ḥajar zāwīya mufarraġ
hollow sections قطاعات أنبوبية مفرغة
qiṭāʿāt unbūbīya mufarraġa
hollow-slab construction إنشاء من البلاط المفرغ
inšāʾ min al-bilāṭ al-mufarraġ
hollow tile آجرة مفرغة
ājurra mufarraġa
hollow-tile floor أرضية من آجر مفرغ
arḍīya min ājurr mufarraġ
hollow wall جدار مجوف
jidār mujawwaf
hollow-web girder عارضة صندوقية
ʿāriḍa ṣundūqīya
homogeneous material مادة متجانسة
mādda mutajānisa
honeycombing تخريم
taḵrīm
honeycomb wall جدار مخرّم
jidār muḵarram
Honigmann method طريقة «هونيجمان»
ṭarīqat hūnīġman
Hooghly bridge جسر مدعم بالحبال
jisr mudʿam bil-hibāl

H

hook bolt	مسمار خطافي الشكل
mismār ḵaṭṭāfī aš-šakl	
Hooke's law	قانون «هوك»
qānūn hūk	
hook gauge	مقياس خطافي
miqyās ḵaṭṭāfī	
hooping	قضبان حلزونية
quḍbān halazūnīya	
hoop stress	إجهاد حلقي
ijhād ḥalaqī	
hopper	قادوس
qādūs	
hopper dredger	كراءة قادوسية
karrā'a qādūsīya	
hopper light	مصباح إضاءة قادوسي
miṣbaḥ iḍā'a qādūsī	
hopper suction dredger	كراءة قادوسية ماصة
karrā'a qādūsīya māṣṣa	
hopper tank	خزان قادوسي الشكل
ḵazzān qādūsī aš-šakl	
horizon	أفق
ufuq	
horizon glass	مرآة الأفق
mir'āt al-ufuq	
horizontal control	التحكم الأفقي
at-taḥakkum al-ufuqī	
horizontal drainage blanket	غطاء صرف أفقي
ḡiṭā' ṣarf ufuqī	
horizontal line	خط أفقي
ḵaṭṭ ufuqī	
horizontal longitudinal load	الحمل الطولي الأفقي
al-ḥiml aṭ-ṭūlī al-ufuqī	
horizontal plane	مستوى أفقي
mustawa ufuqī	
horizontal sliding doors	أبواب منزلقة أفقية
abwāb munzaliqa ufuqīya	
horizontal transverse load	حمل مستعرض أفقي
ḥiml mustaʿraḍ ufuqī	
horsepower	قدرة حصانية
qudra ḥiṣānīya	
horsing-up	إقامة قالب نضوي
iqāmat qālab naḍawī	
hospital	مستشفى
mustašfa	
hostel	بيت الشباب
bait aš-šabāb	
hot-air dryer	مجفف بالهواء الساخن
mujaffif bil-hawā' as-sāḵin	

hot-air heating	التدفئة بالهواء الساخن
at-tadfi'a bil-hawā' as-sāḵin	
hot-air intake	مأخذ الهواء الساخن
ma'ḵaḏ al-hawā' as-sāḵin	
hot-dip coating	طلاء بالغمس الساخن
ṭilā' bil-ḡams as-sāḵin	
hotel	فندق
funduq	
hot flow drying	تجفيف بالهواء الساخن
tajfīf bil-hawā' as-sāḵin	
hot miller	فرازة على الساخن
farrāza ʿala as-sāḵin	
hot rolling	دلفنة على الساخن
dalfana ʿala as-sāḵin	
hot shortness	قصف على الساخن
qaṣf ʿala as-sāḵin	
hot spraying	رش حار
rašš ḥārr	
hot working	تشكيل على الساخن
taškīl ʿala as-sāḵin	
house	منزل
manzil	
house connections	توصيلات منزلية
ṭauṣīlāt manzilīya	
house drain	مصرف المنزل
maṣrif al-manzil	
housing block	بناية سكنية
binaya sakanīya	
Howe truss	جملون «هاو»
jamlūn hau	
humidity	رطوبة
ruṭūba	
humidity of air	رطوبة الهواء
ruṭūbat al-hawā'	
hump yard	ساحة تفريغ
sāḥat tafrīḡ	
humus	دبال
dubāl	
humus tank	خزان الدبال
ḵazzān ad-dubāl	
hundred	مئة . مائة
mi'a	
hundredweight	هندرد ويت
handrad wait	
hung span	باع معلق
bāʿ muʿallaq	
hurdle work	حواجز
ḥawājiz	
hut	كوخ
kūḵ	

hybrid computer حاسبة هجينية
ḥāsiba hajīnīya
hydrant محبس مطافئ
miḥbas maṭāfī'
hydration تميؤ
tamayyu'
hydraulic هيدرولي
hīdrūlī
hydraulic binder محصدة حازمة هيدرولية
miḥṣada ḥāzima hīdrūlīya
hydraulic burster مفجر هيدرولي
mufajjir hīdrūlī
hydraulic cement إسمنت مائي
ismant mā'ī
hydraulic concrete breaker كسارة خرسانة هيدرولية
kassārat k̲arasāna hīdrūlīya
hydraulic dredger كراءة هيدرولية ماصة
karrā'a hīdrūlīya māṣṣa
hydraulic ejector لافظ هيدرولي
lāfiẓ hīdrūlī
hydraulic elements عناصر هيدرولية
'anāṣir hīdrūlīya
hydraulic energy طاقة هيدرولية
ṭāqa hīdrūlīya
hydraulic engineer مهندس هيدروليات
muhandis hīdrūlīyāt
hydraulic engineering الهندسة الهيدرولية
al-handasa al-hīdrūlīya
hydraulic excavation حفر هيدرولي
ḥafr hīdrūlī
hydraulic excavator حفارة هيدرولية
ḥaffāra hīdrūlīya
hydraulic excavator scraper كاشطة حفر هيدرولية
kāšiṭat ḥafr hīdrūlīya
hydraulic fill ترسبات محمولة بالماء المتدفق
tarassubāt maḥmūla bil-mā' al-mutadaffiq
hydraulic fill dam سد هيدرولي
sadd hīdrūlī
hydraulic fluid سائل هيدرولي
sā'il hīdrūlī
hydraulic friction مقاومة التدفق الهيدرولي
muqāwamat at-tadaffuq al-hīdrūlī
hydraulic gradient معدل الانحدار الهيدرولي
mu'addal al-inḥidār al-hīdrūlī

hydraulic jack مرفاع نقال هيدرولي
mirfā' naqqāl hīdrūlī
hydraulic jump قفز هيدرولي
qafz hīdrūlī
hydraulicking إزاحة الترسبات بالماء المتدفق
izāḥat at-tarassubāt bil-mā' al-mutadaffiq
hydraulic lift مصعد هيدرولي
miṣ'ad hīdrūlī
hydraulic lift platform منصة رافع هيدرولي
minaṣṣat rāfi' hīdrūlī
hydraulic machinery آليات هيدرولية
ālīyāt hīdrūlīya
hydraulic main خط مياه رئيسي
k̲aṭṭ miyah ra'īsī
hydraulic mean depth متوسط العمق الهيدرولي
mutawassiṭ al-'umq al-hīdrūlī
hydraulic outrigger ذراع امتداد هيدرولي
d̲irā' imtidād hīdrūlī
hydraulic pile driving دق الخوازيق هيدرولياً
daqq al-k̲awāzīq hīdrūlīyan
hydraulic pipe أنبوب هيدرولي
unbūb hīdrūlī
hydraulic power قدرة هيدرولية
qudra hīdrūlīya
hydraulic pump مضخة هيدرولية
miḍak̲k̲a hīdrūlīya
hydraulic ram مكباس هيدرولي
mikbās hīdrūlī
hydraulic reservoir خزان هيدرولي
k̲azzān hīdrūlī
hydraulics هيدروليات (علم السوائل)
hīdrūlīyāt, 'ilm as-sawā'il
hydraulic structure إنشاء هيدرولي
inšā' hīdrūlī
hydraulic turbine تربين هيدرولي
tūrbīn hīdrūlī
hydraulic works محطة هيدرولية
muḥaṭṭa hīdrūlīya
hydrocarbon binder رباط هيدروكربوني
ribāṭ hīdrūkarbūnī
hydrodynamic pressure ضغط هيدروديناميي
ḍagṭ hīdrūdīnāmī
hydrodynamics هيدرو ديناميك
hīdrūdīnāmīk
hydroelectric power station محطة توليد مائية
muḥaṭṭaṭ taulīd mā'īya

H

hydroelectric scheme مشروع كهرومائي
mašrū‘ kahrumā’ī
hydrofracture تصدع هيدرولي
taṣaddu‘ hīdrūlī
hydrogeological investigation بحث جيولوجية المياه
baḥt jiyūlūjīyat al-miyāh
hydrogeology جيولوجية الماء
jiyūlūjīyat al-mā’
hydrograph خارطة للمياه
ḵāriṭa lil-miyāh
hydrographic survey مسح هيدروغرافي
masḥ hīdrūğrāfī
hydrography هيدروغرافي
hīdrūğrāfī
hydrological cycle دورة المياه الجوفية
daurat al-miyah al-jaufīya
hydrological data بيانات المياه الجوفية
bayānāt al-miyah al-jaufīya
hydrological study دراسة المياه الجوفية
dirāsat al-miyah al-jaufīya
hydrology هيدرولوجية (علم المياه)
hīdrūlūjīya, ‘ilm al-miyah
hydrometer هيدرومتر
hīdrūmītar
hydrophobic cement إسمنت مضاد للماء
ismant muḍād lil-mā’
hydrostatic catenary منحنى سلسلي هيدروستاتي
munhana silsilī hīdrūstātī
hydrostatic excess pressure ضغط زائد هيدروستاتي
ḍağṭ zā’id hīdrūstātī
hydrostatic joint وصلة هيدروستاتية
waṣla hīdrūstātīya
hydrostatic press مكبس هيدروستاتي
mikbas hīdrūstātī
hydrostatic pressure ضغط هيدروستاتي
ḍağṭ hīdrūstātī
hydrostatic pressure ratio نسبة الضغط الهيدروستاتي
nisbat aḍ-ḍağṭ al-hīdrūstātī
hydrostatic valve صمام هيدروستاتي
ṣimām hīdrūstātī
hydroxylated polymers بوليمرات هيدروكسيلية
būlīmrāt hīdrūksīlīya
hygrometer هيجرومتر
hīğrūmitar
hygroscopic coefficient معامل الاسترطاب
mu‘āmil al-istirṭāb

hygroscopic moisture الرطوبة النسبية في الهواء
ar-ruṭūba an-nisbīya fil-hawā’
hypar سطح مكافئ
saṭḥ mukāfi’
hyperbaric chamber غرفة مكافئة
ğurfa mukāfi’a
hysteresis التخلف (المغنطيسي)
at-taḵalluf al-muğnāṭīsī
ignition powder مسحوق الاشعال
masḥuq al-iš‘āl
illuminated sign لافتة مضاءة
lāfita muḍā’a
Imhoff tank خزان تخمير الحمأة
ḵazzān taḵmīr al-ḥama
immersed tube ماسورة مغمورة
māsūra mağmūra
impact تصادم
taṣādum
impact driven pile ركيزة مدقوقة بالصدم
rakīza madqūqa biṣ-ṣadm
impact factor معامل التصادم
mu‘āmil at-taṣādum
impact load حمل صدمي
ḥiml ṣadmī
impact spanner مفتاح ربط آلي
miftaḥ rabṭ ālī
impact study دراسة تأثير الصدم
dirāsat ta’ṯīr aṣ-ṣadm
impact test إختبار الصدم
iḵtibār aṣ-ṣadm
impeller دفاعة مروحية
daffā‘a mirwaḥīya
imperfect frame هيكل ناقص
haikal nāqiṣ
impermeability factor عامل اللاإنفاذية
mu‘āmil al-lā-infāḏīya
impermeable material مادة كتيمة
mādda katīma
impermeable membrane غشاء كتيم
ğišā’ katīm
impervious غير منفذ
ğair munfiḏ
impervious concrete خرسانة كتيمة
ḵarasāna katīma
impervious core قلب كتيم
qalb katīm
import study دراسة الاستيراد
dirāsat al-istīrād

impounding reservoir خزان ماء ضخم
kazzān mā' ḍakm
impregnation تشرب
tašarrub
impulse turbine تربين دفعي
tūrbīn daf'ī
incandescent light ضوء متوهج
ḍau' mutawahhij
incentive-type contract عقد تشجيعي
'aqd tašjī'ī
inch بوصة . إنش
būṣa, inš
incidence rate معدل الحوادث
mu'addal al-ḥawādit
incinerator مرمد
murammid
incline ميل
mail
inclined bar قضيب مائل
qaḍīb mā'il
inclined cableway مصعد كبلي مائل
miṣ'ad kablī mā'il
inclined footway ممر مشاة مائل
mamarr mušāt mā'il
inclined gauge مقياس مائل
miqyās mā'il
inclined lift gate بوابة هويس مائلة
bawwābat hawīs mā'ila
incoming sewage مياه المجاري القادمة
miyāh al-majārī al-qādima
indemnification clause شرط التعويض
šarṭ at-ta'wīḍ
indentation قطع متعرج
qaṭ' muta'arrij
indented bars قضبان محززة
quḍbān muḥazzaza
indented bolt مسمار تثبيت محزز
mismār taṯbīt muḥazzaz
indented wheel دولاب محزز
dūlāb muḥazzaz
indenter أداة ثلم
adāt ṯalm
indenting roller مدحلة تحزيز
midḥalat taḥzīz
indexed addressing تعديل ايجاد المعلومات المسجلة
ta'dīl ījād al-ma'lūmāt al-musajjala
index glass مرآة دليلية
mir'āt dalīlīya
index of liquidity دليل السيولة
dalīl as-suyūla
index of plasticity دليل اللدانة
dalīl al-ladāna
index properties خصائص دليلية
kaṣṣā'iṣ dalīlīya
indirect addressing أمر غير مباشر لايجاد المعلومات
amr ġair mubāšir li-ījād al-ma'lūmāt
indirect heating تدفئة غير مباشرة
tadfi'a ġair mubāšira
indirect lighting إضاءة غير مباشرة
iḍā'a ġair mubāšira
indirect rail fastening تثبيت غير مباشر للسكة
taṯbīt ġair mubāšir lis-sikka
inductive component مكونات حثية
mukawwinat ḥaṯṯīya
industrial building مبنى صناعي
mabna ṣinā'ī
industrial consumption إستهلاك صناعي
istihlāk ṣinā'ī
industrial diamond ألماس صناعي
almās ṣinā'ī
industrial electronics إلكترونيات صناعية
ilikrūnīyat ṣinā'īya
industrialised building طريقة البناء المصنّع
ṭarīqat al-binā' al-muṣanna'
industrial sewage مجارير مياه المصانع
majārīr miyāh al-maṣāni'
industrial switchgear مجموعة مفاتيح صناعية
majmū'at mafatīḥ ṣinā'īya
industrial waste treatment معالجة النفايات الصناعية
mu'ālajat an-nufayāt aṣ-ṣinā'īya
inertia القصور الذاتي
al-quṣūr aḍ-ḏātī
inertial force قوة القصور الذاتي
qūwat al-quṣūr aḍ-ḏātī
inertial surveying system نظام مسح القصور الذاتي
niẓām masḥ al-quṣūr aḍ-ḏātī
infiltration ترشيح
taršīḥ
infiltration capacity سعة الترشيح
si'at at-taršīḥ
influent stream تيار متدفق
tayyār mutadaffiq
infra-red beam أشعة تحت الحمراء
aši''a taḥt al-ḥamrā'

I

infra-red heater
musakkin bil-aši''a taht al-hamrā'
مسخن بالأشعة تحت الحمراء

infra-red photography
taṣwīr bil-aši''a taht al-hamrā'
تصوير بالأشعة تحت الحمراء

ingot
sabīka lit-taškīl
سبيكة للتشكيل

inherent settlement
hubūt bin-naql aḏ-ḏātī
هبوط بالثقل الذاتي

inhibitor
māni' lit-tafā'ul al-kīmāwī
مانع للتفاعل الكيماوي

in-house
dāḵil al-maqarr
داخل المقر

initial setting time
waqt at-taṣallud al-awwalī
وقت التصلد الأولي

initial surface absorption test
iḵtibār imtiṣāṣ as-saṭh al-awwalī
إختبار امتصاص السطح الأولي

injection
ḥaqn
حقن

injection of cement
ḥaqn al-ismant
حقن الاسمنت

injection well
bi'r al-ḥaqn
بئر الحقن

inlaid parquet
arḍīya muzaḵrafa
أرضية مزخرفة

inlet
madḵal
مدخل

inlet culvert
barbaḵ saḥb
بربخ سحب

inlet ventilator
mihwāt al-madḵal
مهواة المدخل

input
daḵl
دخل

input/output interface
saṭh bainī lid-daḵl wal-ḵarj
سطح بيني للدخل والخرج

insert
īlāj
ايلاج

in-situ
fil-mauqi'
في الموقع

in-situ concrete piles
ṣabb al-ḵawāzīq al-ḵarasānīya fil-mauqi'
صب الخوازيق الخرسانية في الموقع

in-situ soil tests
iḵtibārāt at-turba fil-mauqi'
إختبارات التربة في الموقع

inspection chamber
ḥujrat at-taftīš
حجرة التفتيش

inspection gallery
bahū at-taftīš
بهو التفتيش

inspection ladder
sullam al-mu'ayana
سلم المعاينة

inspector
mufattiš
مفتش

inspector of works
mufattiš al-a'māl
مفتش الأعمال

installation
tarkīb
تركيب

instruction cycle
daurat at-tadrīb
دورة التدريب

instruction organisation
tanẓīm at-tadrīb
تنظيم التدريب

instruction type
tadrībī
تدريبي

instrument
jihāz
جهاز

instrumental shaft plumbing
tafdīn al-muhawāt bil-muzawwāt
تفدين المهواة بالمزواة

instrument approach runway
madraj al-iqtirāb bi-ajhizat al-qiyās
مدرج الاقتراب بأجهزة القياس

instrument flight rules
mabādi' liṭ-ṭayarān bi-ajhizat al-qiyās
مبادئ للطيران بأجهزة القياس

instrument landing system
niẓām al-ḥaṭṭ bi-ajhizat al-qiyās
نظام الحط بأجهزة القياس

insulated cable
kabl ma'zūl
كبل معزول

insulating board
lauḥ 'āzil
لوح عازل

insulating lining
biṭāna 'āzila
بطانة عازلة

insulating material
mādda 'āzila
مادة عازلة

insulation
'azl, 'āzil
عزل . عازل

intake
saḥb
سحب

intake tower
burj as-saḥb
برج السحب

integrally-stiffened plating
taṣfīḥ mušakkal bil-baṭq
تصفيح مشكل بالبطق

integrated circuits
dārāt mutakāmila
دارات متكاملة

integrated injection logic
manṭiq al-ḥaqn al-mutakāmil
منطق الحقن المتكامل

integrating meter
'addād mutakāmil
عداد متكامل

intensity of rainfall
šiddat suqūṭ al-amṭār
شدة سقوط الأمطار

intensity of stress
šiddat al-ijhād
شدة الاجهاد

intercepting channel
qanāt i'tirāḍīya
قناة اعتراضية

intercepting drain
maṣrif mu'tarriḍ
مصرف معترض

intercepting sewer
ḡurfat iḥtibās ar-rawā'iḥ
غرفة احتباس الروائح

interception
i'tirāḍ
إعتراض

interceptor drains
maṣārif marfaq al-iḥtibās
مصارف مرفق الاحتباس

interchange
taqāṭu'
تقاطع

interchange layout
muḵaṭṭaṭ at-taqāṭu'
مخطط التقاطع

intercity passenger system
šabakat naql ar-rukkāb bain al-mudun
شبكة نقل الركاب بين المدن

interconnected system
niẓām mutarābiṭ
نظام مترابط

intercooler
mubarrid bainī
مبرّد بيني

interface strength
hubūṭ tadāḵulī
هبوط تداخلي

interfacing microprocessor
mikrūbrūsisar tadāḵulīya
ميكروبروسسور تداخلية

interference-body bolt
mismār taṯbīt iḥtikākī
مسمار تثبيت إحتكاكي

interference settlement
hubūṭ tadāḵulī
هبوط تداخلي

interflow
tadaffuq al-mā' al-bāṭinī
تدفق الماء الباطني

intergranular pressure
aḍ-ḍaḡṭ bain al-ḥubaibāt
الضغط بين الحبيبات

interior span
al-bā' ad-dāḵilī
الباع الداخلي

interlock
tašābuk
تشابك

interlocking
taušīj
توشيج

interlocking piles
ḵawāzīq ta'šīq
خوازيق تعشيق

intermittent sand filter
taršīḥ ramlī mutaqaṭṭi'
ترشيح رملي متقطع

internal area
al-masāḥa ad-dāḵilīya
المساحة الداخلية

internal-combustion engine
muḥarrik dāḵilī al-iḥtirāq
محرك داخلي الاحتراق

internal diameter
quṭr dāḵilī
قطر داخلي

internal environment
bī'a dāḵilīya
بيئة داخلية

internal flow
dafq dāḵilī
دفق داخلي

internal friction
iḥtikāk dāḵilī
إحتكاك داخلي

internal lining
biṭāna dāḵilīya
بطانة داخلية

internal vibration
ihtizāz dāḵilī
إهتزاز داخلي

internal vibrator
hazzāza dāḵilīya
هزازة داخلية

International Civil Aviation Organisation
munaẓẓamat aṭ-ṭayarān al-madanī ad-daulīya
منظمة الطيران المدني الدولية

interpolation
istikmāl
إستكمال

interpretative languages
luḡāt tafsīrīya
لغات تفسيرية

interrupts
muqāṭa'a
مقاطعة

intersection
taqāṭu'
تقاطع

intersection angle
zāwīyat at-taqāṭu'
زاوية التقاطع

intersection point
nuqṭat at-taqāṭu'
نقطة التقاطع

invar
īnfār
إنفار

invert
'aks
عكس

inverted siphon
sīfūn ma'kūs
سيفون معكوس

inverted well
bi'r ma'kūs
بئر معكوس

I

invert level مستوى الانعكاس
mustawa al-in'ikās
ion أيون
īyūn
ion exchange التبادل الأيوني
at-tabādul al-īyūnī
iron حديد
ḥadīd
iron paving رصف حديدي
raṣf ḥadīdī
iron railings قضبان حديدية
quḍbān ḥadīdīya
irradiation تعريض للاشعاع
ta'rīḍ lil-iš'ā'
irrigable area المساحة المروية
al-masāḥa al-marwīya
irrigating head صمام ري
ṣimām rayy
irrigation الري
ar-rayy
irrigation canal قناة الري
qanāt ar-rayy
irrigation requirement متطلبات الري
mutaṭallabāt ar-rayy
isochromatic lines خطوط أيسوكروماتية
ku̱ṭūṭ īsūkrūmātīya
isoclinic lines خطوط الميل المغنطيسي
ku̱ṭūṭ al-mail al-muğnāṭīsī
isolating valve صمام فاصل
ṣimām fāṣil
isolator عازل
'āzil
isometric متناظر
mutanāẓir
isometric projection إسقاط متناظر
isqāṭ mutanāẓir
isotherm خط التحارر
ka̱ṭṭ at-taḥārur
isothermal compression ضغط متحارر
ḍağṭ mutahārir
isotropic نظائر
naẓā'ir
item صنف
ṣanf
ivy اللبلاب
al-lablāb
Izod test إختبار «آيزود»
ik̠tibār īzūd
jack مرفاع السيارة
mirfā' as-sayyārat
jack arch عقد مسطح
'aqd musaṭṭaḥ
jackblock method طريقة البناء المتداخل
ṭarīqat al-binā' al-mutadākil
jacked pile ركيزة مرفوعة
rakīza marfū'a
jacket غلاف . قميص
ğilāf, qamīṣ
jacking pockets جيوب الرفع
jiyūb ar-raf'
jack plane مسحاج
misāj
jack rafter رافدة قصيرة
rāfida qaṣīra
jack rib دعامة قصيرة
du'āma qaṣīra
jack roll مرفاع ملفافي يدوي
mirfā' milfāfī yadawī
jaw breaker كسارة صخور ذات فكين
kassārat ṣuk̠ūr ḏāt fakkain
jet drilling حفر نفثي
ḥafr naftī
jet propulsion دفع نفثي
daf' naftī
jetting تغطيس الركائز دون دق
tağṭīs ar-rakā'iz dūn daqq
jetty حائل أمواج
ḥā'il amwāj
jib ذراع المرفاع
ḍirā' al-mirfā'
jib crane مرفاع ذراعي
mirfā' ḍirā'ī
jib support دعامة ذراع المرفاع
du'āmat ḍirā' al-mirfā'
jib swing control التحكم في تأرجح ذراع المرفاع
at-taḥakkum fi ta'arjuḥ ḍirā' al-mirfā'
jig دليل تشغيل
dalīl tašğīl
jig back خط ترام عكوس
ka̱ṭṭ trām 'akūs
jim crow حانية قضبان
ḥānīyat quḍbān
joggle يوثق
yuwaṯiq
joiner نجار تركيب
najjār tarkīb
joinery نجارة
nijāra

joint filler حشو مفصلي
hašū mifṣalī
jointing سطوح فاصلة
suṭūḥ fāṣila
joints فواصل
fawāṣil
joist رافدة
rāfida
joists روافد
rawāfid
joule جول
jūl
jumbo عربة الحفارة
ʻarabat al-haffāra
jumper قضيب حفر قفاز
qaḍīb ḥafr qaffāz
jump join وصلة تناكب
waṣlat tanāqub
junction ملتقى
multaqa
junction cock محبس ملتقى
miḥbas multaqa
junction design تصميم التوصيل
taṣmīm at-tauṣīl
junction point نقطة الاتصال
nuqṭat al-ittiṣāl
jute fibre ليف الجوت
līf al-jūt
Kaplan turbine توربين «كابلان»
tūrbīn kaplān
keel block كتلة رافدة القص
kutlat rāfidat al-qaṣṣ
kelly جذع سحب مضلع
juḍʻ saḥb muḍallaʻ
kelly bar قضيب حفر مضلع
qaḍīb ḥafr muḍallaʻ
kelvin كلفن
kilfin
kentledge صابورة من نفايات الحديد
ṣāburat nufayāt al-ḥadīd
kerb حافة الرصيف
ḥāffat ar-raṣīf
kerbside طرف الرصيف
ṭarf ar-raṣīf
kerosene كيروسين
kirūsīn
key مفتاح
miftāḥ
keyboard لوحة مفاتيح التنضيد
lauḥat mafātīḥ at-tanḍīd
key operated valve صمام يعمل بمفتاح
ṣimām yaʻmal bi-miftāḥ
keystone حجر واسطة العقد
ḥajar wāsiṭat al-ʻaqd
keyword كلمة دليلية
kalima dalīlīya
kicker وسيلة إرتداد
wasīlat irtidād
kicking piece قطعة الارتداد
qiṭʻat al-irtidād
kiln أتون
atūn
kilogram كيلوغرام
kīlūgrām
kilometre كيلومتر
kīlū mitr
kilowatt hour كيلو واط ساعة
kīlūwāṭ sāʻa
kinematic similarity تشابه حركي
tašābuh ḥarakī
kinetic energy طاقة حركية
ṭāqa ḥarakīya
king pile ركيزة رئيسية
rakīza raʼīsīya
king post دعامة رئيسية
duʻāma raʼīsīya
king tower برج المرفاع
burj al-mirfāʻ
kiosk كشك
kušk
kip كب (كيلو باوند)
kib, kīlūbaund
kitchen مطبخ
maṭbak
kitchen waste disposal رمي فضلات المطبخ
ramī faḍalāt al-maṭbak
knee brace شكال زاوي
šikāl zāwī
kneeler دعامة تثبيت
diʻāmat tatbīt
knife-edge loading حمل سكيني الحد
ḥiml skīnī al-ḥadd
knob قبضة
qabḍa
labels طنف فوق باب أو نافذة
ṭunuf fauq bāb au nāfiḍa
laboratory مختبر
muktabar
labour عمل
ʻamal

labour cost تكلفة العمل
taklifat al-ʿamal
labourer عامل
ʿāmil
labour programming برمجة العمل
barmajat al-ʿamal
laced column عمود مربوط
ʿamūd marbūṭ
lacing ربط
rabṭ
lacing board لوح ربط
lauḥ rabṭ
lacquer لاكيه
lākīya
ladder سلم
sullam
ladder jack مرفاع سلمي
mirfāʿ sullamī
ladder track سكة سلمية
sikka sullamīya
lagging ألواح تثبيت العقد . تغليف
alwaḥ taṯbīt al-ʿaqd, taġlīf
laitance غثاء الخرسانة
ġuṯāʾ al-ḵarasāna
lake بحيرة
buḥaira
lamella roof سطح طويل الباع
saṭḥ ṭawīl al-bāʿ
laminar velocity سرعة الانسياب
surʿat al-insīyāb
laminate صفّح . صفيحي
ṣaffaḥa, ṣafīḥī
laminated glass زجاج رقائقي
zujāj raqāʾiqī
lamp مصباح
miṣbāḥ
lamp holder دواة المصباح
dawāt al-miṣbāḥ
lamphole فتحة المصباح
fatḥat al-miṣbāḥ
lamp standard قاعدة المصباح
qāʿidat al-miṣbāḥ
land drain مصرف أرضي
maṣrif arḍī
land drainage تجفيف الأرض
tajfīf al-arḍ
landing رسو . هبوط
rasū, hubūṭ
landing platform قاعدة هبوط
qāʿidat hubūṭ
landing stage منصة هبوط
minaṣṣat hubūṭ
landing strip مدرج هبوط
madraj hubūṭ
land reclamation إستصلاح الأرض
istiṣlāḥ al-arḍ
landscape gardening بستنة التجميل الهندسي
bastanat at-tajmīl al-handasī
landslide hazard خطر الإنهيار الأرضي
ḵaṭar al-inḥiyār al-arḍī
land surveying مسح أرضي
masḥ arḍī
land surveyor مسّاح
massāḥ
land tie قضيب ربط أرضي
qaḍīb rabṭ arḍī
land treatment معالجة الأرض
muʿālajat al-arḍ
land use إستعمال الأرض
istiʿmāl al-arḍ
lane درب
darb
lane capacity سعة الدرب
siʿat ad-darb
lane loading حمل الدرب
ḥaml ad-darb
Lang lay جديلة «لانج»
jadīlat lānğ
lantern فانوس
fānūs
lantern carriage حاضن الفانوس
ḥāḍin al-fānūs
lap تراكب
tarākub
lap joint وصلة تراكب
waṣlat tarākub
large-diameter sewer مجرور بقطر كبير
majrūr bi-quṭr kabīr
large scale integration تكامل واسع النطاق
takāmul wāsiʿ an-niṭāq
laser أشعة ليزر
ašiʿʿat līzar
latch سقاطة . مزلاج
saqāṭa, mizlāj
lateral bracing شكّال جانبي
šikāl jānibī
lateral buckling hazard خطر التحديب الجانبي
ḵaṭar at-taḥdīb al-jānibī

lateral canal قناة فرعية
qanā far'īya
lateral conduit ماسورة فرعية
māsūra far'īya
lateral-force design تصميم القوة الجانبية
taṣmīm al-qūwa al-jānibīya
lateral load حمل جانبي
ḥiml jānibī
lateral support دعم جانبي
da'm jānibī
lath شريحة خشبية
šarīḥa ḵašabīya
latitude خط العرض
ḵaṭṭ al-'arḍ
lattice شبكية
šabakīya
lattice girder عارضة تشابكية
'āriḍa tašābukīya
lattice jib ذراع مرفاع شبكي
ḏirā' mirfā' šabakī
lavatory basin حوض اغتسال
ḥauḍ iġtisāl
lawn شاش . مرجة
šāš, marja
lay رصة الجدل
raṣṣat al-jidal
lay barge صندل تمديد الأنابيب
ṣandal tamdīd al-anābīb
lay-by موقف إستراحة للسيارات
mauqif istirāḥat lis-sayyārāt
layered map خريطة كنتورية
ḵarīṭa kanṭūrīya
laying-and-finishing machine مكنة فرش وانهاء
makinat farš wa-inhā'
laying plant وحدة فرش
waḥdat farš
layout مخطط
muḵaṭṭaṭ
leach صفى
ṣaffa
leachate ماء يحوي أملاح مذابة
mā' yaḥwī amlāḥ muḏāba
leaching cesspool بالوعة مجاري مسربة
bālū'at majārī musarriba
lead سلك موصل . رصاص
silk muwaṣṣil, raṣāṣ
leading draughtsman رسام رئيسي
rassām ra'īsī
lead line حبل سبر الغور
ḥabl sibr al-ġaur
lead-line soundings قياسات العمق المسبور
qiyāsāt al-'umq al-masbūr
lead poisoning التسمم بالرصاص
at-tasammum bir-rasās
lead rail قضيب رصاصي
qaḍīb raṣāṣ
lead sheath غلاف رصاصي
ġilāf raṣāṣī
lead shot خردق الرصاص
ḵardaq ar-raṣāṣ
leaf springs نوابض ورقية
nawābiḍ waraqīya
leakage تسرب
tasarrub
leaktight joint وصلة مسيكة
waṣla masīka
lean concrete خرسانة ضعيفة
ḵarasāna ḍa'īfa
least count القيمة الصغرى
al-qīma aṣ-ṣuġra
leat مجرى ماء
majra mā'
leech ضلع عمودي
ḍil' 'amūdī
legal and public relations العلاقات القانونية والعامة
al-'alāqāt al-qānūnīya wal-'āmma
legal papers المستندات القانونية
al-mustanadāt al-qānūnīya
legend مفتاح المصطلحات
miftāḥ al-muṣṭalaḥāt
length of dam طول السد
ṭūl as-sadd
length of runway طول المدرج
ṭūl al-madraj
letter-box صندوق الرسائل
ṣunduq ar-rasā'il
level سوّى . مستوي
sawwa, mustawī
level book سجل المسّاح
sijill al-massāḥ
level crossing تقاطع طريق بسكة
taqāṭu' ṭarīq bi-sikka
level line خط مستو
ḵaṭṭ mustawī
levelling تسوية
taswiya

L

levelling rod قضيب تسوية
qaḍīb taswiya
levelling screw برغي تسوية
burġī taswiya
levelling staff قامة تسوية
qāmat taswiya
level-luffing crane مرفاع تسوية
mirfā' taswiya
level of control مستوى التحكم أو المراقبة
mustawa at-taḥakkum au al-murāqaba
level of noise emission مستوى انبعاث الضوضاء
mustawa inbi'āṯ aḍ-ḍauḍā'
level of water منسوب الماء
mansūb al-mā'
level recorder مسجلة المنسوب
musajjilat al-mansūb
level surface سطح مستو
saṭḥ mustawī
level tube ميزان تسوية
mīzān taswīya
lever عتلة . رافعة
'atala rāfi'a
lever arm ذراع الرافعة
dirā' ar-rāfi'a
Lewis bolt مسمار «لويس»
mismār luwīs
library مكتبة
maktaba
lifeline حبل إنقاذ
ḥabl inqāḏ
life linesman مراقب إنقاذ
murāqib inqāḏ
life saving equipment معدات الانقاذ
mu'iddāt al-inqāḏ
lift مصعد
miṣ'ad
lift entrance مدخل المصعد
madḵal al-miṣ'ad
lifter hole ثقب لحشوة النسف
ṯuqb li-ḥašwat an-nasf
lift gate بوابة المصعد
bawwābat al-miṣ'ad
lifting block بكارة رافعة
bakāra rāfi'a
lifting magnet مغنطيس رافع
muġnāṭīs rāfi'
lifting motor محرك رفع
muḥarrik raf'
lifting tackle معدات الرفع
mu'iddāt ar-raf'
lift motor محرك المصعد
muḥarrik al-miṣ'ad
lift power system نظام قدرة الرفع
niẓām qudrat ar-raf'
lift pump مضخة رافعة
miḍakka rāfi'a
lift-slab construction تبليط مزدوج
tablīṭ muzdawij
lift technology تقنية المرافع
taqnīyat al-marāfi'
lift well بئر المصعد
bi'r al-miṣ'ad
light alloys سبائك خفيفة
sabā'ik ḵafīfa
lighthouse منارة
mināra
lighting إنارة . إضاءة
ināra, iḍā'a
light railway سكة حديد ضيقة
sikkat ḥadīd ḍayyiqa
light sensor حساس الضوء
ḥassās aḍ-ḍau'
lightship منارة عائمة
mināra 'ā'ima
lightweight aggregate ركام خفيف الوزن
rukām ḵafīf al-wazn
lightweight concrete خرسانة خفيفة الوزن
ḵarasāna ḵafīfat al-wazn
limb طرف . نصل
ṭaraf, naṣl
lime جير . كلس
jīr, kils
lime-cement mortar ملاط الجير والاسمنت
milāṭ al-jīr wal-ismant
lime concrete خرسانة جيرية
ḵarasāna jīrīya
lime mortar ملاط جيري
milāṭ jīrī
limestone حجر جيري
ḥajar jīrī
limiting gradient الانحدار الحدي
al-inḥidār al-ḥaddī
limit of proportionality حد التناسب
ḥadd at-tanāsub
limit state الحالة الحدية للمبنى
al-ḥāla al-ḥaddīya lil-mabna
limit state design تصميم الحالة الحدية
tasmīm al-ḥāl al-ḥaddīya

limit switch مفتاح كهربائي حدي
miftāḥ kahrabā'ī ḥaddī
limpet قنبلة ملتصقة
qunbula multaṣiqa
linear induction motor محرك حثي خطي
muḥarrik ḥaṯṯī ḵaṭṭī
lined channel قناة مبطنة
qanā mubaṭṭana
line of balance خط الموازنة
ḵaṭṭ al-muwāzana
line of least resistance خط المقاومة الدنيا
ḵaṭṭ al-muqāwama ad-dunya
line of thrust خط الدفع
ḵaṭṭ ad-daf'
liner plate لوح تبطين
lauḥ tabṭīn
liner-plate cofferdam سد تحويل ذو ألواح تبطين
sadd taḥwīl ḏū alwaḥ tabṭīn
line speed limit حد السرعة الخطية
ḥadd as-sur'a al-ḵaṭṭīya
lining بطانة . تبطين
biṭāna, tabṭīn
lining of the ceiling بطانة السقف
biṭānat as-saqf
link حلقة . وصلة
ḥalaqa, waṣla
linking توصيل
tauṣīl
linoleum لينوليوم
līnūlyūm
lintel عتبة الباب العليا
'atabat al-bāb al-'ulya
lip block قطعة الشفة
qiṭ'at aš-šiffa
liquefaction تميع
tamayyu'
liquid asphalt أسفلت سائل
asfalt sā'il
liquidated damages تعويضات نهائية
ta'wīḍāt nihā'īya
liquid cooler مبرِّد سوائل
mubarrid sawā'il
liquid dryer مجفف السائل
mujaffif as-sā'il
liquid effluent نفاية سائلة
nifāya sā'ila
liquid limit حد السيولة
ḥadd as-suyūla
list of materials قائمة المواد
qā'imat al-mawwādd
littoral drift جرف ساحلي
jurf sāḥilī
live load حمل متحرك
ḥiml mutaḥarrik
living room غرفة المعيشة
ğurfat al-ma'īša
load حمل . ثقل
ḥiml, ṯuql
loadbearing pavement رصيف حامل
raṣīf ḥāmil
loadbearing wall جدار حامل
jidār ḥāmil
load carrying capacity السعة الحملية
as-si'a al-ḥamlīya
load change تغير الحمل
tağayyur al-ḥiml
Load Classification Group مجموعة تصنيف الحمل
majmū'at taṣnīf al-ḥiml
Load Classification Number رقم تصنيف الحمل
raqm taṣnīf al-ḥiml
loaded filter مرشح مثقل
muraššiḥ muṯqal
loader حمالة
ḥammāla
load-extension curve منحنى الحمل والامتداد
munḥana al-ḥiml wal-imtidād
load factor عامل الحمل
'āmil al-ḥiml
load-indicating bolt برغي بيان الحمل
burğī bayān al-ḥiml
loading boom ذراع تحميل
ḏirā' taḥmīl
loading gauge مقياس التحميل
miqyās at-taḥmīl
loading shovel مجراف تحميل
mijrāf taḥmīl
load line خط الحمل
ḵaṭṭ al-ḥiml
load spreading properties خصائص توزيع الحمل
ḵaṣṣā'iṣ tauzī' al-ḥiml
load test إختبار الحمل
iḵtibār al-ḥiml
loam تربة رملية طينية
turba ramlīya ṭīnīya
local attraction تجاذب موضعي
rajāḏub mauḍī'ī

L

local ecology علم البيئة المحلية
'ilm al-bī'a al-maḥallīya
locator beacons منارة لتحديد الموقع
mināra li-taḥdīd al-mauqi'
lock قفل . هويس
qifl, hawīs
lockage رسم عبور الهويس
rasm 'ubūr al-hawīs
lock bay حوز الهويس
hauz al-hawīs
locked-coil rope حبل بحلقات متحدة المركز
ḥabl bi-ḥalaqāt muttaḥidat al-markaz
lockers خزانات
kazzānāt
lock gate بوابة الهويس
bawwābat al-hawīs
lock nut صمولة زنق
ṣamūlat zanq
lock paddle قناة الهويس
qanāt al-hawīs
lock sill عتبة الهويس
'atabat al-hawīs
locomotive قاطرة
qāṭira
locomotive crane مرفاع متنقل
mirfā' mutanaqqil
locomotive haulage نقل بالقطارات
naql bil-qiṭārāt
loft سقيفة
saqīfa
log زند خشب
zand kašab
logic circuits دوائر منطقية
dawā'ir manṭiqīya
long column عمود متطاول
'amūd mutaṭāwil
longitudinal bar قضيب طولي
qaḍīb ṭūlī
longitudinal gradient إنحدار طولي
inḥidār ṭūlī
longitudinal reinforcement تسليح طولي
taslīḥ ṭūlī
longitudinal section قطاع طولي
qiṭa' ṭūlī
longitudinal system نظام طولاني
niẓām ṭūlānī
long span باع طولي
bā' ṭūlī
long-span roof سطح ذو باع طولي
saṭḥ ḏū bā' ṭūlī
long ton طن انجليزي
ṭann inglīzī
loose-boundary hydraulics هيدروليات الحدود السائبة
hīdrūlīyāt al-ḥudūd as-sā'iba
loose core قلب سائب
qalb sā'ib
loose ground تربة حبيبية
turba ḥubaibīya
loose material مادة سائبة
mādda sā'iba
lorry سيارة شحن
sayyārat šaḥn
loss of ground فقد التربة المحفورة
faqd at-turba al-maḥfūra
loss of head فقد الضغط
faqd aḍ-ḍaǧṭ
loss of prestress فقد الاجهاد السابق
faqd al-ijhād as-sābiq
lost formwork الهيكل المفقود
al-haikal al-mafqūd
lost head الطاقة الضغطية المفقودة
aṭ-ṭāqa aḍ-ḍaǧṭīya al-mafqūda
loudspeaker مكبر الصوت
mukabbir aṣ-ṣaut
lounge ردهة
radha
low accident rate معدل حوادث منخفض
mu'addal ḥawādiṯ munkafiḍ
low-alloy carbon steel فولاذ كربوني منخفض الخليط المعدني
fūlāḏ kārbūnī munkafiḍ al-kaliṭ al-ma'danī
low-carbon steel فولاذ منخفض الكربون
fūlāḏ munkafiḍ al-kārbūn
low-cost construction إنشاء منخفض التكاليف
inšā' munkafiḍ at-takālīf
low cost road طريق منخفض التكاليف
ṭarīq munkafiḍ at-takālīf
low density airport مطار ذو كثافة منخفضة
maṭār kafīḍ al-kaṯāfa
lower bar القضيب السفلي
al-qaḍīb as-suflī
lower plate طرف المزواة
ṭaraf al-mizwāt
lower reservoir الخزان السفلي
al-kazzān as-suflī

lower surface *as-saṭḥ as-suflī*	السطح السفلي
low heat cement *ismant munḵafiḍ al-ḥarāra*	إسمنت منخفض الحرارة
low initial cost *takālif awwalīya munḵafiḍa*	تكاليف أولية منخفضة
low level *mustawa munḵafiḍ*	مستوى منخفض
low maintenance cost *takālīf ṣiyāna munḵafiḍa*	تكاليف صيانة منخفضة
low-pressure sodium vapour lamp *miṣbaḥ buḵār sūdyūm munḵafiḍ aḍ-ḍaḡṭ*	مصباح بخار صوديوم منخفض الضغط
low productivity *intājīya munḵafiḍa*	إنتاجية منخفضة
low tolerance *tafāwut munḵafiḍ*	تفاوت منخفض
low water fuel cut-off *inqiṭāʿ li-inḵifāḍ al-māʾ*	إنقطاع لانخفاض الماء
low-water valve *ṣimām inḵifāḍ mansūb al-māʾ*	صمام إنخفاض في منسوب الماء
LPG cylinder *jarrat ḡāz*	جرة غاز
L-shaped pier *rakīza bi-zāwiya qāʾima*	ركيزة بزاوية قائمة
luffing cableway mast *burj miṣʿad ḥablī muqawwa*	برج مصعد حبلي مقوى
luffing jib crane *mirfāʿ bi-ḏirāʿ suflī*	مرفاع بذراع سفلي
lumen output *ḵarj «lūmin»*	خرج «لومن»
luminance level *mustawa an-nuṣūʿ*	مستوى النصوع
lumping *taḡyīr sarīʿ lis-sikkak al-ḥadīdīya*	تغيير سريع للسكك الحديدية
lump-sum contract *ʿaqd bi-mablaḡ ijmālī*	عقد بمبلغ إجمالي
lurching allowance *at-tamāyul al-masmūḥ*	التمايل المسموح
lux *luks*	لكس
macadam *ḥaṣa ar-raṣf*	حصى الرصف
macadam road *ṭarīq marṣūf bil-ḥaṣbāʾ*	طريق مرصوف بالحصباء
macadam spreader *makina li-raṣf aṭ-ṭuruq*	مكنة لرصف الطرق
machine *makina, āla*	مكنة . آلة
machine code programming *barmajat ramz al-āla*	برمجة رمز الآلة
machine cycle *daurat al-āla*	دورة الآلة
machine foundations *qāʿidat tatbīt al-āla*	قاعدة تثبيت الآلة
machine guard *wāqī al-āla*	واقي الآلة
machine tools *ālāt makinīya*	آلات مكنية
machine tunnelling *ḥafr al-anfāq bil-makina*	حفر الانفاق بالمكنة
macro assembler *barnāmij tajmīʿ al-kumbyūtar*	برنامج تجميع الكمبيوتر
magazine *makzan, mustaudaʿ*	مخزن . مستودع
magnesite *maḡnasīt*	مغنسيت
magnetic bearing *al-ittijāh az-zāwī al-muḡnaṭīsī*	الاتجاه الزاوي المغنطيسي
magnetic compass *būsala maḡnaṭīsīya*	بوصلة مغنطيسية
magnetic declination *al-mail al-maḡnaṭīsī*	الميل المغنطيسي
magnetic flow meter *ʿaddād dafq maḡnaṭīsī*	عداد دفق مغنطيسي
magnetic levitation vehicle *ʿarabat ḵiffa maḡnaṭīsīya*	عربة خفة مغنطيسية
magnetic north pole *quṭb aš-šamāl al-maḡnāṭīsī*	قطب الشمال المغنطيسي
magnetic south pole *quṭb al-janūb al-maḡnāṭīsī*	قطب الجنوب المغنطيسي
magnetic tape *šarīṭ muḡnaṭīsī*	شريط مغنطيسي
magnetic variation *at-taḡayyur al-maḡnaṭīsī*	التغير المغنطيسي
magneto *maḡnīṭ, muwallid fulṭīya*	مغنيط . مولد فلطية
magnetometer *maḡnītūmitar*	مغنيطومتر
magnitude of loading *miqdār al-ḥiml*	مقدار الحمل
magnox reactor *mufāʿil maḡnūks*	مفاعل «ماجنوكس»

M

MAG welding لحام بالأكسجين
liḥām bil-aksijīn
mahogany خشب الماهوغوني
kašab al-māhūgūnī
main خط رئيسي
katt ra'īsī
main bar قضيب رئيسي
qaḍīb ra'īsī
main beam عارضة رئيسية
'āriḍa ra'isīya
main cable كبل رئيسي
kabl ra'īsī
main canal قناة رئيسية
qanāt ra'īsīya
main contractor مقاول رئيسي
muqāwil ra'īsī
main door باب رئيسي
bāb ra'īsī
main drain مصرف رئيسي
maṣrif ra'īsī
mainframe هيكل إطار رئيسي
haikal iṭār ra'īsī
main memory ذاكرة رئيسية
ḏakira ra'īsīya
main program برنامج رئيسي
barnāmij ra'īsī
main road طريق رئيسي
ṭarīq ra'īsī
main rod عمود الاتصال الرئيسي
'amūd al-ittiṣāl ar-ra'īsī
main sewer مجرور رئيسي
majrūr ra'īsī
main support دعامة رئيسية
di'āma ra'īsīya
main switch مفتاح كهربائي رئيسي
miftāḥ kahrabā'ī ra'īsī
maintenance صيانة
ṣīyāna
maintenance buildings منشآت الصيانة
munšā'āt aṣ-ṣīyāna
maintenance costs تكاليف الصيانة
takālīf aṣ-ṣīyāna
maintenance hangars عنابر الصيانة
'anābir aṣ-ṣīyāna
maintenance interval فترات الصيانة
fatrāt aṣ-ṣīyāna
maintenance manual دليل إرشادات الصيانة
dalīl iršādāt aṣ-ṣīyāna
maintenance period مدة الصيانة
muddat aṣ-ṣīyāna
maintenance records سجلات الصيانة
sijillāt aṣ-ṣīyāna
main tie شدادة رئيسية
šiddāda ra'īsīya
maisonette شقة ذات طابقين
šuqqa ḏāt ṭabiqain
making good إنهاء جيد
inhā' jayyid
malleability قابلية التطريق
qābilīyat at-taṭrīq
malleable cast iron حديد مطروق
ḥadīd maṭrūq
mammoth pump مضخة ضخمة
miḍakka ḍakma
management contract عقد إدارة
'aqd idāra
managing surveyor مسّاح إداري
massāḥ idārī
mandrel ممسك العدة
mamsak al-'udda
manganese steel فولاذ منغنيزي
fūlāḏ mangānīzī
man handling يحرك أو يدير بالقوة البدنية
yuḥarrik au yudīr bil-qūwa al-badanīya
manhole حفرة تفتيش المجاري
ḥufrat taftīš al-majārī
manhole cover غطاء حفرة التفتيش
giṭā' ḥufrat at-taftīš
man hours ساعات عمل
sā'āt 'amal
manifold متعدد . متشعب
muta'addid, mutašā''ib
manipulator ذراع اللحام
ḏirā' al-liḥām
man-lock غلق المدخل لحجرة مضغوطة
gālq al-madkal li-ḥujra madgūṭa
man-machine interface علاقة الرجل مع الماكينة
'alāqat ar-rajul ma' al-mākina
manmade fibre ليف إصطناعي
līf iṣṭinā'ī
manometer مانومتر (مقياس ضغط)
mānūmitar, miqyās ḍagṭ
manual-block system نظام الاشارات اليدوي
niẓām al-išārāt al-yadawī
manual metal-arc welding لحام يدوي بالقوس المعدني
liḥām yadawī bil-qaus al-ma'danī

manuals of procedures — كتب الاجراءات
kutub al-ijrā'āt

manuals of standing instructions — كتب الارشادات الدائمة
kutub al-iršādāt ad-dā'ima

manual reset — ضبط يدوي
ḍabṭ yadawī

manufacturer's advice — نصيحة المُصنع
nasīhat al-muṣanni'

map — خريطة
ḵarīṭa

mapping — رسم خرائط
rasm ḵarā'iṭ

mapping institute — جمعية وضع الخرائط
jam'īyat waḍ' al-ḵarā'iṭ

maraging steel — فولاذ «ماراجين»
fūlāḏ marājīn

marble — رخام . مرمر
ruḵām, marmar

marbled tiles — بلاط رخامي
bilāṭ ruḵāmī

marble facing — تلبيس بالرخام
talbīs bir-ruḵām

marble gravel — حصى رخامي
ḥaṣa ruḵāmī

marbling — تجزيع كالرخام
tajzī' kar-ruḵām

marigraph — مقياس المد والجزر
miqyās al-madd wal-jazr

marina — حوض لرسو السفن
ḥauḍ li-rasū as-sufun

marine borers — مثاقب بحرية
maṯāqib baḥrīya

marine mooring structure — إنشاء لإرساء بحري
inšā' li-irsā' baḥrī

marine surveying — مسح بحري
masḥ baḥrī

marine terminal facilities — مرافق المحطة البحرية
marāfiq al-muḥaṭṭa al-baḥrīya

maritime plants — معدات بحرية
mu'iddāt baḥrīya

market — سوق
sūq

market study — دراسة السوق
dirāsat as-sūq

masonry — حرفة البناء
ḥarfat al-binā'

masonry bond — ترابط بنائي
tarābuṭ binā'ī

masonry cement — إسمنت بورتلاندي
ismant būrtlānd

mass concrete — خرسانة عادية
ḵarasāna 'ādīya

mass concrete fill — حشو بالخرسانة العادية
hašu bil-ḵarasāna al-'ādīya

mass curve — منحنى إجمالي
munhana ijmālī

mass flow meter — عداد الدفق الكمي
'addād ad-dafq al-kammī

mass-haul curve — منحنى بيان كمية الحفر
munhana bayān kammīyat al-ḥafr

massive masonry pier — ركيزة حجر
rakīzat ḥajar

mast — سارية
sārīya

master plan — خطة رئيسية
ḵuṭṭa ra'īsīya

master specification — مواصفات رئيسية
muwāṣafāt ra'īsīya

masthead gear — تجهيزات قمة الصاري
tajhīzāt qummat aṣ-ṣārī

mastic — مصطكاء
muṣṭaka'

mastic asphalt — إسفلت صمغي
isfalt samḡī

mat — حصيرة . طلية عاتمة
ḥaṣīra, ṭulya 'ātima

material hoist — مرفاع المواد
mirfā' al-mawādd

material procurement specification — مواصفات الحصول على المواد
muwāṣafāt al-ḥuṣūl 'ala al-mawādd

materials cost — تكلفة المواد
taklifat al-mawādd

materials handling — مناولة المواد
munāwalat al-mawādd

materials lock — غلق هوائي للمواد
ḡalq hawā'ī lil-mawādd

materials management — إدارة المواد
idārat al-mawādd

materials programming — برمجة المواد
barmajat al-mawādd

materials progressing — معالجة المواد
mu'ālajat al-mawādd

materials scheduling — برنامج تسليم المواد
barnāmij taslīm al-mawādd

M

materials specification مواصفات المواد
muwāṣafāt al-mawādd
material testing laboratory مختبر إختبار المواد
muktabar iktibār al-mawādd
mat foundation أساس حصيري
asās ḥaṣīrī
matrix قالب . مادة الترابط
qālab, māddat at-tarābuṭ
matrix system نظام مادة الترابط
niẓām māddat at-tarābuṭ
mattress فرشة . حشية
farša, hašya
maturing إنضاج الاسمنت بالترطيب
inḍāj al-ismant bit-tarṭīb
maximum cant الانحدار الأقصى
al-inḥidār al-aqṣā
maximum cement content محتوى الاسمنت الأقصى
muḥtawa al-ismant al-aqṣa
maximum dry density كثافة الجفاف القصوى
kaṯāfat al-jafāf al-quṣwa
maximum gradient معدل الانحدار الأقصى
muʿaddal al-inḥidār al-aqṣa
maximum intensity of rainfall المعدل الأقصى لهطول الأمطار
al-muʿaddal al-aqṣa li-huṭūl al-amṭār
maximum load الحمل الأقصى
al-ḥiml al-aqṣa
maximum permissible speed السرعة القصوى المسموح بها
as-surʿa al-quṣwa al-masmūḥ bi-ha
maximum speed السرعة القصوى
as-surʿa al-quṣwa
maximum water level منسوب الماء الأقصى
mansūb al-māʾ al-aqṣa
mean متوسط
mutawassiṭ
mean depth متوسط العمق
mutawassiṭ al-ʿumq
mean high water متوسط إرتفاع الماء
mutawassiṭ irtifāʿ al-māʾ
mean low water متوسط إنخفاض الماء
mutawassiṭ inkifāḍ al-māʾ
mean radiant temperature متوسط الحرارة المشعة
mutawassiṭ al-ḥarāra al-mušiʿʿa

mean sea level متوسط مستوى سطح البحر
mutawassiṭ mustawa saṭḥ al-baḥr
means of excavation وسائل الحفر
wasāʾil al-ḥafr
mean velocity متوسط السرعة
mutawassiṭ as-surʿa
measured value قيمة مقاسة
qīma muqāsa
measurement قياس
qiyās
measurement of quantity قياس الكميات
qiyās al-kammīyāt
measuring chain سلسلة قياس
silsilat qiyās
measuring element عنصر قياس
ʿunṣur qiyās
measuring tape شريط قياس
šarīṭ qiyās
measuring weir سد صغير للقياس
sadd saġīr lil-qiyās
mechanic ميكانيكي
mikānīkī
mechanical advantage الفائدة الآلية
al-fāʾida al-ālīya
mechanical analysis تحليل آلي
taḥlīl ālī
mechanical bond ترابط ميكانيكي
tarābuṭ mīkānīkī
mechanical construction تركيب ميكانيكي
tarkīb mīkānīkī
mechanical conveyor belt سير ناقلة ميكانيكية
sair nāqila mīkānīkīya
mechanical dredger كراءة ميكانيكية
karrāʾa mīkānīkīya
mechanical efficiency فعالية ميكانيكية
faʿālīya mīkānīkīya
mechanical engineering work أعمال هندسية ميكانيكية
aʿmāl handasīya mīkānīkīya
mechanically operated dredger كراءة تعمل ميكانيكياً
karrāʾa taʿmal mīkānīkīyan
mechanical rammer مدك ميكانيكي
midakk mīkānīkī
mechanical shovel مجرفة ميكانيكية
mijrafa mīkānīkīya
mechanical vibration إهتزاز ميكانيكي
ihtizāz mīkānīkī

M

mechanical washer *falaka mīkānīkīya*	فلكة ميكانيكية
medical examination *fuḥūṣ ṭubbīya*	فحوص طبية
medium density airport *maṭār mutawassiṭ al-kaṯāfa*	مطار متوسط الكثافة
medium-span roof *saṭḥ mutawassiṭ al-bāʿ*	سطح متوسط الباع
medium tolerance *tafāwut mutawassiṭ*	تفاوت متوسط
meeting post *ʿamūd iltiqāʾ*	عمود إلتقاء
mekometer *miqyās masāfāt*	مقياس مسافات
member *juzʾ*	جزء
membrane *ġišāʾ*	غشاء
membrane electric separation *faṣl kahrabāʾī ġišāʾī*	فصل كهربائي غشائي
membrane pressure separation *faṣl bi-ḍaġṭ al-ġišāʾī*	فصل بالضغط الغشائي
membrane process *ʿamalīyat taḥlīyat al-māʾ al-ġišāʾīya*	عملية تحلية الماء الغشائية
membrane waterproofing *aṣ-ṣumūd al-ġišāʾī lil-māʾ*	الصمود الغشائي للماء
memory *ḏākira*	ذاكرة
memory management *idārat aḏ-ḏākira*	إدارة الذاكرة
memory system *niẓām aḏ-ḏākira*	نظام الذاكرة
mercury-vapour fluorescent lamp *miṣbāḥ flūrī ziʾbaqī*	مصباح فلوري زئبقي
mercury-vapour lamp *miṣbāḥ ziʾbaqī*	مصباح زئبقي
mercury vapour with halides *buḵār ziʾbaqī maʿa hālīd*	بخار زئبقي مع هاليد
merging section *qiṭāʿ ad-damj*	قطاع الدمج
meridian *ḵaṭṭ az-zawāl*	خط الزوال
mesh *šabaka*	شبكة
mesh reinforcement *at-tadʿīm aš-šabakī lil-ḵarasāna*	التدعيم الشبكي للخرسانة
metal *maʿdan*	معدن
metal-arc welding *al-liḥām bil-qaus al-maʿdanī*	اللحام بالقوس المعدني
metal bearing *maḥmal maʿdanī*	محمل معدني
metal-deck roof *saṭḥ maʿdanī*	سطح معدني
metalled road *ṭarīq marṣūf*	طريق مرصوف
metalling *saṭḥ aṭ-ṭarīq*	سطح الطريق
metallurgical cement *ismant maʿādan*	إسمنت معادن
metal oxide *aksīd al-maʿdan*	أكسيد المعدن
meteorological equipment *ajhizat ar-raṣd*	أجهزة الرصد
meteorology *ʿilm al-arṣād al-jawwīya*	علم الأرصاد الجوية
meter *ʿaddād*	عداد
metered supply *imdād muqās*	إمداد مُقاس
methane *ġāz al-mīṯān*	غاز الميثان
metre *matr*	متر
metric system *niẓām matrī*	نظام متري
mezzanine *ṭabiq mutawassiṭ*	طابق متوسط
microcomputer *mīkrū-kumbyūtar*	ميكروكمبيوتر
micrometer *mīkrūmitar*	ميكرومتر
micrometer gauge *miqyās mīkrūmitrī*	مقياس ميكرومتري
micron *mīkrūn*	ميكرون
microphone *mīkrūfūn*	ميكروفون
microprocessor *mīkrūbrūsisūr*	ميكروبروسسور
microprocessor operation *tašġīl al-mīkrūbrūsisūr*	تشغيل الميكروبروسسور

M

microptic theodolite — مزاواة (تيودوليت)
mizwāt, tīyūdūlīt
microscope — ميكروسكوب
mīkrūskūb
micro-strainer — مصفاة دقيقة
miṣfat daqīqa
middle third — ثلث الوسط
ṯulṯ al-wasaṭ
middling-board — لوح تثبيت
lauḥ taṯbīt
middling frame — إطار تثبيت
iṭār taṯbīt
mid-ordinate — الاحداثي الأوسط
al-iḥdāṯī al-ausaṭ
midpoint — نقطة الوسط
nuqṭat al-wasaṭ
midspan — الباع الأوسط
al-bā' al-ausaṭ
midspan clearance — خلوص الباع الأوسط
ḵulūṣ al-bā' al-ausaṭ
MIG welding — لحام بسلك معدني
liḥām bi-silk ma'danī
mild steel — فولاذ طري
fūlāḏ ṭarī
mild steel bar — قضيب فولاذ طري
qaḍīb fūlāḏ ṭarī
mile — ميل
mīl
military airport — مطار عسكري
maṭār 'askarī
military harbour — ميناء عسكري
mīnā' 'askarī
mill — طاحونة
ṭāḥūna
millimetre — مليمتر
milimītar
millimicron — مليميكرون
milimīkrūn
milling — طحن . تفريز
ṭaḥn, tafrīz
milling machine — مكنة تفريز
makinat tafrīz
million — مليون
milyūn
miner's dip needle — إبرة الميل لعامل المنجم
ibrat al-mail li-'āmil al-manjam
mineshaft — مهواة المنجم
mihwāt al-manjam
minicomputer — كمبيوتر صغير
kumbyūtar ṣaġīr
minimum air velocity — سرعة الهواء الدنيا
sur'at al-hawā' ad-dunya
minimum cement content — محتوى الاسمنت الأدنى
muḥtawa al-ismant al-adna
mining — تعدين . زرع الالغام
ta'dīn, zar' al-alġām
mini-roundabout — دائرة إلتفاف صغيرة
dā'irat iltifāf ṣaġīra
minor road — طريق فرعي
ṭarīq far'ī
minus sight — مهداف التسوية السالب
mihdāf at-taswiya as-sālib
misappropriation — سوء الاستعمال
sū' al-isti'māl
misfires — إخفاق الاشعال
iḵfāq al-iš'āl
mitre post — عمود مشطوب
'amūd maštūb
mitre sill — عتبة مشطوبة
'ataba maštūba
mix — خلط
ḵalṭ
mix design — تصميم الخلط
taṣmīm al-ḵalṭ
mixed-flow turbine — تربين الدفق المختلط
tūrbīn ad-dafq al-muḵtalaṭ
mixer — خلاط . خلاطة
ḵallāṭ, ḵallāṭa
mix-in-place — خلط في الموقع
ḵalṭ fil-mauqi'
MMA welding — لحام يدوي بالقوس المعدني
liḥam yadawi bil-qaus al-ma'danī
mobile belt loader — حمالة متحركة بالسير
ḥammāla mutaḥarrika bis-sair
mobile concrete pump — مضخة خرسانة متحركة
miḍakkat ḵarasāna mutaḥarrika
mobile crane — ونش متنقل
winš mutanaqqil
mobile hoist — مرفاع متنقل
mirfā' mutanaqqil
model — نموذج
namūḏij
model analysis — تحليل النماذج
taḥlīl an-namāḏij

M

modular ratio النسبة المعيارية
an-nisba al-maʿyārīya
modulator مضمن
muḍammin
modulus of elasticity مُعامل المرونة
muʿāmil al-murūna
modulus of incompressibility مُعامل اللاإنضغاطية
muʿāmil al-lā-indiġāṭīya
modulus of resilience مُعامل الرجوعية
muʿāmil at-rujūʿīya
modulus of rigidity مُعامل الصلابة
muʿāmil aṣ-ṣalāba
modulus of rupture مُعامل التصدع
muʿāmil at-taṣaddu'
modulus of section مُعامل المقطع
muʿāmil al-maqṭaʿ
Mohr's circle دائرة «موهر»
dā'ira muhar
Mohr's circle of stress دائرة «موهر» للاجهاد
dā'irat muhar lil-ijhād
moisture content المحتوى الرطوبي
al-muḥtawa ar-ruṭūbī
moisture meter مقياس الرطوبة
maqāyīs ar-ruṭūba
moisture movement تحرك الرطوبة
taḥarruk ar-ruṭūba
mole حفارة أنفاق
ḥaffārat anfāq
mole drain فتحة تصريف
fatḥat taṣrīf
mole plough حفارة جارفة
ḥaffāra jārifa
moling حفر أنفاق الأنابيب
ḥafr anfāq al-anābīb
moment distribution توزيع العزم
tauzīʿ al-ʿazm
moment of a force عزم القوة
ʿazm al-qūwa
moment of inertia عزم العطالة
ʿazm al-ʿaṭāla
moment of resistance عزم المقاومة
ʿazm al-muqāwama
momentum كمية التحرك
kammīyat at-taḥarruk
monkey رأس مدق الخوازيق
ra's midaqq al-ḵawāzīq
monoblock concrete sleeper عارضة خرسانية أحادية
ʿāriḍa ḵarasānīya uḥādīya
monocable أحادي الكبل
uḥādī al-kabl
monolith عمود منفصل
ʿamūd munfaṣil
monolithic متآلف
muta'ālif
monorail سكة أحادية
sikka uḥādīya
monotower crane مرفاع أحادي البرج
mirfāʿ uḥādī al-burj
monument نصب تذكاري
nuṣb tiḏkārī
mooring buoy عوامة إرساء
ʿawwāmat irsā'
mooring-buoy anchor مثبتات عوامة الارساء
muṯabbitāt ʿawwāmat al-irsā'
mooring dolphins مراسي إرساء
marāsī irsā'
mooring line حبل الارساء
ḥabl al-irsā'
mooring post وتد الارساء
watad al-irsā'
mooring wire rope حبل سلكي للارساء
ḥabl lil-irsā'
mortar ملاط
milāṭ
mortise and tenon joint وصلة نقرة ولسان
waṣlat naqra wa-lisān
mortise lock قفل مبيت
qufl mubayyit
mosaic فسيفساء
fusaifisā'
most probable value القيمة الأكثر إحتمالا
al-qīma al-akṯar iḥtimālan
motion study دراسة حركة العامل
dirāsat ḥarakat al-ʿāmil
motor grader مدرجة آلية
mudarrija ālīya
motor truck شاحنة
šāḥina
motorway طريق سريع
ṭarīq sarīʿa
mould قالب الصب
qālab aṣ-ṣabb
mouldboard لوح التشكيل
lauḥ at-taškīl
moulding machine مكنة تشكيل
makinat taškīl

M

moulding press مكبس تشكيل
mikbas taškīl
moulding sand رمل السبك الطفالي
raml as-sabk aṭ-ṭufālī
mound هضبة صغيرة
haḍaba ṣaǧīra
mound breakwater مصد أمواج ركامي
miṣadd amwāj rukāmī
mountain railway سكة حديد جبلية
sikkat ḥadīd jabalīya
mouth فوهة . فم
fauha, famm
movable bridge جسر متحرك
jisr mutaḥarrik
movable dam سد متحرك
sadd mutaḥarrik
movable scaffold سقالة متحركة
saqāla mutaḥarrika
move instruction تعليمات الحركة
taʿlīmāt al-ḥaraka
movement joints وصلات الحركة
waṣlāt al-ḥaraka
moving head disc قرص متحرك الرأس
qurṣ mutaḥarrik ar-raʾs
moving load حمل متحرك
ḥiml mutaḥarrik
muck نفاية
nufāya
muck excavation حفر الطين
ḥafr aṭ-ṭīn
mucking تعزيل الصخر
taʿzīl aṣ-ṣak̲r
mudcapping تغطية الوحل
taǧtiyat al-waḥl
mudguard واقية الوحل
wāqīyat al-waḥl
mud jacking رفع الوحل
rafʿ al-waḥl
mud soil تربة طينية
turba ṭīnīya
mud trap مصيدة الوحل
miṣyadat al-waḥl
multi-bucket excavator حفارة متعددة القواديس
haffāra mutaʿaddidat al-qawādīs
multicore cable كبلات متعددة الأسلاك
kablāt mutaʿaddidat al-aslāk
multicore power cable كبلات طاقة متعددة الأسلاك
kablāt ṭāqa mutaʿaddidat al-aslāk
multi-flue chimney مدخنة متعددة المجاري
midk̲ana mutaʿaddidat al-majārī
multipair cables كبلات متعددة الأزواج
kablāt mutaʿaddidat al-azwaj
multi-phase flow دفق متعدد المراحل
dafq mutaʿaddid al-marāḥil
multiple arch concrete dam سد خرسانة متعدد القناطر
sadd k̲arasāna mutaʿaddid al-qanāṭir
multiple-arch dam سد متعدد القناطر
sadd mutaʿaddid al-qanāṭir
multiple-expansion engine محرك متعدد المراحل
muḥarrik mutaʿaddid al-marāḥil
multiple shear قص متعدد
qaṣṣ mutaʿaddid
multi-register processor بروسسور متعدد المسجلات
brūsisūr mutaʿaddid al-musijjilāt
multistage centrifugal compressor ضاغط نابذ متعدد المراحل
ḍāǧiṭ nābiḏ mutaʿaddid al-marāhil
multistage flash distillation تحلية الماء بالبثق متعدد المراحل
taḥliya al-māʾ bil-baṭq mutaʿaddid al-marāḥil
multi-step control مضبط مدرّج
maḍbaṭ mudarraj
multi-storey building بناية متعددة الطوابق
bināya mutaʿaddidat aṭ-ṭawābiq
multi-storey tower برج متعدد الطوابق
burj mutaʿaddid aṭ-ṭawābiq
multi-wheel roller دلفين بعجلات هوائية
dalfīn bi-ʿajalāt hawāʾīya
multizone متعدد المناطق
mutaʿaddid al-manāṭiq
municipal consumption إستهلاك بلدي
istihlāk baladī
municipal engineering الهندسة البلدية
al-handasa al-baladīya

municipal works الأشغال البلدية
al-ašġāl al-baladīya
museum متحف
mathaf
mushroom construction إنشاء من السطوح والأعمدة
inšā' min as-suṭūḥ wal- a'mida
nappe صخر مغترب
ṣakr muġtarib
narrow-base tower برج ضيق القاعدة
burj ḍayyiq al-qā'ida
narrow gauge سكة حديدية ضيقة
sikka ḥadīdīya ḍayyiqa
national standards المواصفات الوطنية
al-muwāṣafāt al-waṭanīya
natural asphalt أسفلت طبيعي
isfalt ṭabī'ī
natural cement إسمنت طبيعي
ismant ṭabī'ī
natural cooling تبريد طبيعي
tabrīd ṭabī'ī
natural draught سحب طبيعي
saḥb ṭabī'ī
natural-draught cooling tower برج تبريد بالسحب الطبيعي
burj tabrīd bis-saḥb aṭ-ṭabī'ī
natural foundation أساس طبيعي
asās ṭabī'ī
natural frequency تردد طبيعي
tarradud ṭabī'ī
natural gas الغاز الطبيعي
al-ġāz aṭ-ṭabī'ī
natural ground الأرض الطبيعية
al-arḍ aṭ-ṭabī'īya
natural harbour ميناء طبيعي
mīnā' ṭabī'ī
natural rock صخر طبيعي
ṣakr ṭabī'ī
natural scale مقياس رسم طبيعي
miqyās rasm ṭabī'ī
natural slope إنحدار طبيعي
inḥidar ṭabī'ī
natural ventilation تهوية طبيعية
tahwiya ṭabī'īya
Nautical Almanac تقويم بحري
taqwīm baḥrī
nautical mile ميل بحري
mīl baḥrī
Navier's hypothesis إفتراض «نافيير»
iftirāḍ nāfīr
navigable channel قناة صالحة للملاحة
qanāt ṣāliḥa lil-milāḥa
navigable waterway ممر مائي صالح للملاحة
mamarr mā'ī ṣāliḥ lil-milāḥa
navigation ملاحة
milāḥa
navigation aids معينات ملاحية
mu'īnāt milāḥīya
navvy عامل حفر. مجرفة آلية
'āmil ḥafr, mijrafa ālīya
N-channel metal oxide semiconductor القناة الحيادية لأكسيد المعدن شبه الناقل
al-qanāt al-ḥīyādīya li-aksīd al-mā'dan šubh an-nāqil
NDT إختبار غير هدام
ikṭibār ġair haddām
neap tide مد أو جزر ناقص
madd au jazr nāqiṣ
neat lines حدود الحفر المأجور
ḥudūd al-ḥafr al-ma'jūr
necking تخصر
takaṣṣur
needle إبرة . مسلة
ibra, misalla
needle beam قدة رفيعة معترضة
qudda rafī'a mu'tariḍa
needle instrument مقياس بإبرة مؤشرة
miqyās bi-ibra mu'aššira
needle jet منفث إبري
munaffiṯ ibrī
needle nozzle فوهة إبرية
fauha ibrīya
needle roller bearing محمل أسيطينات إبري
maḥmal usīṭīnāt ibrī
needle traverse مسح إجتيازي بالبوصلة
masḥ ijtiyāzī bil-būṣala
needle valve صمام إبري
ṣimām ibrī
needle weir قنطرة إحتجاز بهيكل ثابت
qanṭarat iḥtijāz bi-haikal ṯābit
needling تركيب جائز في الجدار
tarkīb jā'iz fil-jidār
negative camber تقوس سالب
taqawwus sālib
negative skin friction إحتكاك سطحي سالب
iḥtikāk saṭḥī sālib

N

negotiated contract عقد مفاوَض
ʿaqd mufāwaḍ
net duty رسم الماء للمزرعة
rasm al-māʾ lil-mazraʿa
net loading intensity شدة التحميل الصافي
šiddat at-taḥmīl aṣ-ṣāfī
net ton الطن الأمريكي
aṭ-ṭann al-amrīkī
network analysis تحليل للشبكة
taḥlīl liš-šabaka
network of pipes شبكة أنابيب
šabakat anābīb
neutral axis محور التعادل
miḥwar at-taʿādul
neutral pressure ضغط متعادل
ḍaġṭ mutaʿādil
nibbling machine مقرضة (للألواح المعدنية)
miqraḍa lil-alwaḥ al-maʿdanīya
nine تسعة
tisʿa
nineteen تسعة عشر
tisʿataʿašar
ninety تسعون
tisʿūn
nip قرص . قرض
qarasa, qaraḍa
nipple ثقب تزييت
ṯuqb tazyīt
nitrocellulose نتروسليلوز
nītrū sīlīlūz
nitroglycerin نتروغليسرين
nītrūġlīsrīn
node عقدة . عجرة
ʿuqda, ʿajra
Noise and Number Index دليل الضوضاء والرقم
dalīl aḍ-ḍauḍāʾ war-raqm
noise control التحكم في الضوضاء
at-taḥakkum fiḍ-ḍauḍāʾ
noise eliminator مانع الضجيج
māniʿ aḍ-ḍajīj
noiseless engine محرك صامت
muḥarrik ṣāmit
noise level مستوى الضجيج
mustawa aḍ-ḍajīj
nominal mix الخليط الإسمي
al-ḵalīṭ al-ismī
nominal output الخرج الإسمي
al-ḵarj al-ismī

nomogram نوموچرام : رسم بياني
nūmūğrām, rasm bayānī
non-bearing wall جدار غير داعم
jidār ġair dāʿim
non-classified road طريق غير مصنف
ṭarīq ġair muṣannaf
non-clogging centrifugal pump مضخة نابذة عديمة الانسداد
miḍakka nābiḍa ʿadīmat al-insidād
non-cohesive soil تربة عديمة التماسك
turba ʿadīmat at-tamāsuk
non-collusion affidavit شهادة عدم التآمر
šahādat ʿadam at-taʾāmur
non-destructive testing إختبار غير هدام
iḵtibār ġair haddām
non-illuminated signs لافتات غير مضاءة
lāfitāt ġair muḍāʾa
non-instrument runway مدرج بدون أجهزة
madraj bidūn ajhiza
non-load-bearing wall جدار غير حامل
jidār ġair ḥāmil
non-metallic minerals معادن لافلزية
maʿādin lā-fillizīya
non-metallic sheathed cable كبل مدرع لافلزي
kabl mudarraʿ lā-fillizī
non-return valve صمام لا رجعي
ṣimām lā-rajʿī
non-reversible clutch قابض لا إنعكاسي
qābiḍ lā-inʿikāsī
non-skid surface سطح مضاد للانزلاق
saṭḥ muḍādd lil-inzilāq
non-slip floor أرضية لا إنزلاقية
arḍīya lā-inzilāqīya
non-slip surface سطح عديم الانزلاق
saṭḥ ʿadīm al-inzilāq
non-tilting drum machine مكنة دوارة غير قلاّبة
makina dawwāra ġair qallāba
non-tilting mixer خلاطة غير قلاّبة
ḵallāṭa ġair qallāba
normal flow pattern نمط الدفق العادي
namaṭ ad-dafq al-ʿādī
normalizing معالجة بالحرارة
muʿālaja bil-ḥarāra
normal law of error قانون الخطأ العادي
qānūn al-ḵaṭaʾ al-ʿādī
normal lift مصعد عادي
miṣʿad ʿādī

normal maintenance صيانة عادية
ṣīyāna ʿādīya
normal minimum flight path مسار الطيران الأدنى والعادي
masār aṭ-ṭayarān al-adna al-ʿādī
normal stress الاجهاد العادي
al-ijhād al-ʿādī
normal working conditions ظروف العمل العادية
ẓurūf al-ʿamal al-ʿādīya
nose devices أدوات النتوء
adawāt an-nutūʾ
nose of a step الجزء الناتئ من الدرجة
al-juzʾ an-natiʾ min ad-daraja
nosing شفة الدرجة
šaffat ad-daraja
notch ثلمة . حز. نقر
ṯalma, ḥazz, naqr
notched bar test إختبار التصادم بالقضيب المحزز
iḳtibār at-tasādum bil-qaḍīb al-muḥazzaz
notched weir سد صغير محزز
sadd ṣaġīr muḥazzaz
notch effect تأثير التحزيز
taʾṯīr at-taḥzīz
notcher مكنة نقر وتحزيز
makinat naqr wa-taḥzīz
notching وصل بالنقر
waṣl bin-naqr
not negotiable لا يقبل المفاوضة
la-yaqbil al-mufāwaḍa
not to scale بغير مقياس الرسم
bi-ġair miqyās ar-rasm
nozzle فوهة . صنبور
fauha, ṣanbūr
nozzle orifice فتحة الفوهة
fatḥat al-fauha
nuclear energy الطاقة النووية
aṭ-ṭāqa an-nawawīya
nuclear energy plant محطة طاقة نووية
mahaṭṭat ṭāqa nawawīya
nuclear fuel وقود نووي
wuqūd nawawī
nuclear power station محطة توليد بالطاقة النووية
muḥaṭṭaṭ taulīd biṭ-ṭāqa an-nawawīya
nuclear reaction تفاعل نووي
tafāʿul nawawī
number of lifts عدد المصاعد
ʿadad al-maṣāʿid
nut صمولة
ṣamūla
nutating-disk meter مقياس الترنح
miqyās at-tarannuḥ
nut wrench مفتاح ربط الصمولات
miftāḥ rabṭ aṣ-ṣamūlāt
oak بلوط . سنديان
ballūṭ, sindīyān
object libraries مكتبات خرج الكمبيوتر
maktabāt ḳarj al-kumbyūtar
object library editing تحرير مكتبة خرج الكمبيوتر
taḥrīr maktabat ḳarj al-kumbyūtar
object program برنامج نتائج الكومبيوتر
barnāmij natāʾij al-kumbyūtar
oblique مائل
māʾil
oblique aerial photograph صورة جوية مائلة
ṣūra jawwīya māʾila
oblique arch قوس مائل المحور
qaus māʾil al-miḥwar
oblique offset بعد أفقي مائل
buʿd ufuqi māʾil
oblique projection إسقاط مائل
isqāṭ māʾil
oblique section قطاع مائل
qiṭāʿ māʾil
oblique valve صمام مائل
ṣimām māʾil
obliterating paint دهان طمس
dihān ṭams
obscure glass زجاج معتم
zujāj muʿattam
obscure wired glass زجاج معتم مقوى بالأسلاك
zujāj muʿattam muqawwa bil-aslāk
observatory مرصد
marṣad
obstruction-clearance line خط خلوص الحواجز
ḳaṭṭ ḳulūṣ al-ḥawājiz
obstruction criteria معايير الحواجز
māʿyīr al-ḥawājiz
obstruction lights أضواء الحواجز
aḍwāʾ al-ḥawājiz
obstruction warning device أداة التحذير من الحواجز
adāt at-taḥḏīr bil-hawājiz

O

obtuse angle crossing — تقاطع منفرج الزاوية
taqāṭuʿ munfarij az-zāwīya

occupational hazard — خطر مهني
ḵaṭar mihanī

octal — مثمن الأضلاع
muṯamman al-aḏlāʿ

office — مكتب
maktab

office building — بناية مكاتب
bināyat makātib

off-peak load — حمل دون الذروة
ḥiml dūn aḏ-ḏurwa

offset — زيحان
zayaḥān

offset scale — مقياس مدرج قصير
miqyās mudarraj qaṣīr

offset spanner — مفتاح ربط مجنب
miftāḥ rabṭ mujannab

offset yield strength — قوة الخضوع الجانبية
quwwat al-ḵuḍuʿ al-janibīya

off-shore dock — حوض بعيد عن الشاطئ
ḥauḍ baʿīd ʿan aš-šāṭiʾ

off-shore mooring — إرساء بعيد عن الشاطئ
irsāʾ baʿīd ʿan aš-šāṭiʾ

ogee — حلية معمارية مموجة
hilya miʿmārīya mumawwaja

ogee joint — وصلة ذكر وأنثى
waṣlat ḏakar wa-unṯa

Ohio cofferdam — سد مزدوج الجدار
sadd muzdawaj al-jidār

ohm — أوم
ūm

oil bath filter — مرشح ذو حمام زيتي
muraššiḥ ḏū ḥammām zaitī

oil cooler — مبرِّد الزيت
mubarrid az-zait

oil discharge duct — قناة تصريف الزيت
qanāt taṣrīf az-zait

oil drain — مصرف الزيت
maṣrif az-zait

oil-fired power station — محطة طاقة تعمل بالمازوت
muḥaṭṭat ṭāqa taʿmal bil-māzūt

oil inlet pipe — أنبوب دخول الزيت
unbūb duḵūl az-zait

oil outlet — مخرج الزيت
maḵraj az-zait

oil pressure indicator — مؤشر ضغط الزيت
muʾaššir ḍaġṭ az-zait

oil pressure regulator — مُنظم ضغط الزيت
munaẓẓim ḍaġṭ az-zait

oil storage tank — خزان الزيت
ḵazzān az-zait

oil-well cement — إسمنت بطيء الشك
ismant baṭīʾ aš-šakk

oil-well derrick — برج بئر النفط
burj biʾr an-nafṭ

on a large scale — على نطاق واسع
ʿala niṭāq wāsiʿ

on centre — في المركز
fil-markaz

one — واحد
wāḥid

one coat — طلية واحدة
ṭalya wāḥida

one-pipe system — نظام أحادي الماسورة
niẓām uḥādī al-māsūra

one way traffic — مرور باتجاه واحد
murūr bi-ittijāh wāḥid

on-off control — مضبط الوصل والقطع
miḍbaṭ al-waṣl wal-qaṭʿ

onshore marine-terminal facilities — مرافق المحطة البحرية الساحلية
marāfiq al-muḥaṭṭa al-baḥrīya as-sāhilīya

on-site concrete — صب الخرسانة بالموقع
ṣabb al-ḵarasāna bil-mauqiʿ

on-site incineration — حرق القمامة بالموقع
ḥarq al-qumāma bil-mauqiʿ

on-track ballast cleaner — آلة تنظيف الحصى عن السكة
ālat tanẓīf al-ḥaṣa ʿan as-sikka

on-track equipment — معدات السكة الحديدية
muʿiddāt as-sikka al-ḥadīdīya

on-track maintenance work — أعمال صيانة السكة الحديدية
aʿmāl ṣīyanat as-sikka al-ḥadīdīya

on-track tamping machine — آلة الدك على السكة
ālat ad-dakk ʿala as-sikka

opaque glass — زجاج غير شفاف
zujāj ġair šaffāf

OP CODE — رمز أمر التشغيل
ramz amr at-tašġīl

open-bottom shaft — بئر مفتوح القاع
biʾr maftūḥ al-qāʿ

open caisson قيسون مفتوح
qaisūn maftūḥ
open channel قناة مكشوفة
qanāt makšūfa
open cut حفر مكشوف
ḥafr makšūf
open-divided scale مقياس مفتوح التدريج
miqyās maftūḥ at-tadrīj
open drain مصرف مفتوح
maṣrif maftūḥ
open floor أرضية مكشوفة
arḍīya makšūfa
open-frame girder عارضة «ڤيرنديل»
ʿāriḍat fīrndīl
open-grid flooring أرضية شبكية
arḍīya šabakīya
open gutter مزراب مفتوح
mizrāb maftūḥ
opening فتحة . إفتتاح
fatḥa, iftitāḥ
open jetty رصيف مكشوف
raṣīf makšūf
open sheeting ألواح متباعدة
alwaḥ mutabāʿida
open stairway درج مفتوح
darraj maftūḥ
open-tank treatment علاج الخشب في الخزان المفتوح
ʿilāj al-ḵašab fil-ḵazzān al-maftūḥ
open traverse عارضة مفتوحة
ʿāriḍa maftūḥa
open-type construction إنشاء مفتوح
inšāʾ maftūḥ
open-web girder عارضة تشابكية
ʿāriḍa tašābukīya
open-web steel joist رافدة فولاذية تشابكية
rāfida fūlāḏīya tašābukīya
open-well stair درج ببئر مفتوح
darraj bi-biʾr maftūḥ
open wharf رصيف بحري مفتوح
raṣīf baḥrī maftūḥ
operand الكمية المتأثرة
al-kammīya al-mutaʾaṯira
operating costs تكاليف التشغيل
takālīf at-tašġīl
operating hand wheel دولاب التشغيل اليدوي
dūlāb at-tašġīl al-yadawī
operating pressure ضغط التشغيل
ḍaġṭ at-tašġīl
operating speed سرعة التشغيل
surʿat at-tašġīl
operating temperature درجة حرارة التشغيل
darajat ḥarārat at-tašġīl
operation code مبادئ التشغيل
mabādiʾ at-tašġīl
operation manual دليل إرشادات التشغيل
dalīl iršādāt at-tašġīl
operation waste فضلات التشغيل
faḍalāt at-tašġīl
operator موظف التشغيل
muwaẓẓaf at-tašġīl
optical coincidence bubble فقاعات تطابق موشوري
fuqāʿāt tatābuq mūšūrī
optical compensation تعويض بصري
taʿwīḍ baṣarī
optical distance measurement قياس البعد البصري
qiyās al-buʿd al-baṣarī
optical plummet ثقل الفادن البصري
ṯuql al-fādin al-baṣarī
optical-reading theodolite مزواة بقراءة بصرية
muzwāt bi-qirāʾa baṣarīya
optical sound recording تسجيل الصوت الموشوري
tasjīl aṣ-ṣaut al-mūšūrī
optical wedge موشور بصري
mūšūr baṣarī
optimum moisture content محتوى الرطوبة المثلى
muḥtawa ar-ruṭūba al-muṯla
orange-peel bucket قادوس نصف كروي
qādūs niṣf kurawī
orchard بستان
bustān
order of precedence ترتيب الأقدمية
tartīb al-aqdamīya
ordinary lay تسوية عادية
taswiya ʿādīya
ordinary Portland cement إسمنت بورتلاند عادي
ismant būrtlānd ʿādī
ordinates الاحداثيات الرأسية
al-iḥdāṯīyāt ar-raʾsīya
ordnance survey مصلحة المساحة
maṣlaḥat al-massāha
organic content المحتوى العضوي
al-muḥtawa al-ʿuḍwī
organic silt غرين عضوي
ġarīn ʿuḍwī

O

oriented-core barrel	*istiwānat al-ḥafr al-muwajjah*	أسطوانة الحفر الموجَّه
orifice	*fauha, fatḥa*	فوهة . فتحة
orifice meter	*miqyās at-tadaffuq al-fauhī*	مقياس التدفق الفوهي
orifice plate	*lauḥ at-tadaffuq al-fauhī*	لوح التدفق الفوهي
orifice speed	*sur'at at-tadaffuq al-fauhī*	سرعة التدفق الفوهي
origin	*aṣl, maṣdar, manša'*	أصل . مصدر . منشأ
orthographic projection	*isqāṭ muta'āmid*	إسقاط متعامد
orthophoto	*ṣūra 'amūdīya*	صورة عمودية
orthotropic	*musta'mad*	مستعمد
orthotropic-deck bridge	*jisr bi-saṭḥ musta'mad*	جسر بسطح مستعمد
orthotropic plate floor	*arḍīyat alwāḥ musta'mada*	أرضية ألواح مستعمدة
oscillating piston	*kabbās mutaḏabḏib*	كباس متذبذب
osmosis	*at-tanāḍuḥ*	التناضح
osmotic pressure	*aḍ-ḍaġṭ at-tanāḍuḥī*	الضغط التناضحي
outburst bank	*hayajān aḍ-ḍaffa*	هيجان الضفة
outcrop	*nutū' aṣ-ṣaḵr*	نتوء الصخر
outer door	*al-bāb al-ḵārijī*	الباب الخارجي
outer zone	*al-minṭaqa al-ḵārijīya*	المنطقة الخارجية
outfall	*maḵraj at-taṣrīf*	مخرج التصريف
outfall culvert	*majrūr maḵraj at-taṣrīf*	مجرور مخرج التصريف
outfall drain	*maḵraj at-taṣrif*	مخرج التصريف
outfall sewer	*majrūr ḵārijī ra'īsī*	مجرور خارجي رئيسي
outlet	*maḵraj*	مخرج
outlet capacity	*si'at al-maḵraj*	سعة المخرج

outlet channel	*qanāt al-ḵurūj*	قناة الخروج
outlet control unit	*waḥdat at-taḥakkum fil-maḵraj*	وحدة التحكم في المخرج
outlet culvert	*majra al-ḵurūj as-sufli*	مجرى الخروج السفلي
outlet ventilator	*mihwāt al-maḵraj*	مهواة المخرج
outline drawing	*rasm ijmālī*	رسم إجمالي
outline plan	*ḵuṭṭa ijmālīya*	خطة إجمالية
out of phase	*ḵārij aṭ-ṭaur*	خارج الطور
output	*nitāj, mardūd*	نتاج . مردود
output device	*wasīlat al-ḵarj*	وسيلة الخرج
outrigger	*ḏirā' imtidād, misnad*	ذراع إمتداد . مسند
outrigger scaffold	*minaṣṣat imtidād*	منصة إمتداد
outside view	*manẓar ḵārijī*	منظر خارجي
overall depth	*al-'umq al-kullī*	العمق الكلي
overall efficiency	*al-kifāya al-ijmālīya*	الكفاية الاجمالية
overall height	*al-irtifā' al-ijmālī*	الارتفاع الاجمالي
overall length	*aṭ-ṭūl al-ijmālī*	الطول الاجمالي
overbreak	*al-ḥafr az-zā'id*	الحفر الزائد
overburden	*taḥmīl mufriṭ*	تحميل مفرط
overburden pressure	*ḍaġṭ at-taḥmīl al-mufriṭ*	ضغط التحميل المفرط
over-consolidated clay	*ṣalṣāl mufriṭ al-indimāj*	صلصال مفرط الاندماج
overfall	*masqaṭ miyāh as-sadd*	مسقط مياة السد
overflow dam	*sadd aṭ-ṭafḥ*	سد الطفح
overflow pipe	*māsūrat aṭ-ṭafḥ*	ماسورة الطفح
overflow pit	*ḥufrat aṭ-ṭafḥ*	حفرة الطفح

overflow spillway قناة تصريف الفائض
qanāt taṣrīf al-fā'iḍ
overflow stand ماسورة قائمة لتصريف الفائض
māsūra qā'ima li-taṣrīf al-fā'id
overflow weir سد الفائض
sadd al-fā'iḍ
overhanging متدلي
mutadallī
overhaul ترميم
tarmīm
overhead conductor موصل معلق
muwaṣṣil mu'allaq
overhead conveyor ناقل معلق
nāqil mu'allaq
overhead costs نفقات إضافية
nafaqāt iḍāfīya
overhead crane مرفاع معلق
mirfā' mu'allaq
overhead lines خطوط معلقة
kuṭūṭ mu'allaqa
overhead pipes أنابيب معلقة
anābīb mu'allaqa
overhead ropeway طريق حبلي معلق
ṭarīq ḥablī mu'allaq
overheads مصروفات عامة
maṣrūfāt 'āmma
overhead travelling crane مرفاع علوي متحرك
mirfā' 'ulwī mutaḥarrik
overhead valve صمام معلق
ṣimām mu'allaq
overheating فرط الاحماء
farṭ al-iḥmā'
overhung beam عارضة معلقة
'āriḍa mu'allaqa
overlap block signalling الاشارة الهادية المتداخلة
al-išāra al-hadīya al-mutadākila
overlapping joint وصلة متراكبة
waṣla mutarākiba
overload حمل زائد
ḥiml zā'id
overload capacity سعة التحمیل
si'at at-taḥammul
overload conditions ظروف التحميل
ẓurūf at-taḥammul
overloading تجاوز التحميل
tajāwuz at-taḥmīl
overload tests إختبارات فرط التحميل
iktibārāt farṭ at-taḥmīl

overrun brake مكبح العربة المقطورة
mikbaḥ al-'araba al-maqṭūra
oversize ضخم
ḍakm
overtopping تعلية
ta'līya
overturning إنقلاب
inqilāb
owner مالك
mālik
owner's protective public liability تأمين المسؤولية العامة لحماية المالك
ta'mīn al-mas'ūliya al-'āmma li-ḥimāyat al-mālik
oxidation أكسدة . تأكسد
aksada, ta'aksud
oxidation pond بركة الأكسدة
burkat al-aksada
oxy-acetylene flame لهب الأكسجين والاستلين
lahab al-aksijīn wal-astalīn
oxy-acetylene welding لحام بالأكسجين والاستلين
liḥām bil-aksijīn wal-astalīn
oxychloride cement إسمنت المغنسيت
ismant al-magnasīt
ozonizing تشبيع بالأوزون
tašbī' bil-ūzūn
pack حزمة . حشوة
ḥuzma, hašwa
packed معبأ . محشو
mu'aba', maḥšū
packer بطانة تقوية
biṭānat taqwīya
packing تعبئة . حشوة
ta'bi'a, hašwa
paddle مغدف . محراك
magdaf, miḥrāk
paddle hole فتحة التغديف
fatḥat at-tagdīf
pad foundation حصيرة دعم
sīrat da'm
padlocked plug سدادة قفل
sidādat qifl
page صفحة
ṣafḥa
paged memory ذاكرة مرقمة
ḏākira muraqqama
paint دهان طلاء
dihān, ṭilā'

P

painter دهّان
dahhān
paint spray gun مسدس رش الدهان
musaddas rašš ad-dihān
paint work أعمال الدهان
a'māl ad-dihān
palace قصر
qaṣr
pallet منصة نقالة
minaṣṣa naqāla
pallet conveyor ناقل بمنصة
nāqil bi-minaṣṣa
panel لوح . لوحة
lauḥ, lauḥa
panel point عقدة اللوح
'uqdat al-lauḥ
panel system نظام الألواح
niẓām al-alwāḥ
pan head رأس مخروطي
ra'suma makrūṭī
pan mixer خلاط ملاط
kallāṭ milāṭ
pannier سلة كبيرة
salla kabīra
pantograph بانتوغراف . منساخ
bantūgrāf, minsāk
paperhanging لصق ورق الجدران
laṣq waraq al-judrān
parabola قطع مكافئ
qaṭ' mukāfi'
paraffin برافين . شمع معدني
barāfīn
parallel-flanged beam ركيزة عامة
rakīza 'āmma
parallel heading واجهة متوازية
wājiha mutawāzīya
parallel interface سطح بيني متواز
saṭḥ baini mutawāzī
parallel-motion equipment معدات بحركة متوازية
mu'iddāt bi-ḥaraka mutawāzīya
parallel runways مدارج متوازية
madārij mutawāzīya
parallel tracks سكة متوازنة
sikka mutawāzina
parameters بارامترات . معالم
baramitrāt, ma'ālim
parapet سور منخفض
sūr munkafiḍ
parapet wall جدار حاجز
jidār ḥājiz
park أوقف . متنزه
auqafa, mutanazzah
parking area منطقة إنتظار السيارات
minṭaqat intiẓār as-sayyārāt
parquet floor أرضية خشبية مزخرفة
arḍīya kašabīya muzakrafa
partially fixed مثبت جزئياً
muṯabbat juz'īyan
partially-separate system نظام منفصل جزئياً
niẓām munfaṣil juz'īyan
partial overhaul إصلاح جزئي
iṣlāḥ juz'ī
partial prestressing إجهاد جزئي مسبق
ijhād juz'ī musbaq
particle size حجم الجسيمات
ḥajm al-jusaimāt
particle-size distribution توزيع حجم الجسيمات
tauzī' ḥajm al-jusaimāt
parties concerned الأطراف المعنية
al-aṭrāf al-ma'nīya
parting agent مركب إستخلاص بالفصل
murakkab istiklāṣ bil-faṣl
partition فاصل . حاجز
fāṣil, ḥājiz
partition wall جدار حاجز أو فاصل
jidār ḥājiz au fāṣil
party wall جدار مشترك
jidār muštarak
passage إجتياز . مرور
ijtiyāz, murūr
passenger catchment area منطقة تجمع الركاب
minṭaqat tajammu' ar-rukkāb
passenger coach حافلة ركاب
ḥāfilat rukkāb
passenger conveyor ناقلة ركاب
nāqilat rukkāb
passenger lift مصعد ركاب
miṣ'ad rukkāb
passenger-loading device وسيلة تحميل الركاب
wasīlat taḥmīl ar-rukkāb
passenger terminal محطة الركاب
muḥaṭṭat ar-rukkāb
passing place مكان العبور
makān al-'ubūr
passive resistance مقاومة سلبية
muqāwama salbīya
password كلمة السر
kalimat as-sirr

P

paternoster رافعة بالقواديس
rāfi'a bil-qawādīs
path مسار
masār
paved apron مدرج مرصوف
madraj marṣūf
paved surface سطح مرصوف
saṭḥ marṣūf
paved surround محيط مرصوف
muḥīṭ marṣūf
pavement رصيف
raṣīf
pavement overlays أغطية الرصيف
aḡṭīyat ar-raṣīf
pavement thickness ثخانة الرصف
ṯukānat ar-raṣf
pavilion جناح
janāḥ
paving الرصف
ar-raṣf
paving asphalt أسفلت الرصف
isfalt ar-raṣf
paving brick طوب الرصف
ṭaub ar-raṣf
paving flag حجر رصف لوحي
ḥajar raṣf lauḥī
paving stones حجر الرصف
ḥajar ar-raṣf
payment دفع . دفعة
daf', daf'a
payment by instalments دفع بالتقسيط
daf' bit-taqsīṭ
payroll كشف الرواتب
kašf ar-rawātib
P-channel metal oxide semiconductor القناة الموجبة لأكسيد المعدن شبه الناقل
al-qanāt al-mūjaba li-aksīd al-ma'dan šubh an-nāqil
peak power القدرة الذروية
al-qudra aḏ-ḏarwīya
peat blasting سفع خثي
safḥ kaṯṯī
pedestrian-controlled dumper قلابة يتحكم فيها عامل سيار
qallāba yataḥakam fihā 'āmil sayyār
pedestrian crossing منطقة عبور المشاة
minṭaqat 'ubūr al-mašāt
peg وتد . سدادة
watad, sidāda
peg-top paving رصف بحجارة صوانية
raṣf bi-ḥajara ṣuwānīya
pellicular water ماء لاصق
mā' lāṣiq
Pelton wheel تربين دفعي للتصادم
tūrbīn daf'ī lit-taṣādum
pencil قلم رصاص . حزمة
qalam riṣaṣ, ḥuzma
Pencoyd formula for impact معادلة «بنكويد» للتصادم
mu'ādalat binkwīd
penetration إختراق . تغلغل
ikṯirāq, taḡalḡul
penetration test إختبار الاختراق
ikṯibār al-ikṯirāq
penetrometer مقياس الاختراقية
miqyās al-ikṯirāqīya
penning gate بوابة تحكم عمودية
bawwābat taḥakkum 'amūdīya
penstock بربخ : قناة ضبط جريان الماء
barbak, qanāt ḍabṭ jarayān al-mā'
penthouse طابق إضافي
ṭābiq iḍāfī
peptizing agents مواد خفض سرعة السوائل
mawādd kafḍ sur'at as-sawā'il
perch قياس تكعيبي
qiyās tak'ībī
perched water table سطح الماء الباطني
saṭḥ al-mā' al-bāṭinī
percolating filter طبقة مرشحة
ṭabaqa muraššiḥa
percolating-filter treatment المعالجة بمرشح تخلل
al-mu'ālaja bi-muraššiḥ tušīl
percolation تخلل . ترشيح
takallul, taršīḥ
percussion صدم . قدح . دق
ṣadm, qadḥ, daqq
percussion drill حفارة بالدق
ḥaffāra bid-daqq
percussion tools ادوات الحفر بالدق
adawāt al-ḥafr bid-daqq
percussive-rotary drilling حفر دوار بالدق
ḥafr dawwār bid-daqq
perfect frame إطار كامل
iṭār kāmil

P

performance أداء
idā'
performance curve منحنى الأداء
munhana al-idā'
performance specification مواصفات الأداء
muwāṣafat al-idā'
performance testing إختبار الأداء
iḵtibār al-idā'
periodical maintenance الصيانة الدورية
aṣ-ṣīyāna ad-daurīya
period of construction فترة الانشاء
fatrat al-inšā'
peripheral محيطي
muḥīṭī
permafrost أرض دائمة التجمد
arḍ dā'imat at-tajammud
permanent adjustment ضبط دائم
ḍabṭ dā'im
permanent frame هيكل دائم
haikal dā'im
permanent labour عمل دائم
'amal dā'im
permanent set أثر دائم
aṯar dā'im
permanent shuttering تغليف دائم
taġlīf dā'im
permanent staff موظفون دائمون
muwaẓẓafūn dā'imūn
permanent way طريق ثابت
ṭarīq ṯābit
permeability قابلية النفاذ
qābilīyat an-nafāḏ
permeability coefficient معامل النفاذية
mu'āmil an-nafāḏīya
permeameter مقياس الانفاذية
miqyās al-infāḏīya
permissible overload تجاوز الحمل المسموح
tajāwuz al-ḥiml al-masmūḥ
perpendicular axes محاور متعامدة
mahāwir muta'āmida
perpetual screw لولب دودي
laulab dūdī
personal equation معادلة شخصية
mu'ādala šaḵṣīya
personal rapid transit النقل الشخصي السريع
an-naql aš-šaḵṣī as-sarī'
personnel hoist مصعد أفراد
miṣ'ad afrād
perspective منظور
manẓūr

Perspex (TM) برسبكس
birsbiks
pervibration هزهزة الخرسانة
hazhazat al-ḵarasāna
pervious base-course layer طبقة الرصف السابقة
ṭabaqat ar-raṣf as-sābiqa
PETN بنتاريثريت تترانيترات
bintarīṯrīt titrānītrāt
petrographic microscope ميكروسكوب بتروجرافي
mīkrūskūb bitrūġrāfī
petrol بنزين
banzīn
petroleum products منتوجات بترولية
mantūjāt bitrūlīya
petrol interceptor مصيدة البنزين
miṣyadat al-banzīn
pH control التحكم في الرقم الهيدروجيني
at-taḥakkum fir-raqm al-hīdrūjīnī
photocell خلية ضوئية
ḵalīya ḍau'īya
photoclinometer مقياس ميل ضوئي
miqyās mail ḍau'ī
photoconductive cell خلية ذات موصلية ضوئية
ḵalīya ḏāt mauṣilīya ḍau'īya
photocopy نسخة مصوّرة
nusḵa musawara
photo diode دايود ضوئي
dayūd ḍau'ī
photoelasticity مرونة ضوئية
muruna ḍau'īya
photoelectric cell خلية كهرضوئية
ḵalīya kahrḍau'īya
photoelectric detector مكشاف كهرضوئي
mikšāf kahrḍau'ī
photoelectric device وسيلة كهرضوئية
wasīla kahrḍau'īya
photoelectric effect تأثير كهرضوئي
ta'ṯīr kahrḍau'ī
photogrammetric survey مسح تصويري
masḥ taṣwīrī
photogrammetry تصوير جوي
taṣwīr jawwī
photomicrograph صور مجهرية
ṣūra mijharīya
photostat نسخ
fūtūstāt
phototransistor ترانزستور ضوئي
trānzistūr ḍau'ī

P

photovoltaic cell خلية فلطائية ضوئية
kalīya fulṭā'īya ḍau'īya
physical properties خواص طبيعية
kawāṣ ṭabī'īya
pick صاقور . نقر
ṣāqūr, naqr
pickling of metal نقر المعدن
naqr al-mā'dan
picture window نافذة بابية
nāfiḏa bābīya
piece work شغل بالقطعة
šuḡl bil-qiṭ'a
pier ركيزة
rakīza
pier cap العارضة الافقية للركيزة
al-'āriḍa al-ufuqīya lir-rakīza
pier construction إنشاء الركائز
inšā' ar-rakā'iz
pier of a wall ركيزة الجدار
rakīzat al-jidār
piezometer tube مقياس الضغط العالي
miqyās aḍ-ḍaḡṭ al-'ālī
piezometric surface مستوى الماء الباطني
mustawa al-mā' al-bāṭinī
pig كتلة معدن خام . مكشطة
kutlat ma'dan kām mukaššata
pigment خضب . صبغ
kaḍb, ṣabḡ
pilaster عمود جداري ناتئ
'amūd jidārī nāti'
pile كومة . ركيزة أساس
kūma, rakīzat asās
pile bridge جسر على ركائز
jisr 'ala rakā'iz
pile cap عرقة رأسية
'araqa ra'sīya
pile core قلب الركيزة
qalb ar-rakīza
pile cushion وسادة الركيزة
wisādat ar-rakīza
piled foundation أساس على ركائز
asās 'ala rakā'iz
pile-drawer منزعة الركائز
munazzi'at ar-rakā'iz
pile-driver مدق الركائز
midaqq ar-rakā'iz
pile driving دق الركائز
daqq ar-rakā'iz
pile extractor نازع الركائز
nāzi' ar-rakā'iz
pile foundation work أعمال أساسات الركائز
a'māl asāsāt ar-rakā'iz
pile foundations أساسات الركائز
asāsāt ar-rakā'iz
pile frame هيكل ركائز
haikal rakā'iz
pile group مجموعة ركائز
majmū'at rakā'iz
pile hammer مطرقة ركائز
miṭraqat rakā'iz
pile head رأس الركيزة
ra's ar-rakīza
pile helmet خوذة الركيزة
kūḏat ar-rakīza
pile hoop طوق الركيزة
ṭauq ar-rakīza
pile pier رصيف بحري على ركائز
raṣīf baḥrī 'ala rakā'iz
pile-placing methods طرق وضع الركائز
ṭuruq waḍ' ar-rakā'iz
pile plan خطة الركائز
kuṭṭaṭ ar-rakā'iz
pile planking تغطية الركيزة
taḡṭīyat ar-rakā'iz
pile ring حلقة ركيزة
ḥalaqat rakīza
pile settlement استقرار الركيزة
istiqrār ar-rakīza
pile shoe نعل الركيزة
na'l ar-rakīza
pile underpinning تدعيم الأساس بركائز
tad'īm al-asās bir-rakā'iz
piling إنشاء مدعم بركائز
inšā' mud'am bir-rakā'iz
piling board لوح الركائز
lauḥ ar-rakā'iz
pillar عمود . دعامة
'amūd, di'āma
pillar crane مرفاع عمودي
mirfā' 'amūdī
pillar drill ثقابة عمودية
ṯuqāba 'amūdīya
pilot circuit دائرة دليلية
dā'ira dalīlīya
pilot lamp مصباح دليلي
miṣbāḥ dalīlī
pilot plant وحدة صناعية تجريبية
waḥda ṣinā'īya tajrībīya

P

pilot shaft عمود دليلي
ʿamūd dalīlī
pilot tunnel نفق دليلي
nafaq dalīlī
pinchers كماشة
kammāša
pin connection لسان توصيل ذكر
lisān tauṣīl ḏakar
pin drill مثقب توسيع
miṯqab tausiʿ
pin joint وصلة مسمارية
waṣla mismārīya
pinned تثبيت بالمسامير
taṯbīt bil-masāmīr
pipe أنبوب . ماسورة
unbūb · māsūra
pipe bedding غرس الأنابيب
ġars al-anābīb
pipe bender حانية أنابيب
hānīyat anābīb
pipe casing تغليف الأنابيب
taġlīf al-anābīb
pipe drains مصارف أنبوبية
maṣārif unbūbīya
pipe fitter عامل أنابيب
ʿāmil anābīb
pipe fittings قطع تركيب الأنابيب
qiṭaʿ tarkīb al-anābīb
pipe flange شفة الأنبوب
šaffat al-unbūb
pipe friction إحتكاك الأنابيب
iḥtikāk al-anābīb
pipelayer مدّاد أنابيب
maddād anābīb
pipe liners بطانات الأنابيب
biṭānāt al-anābīb
pipe-pushing مد الأنابيب بالدفع
madd al-anābīb bid-dafʿ
pipe sewer مجارير أنبوبية
majārīr unbūbīya
pipe welding لحام الأنابيب
liḥām al-anābīb
piping أنابيب
anābīb
piping network شبكة أنابيب
šabakat anābīb
piston engine محرك ذو كباس
muḥarrik ḏū kabbās
piston-type ventilation تهوية بالكباس
tahwiya bil-kabbās

pit منجم
manjam
pitch-bitumen قطران
quṭrān
pitched roof سطح مائل
saṭḥ māʾil
pitched work جدار مائل
jidār māʾil
pitcher حجر رصف غرانيتي
ḥajar raṣf ġrānītī
pitch fibre ألياف زفتية
alyāf ziftīya
pitching تبطين بالزفت
tabṭīn biz-zift
pitching ferrules أنابيب ثقوب الرفع
anābīb ṯuqūb ar-rafʿ
pitting تنقر
tanaqqur
pit type toilet مرحاض ذو حفرة
mirhāḍ ḏū ḥufra
pivot bridge جسر محوري
jisr miḥwarī
pivoted window نافذة محورية
nāfiḏa miḥwarīya
pivot joint وصلة محورية
waṣla miḥwarīya
placing boom ذراع رفع الاسمنت
ḏirāʿ rafʿ al-ismant
placing plant سقالة رفع الاسمنت المجبول
saqālat rafʿ al-ismant al-majbūl
plain concrete خرسانة عادية
ḵarasāna ʿādīya
plain concrete wall جدار خرسانة عادية
jidār ḵarasāna ʿādīya
plain line خط عادي
ḵaṭṭ ʿādī
plain suction dredger كراءة سحب عادي
karrāʾat saḥb ʿadī
plan خطة
ḵuṭṭa
plane frame هيكل مستو
haikal mustawī
plane of rupture مستوى التصدع
mustawa at-taṣadduʿ
plane of saturation مستوى التشبع
mustawa at-tašabbuʿ
planer مقشطة
miqšaṭa

plane surveying — المساحة المسطحة
al-misāḥa al-musaṭṭaḥa

plane table — لوحة مسح مستوية
lauḥat masḥ mustawīya

plane-table surveying — مسح باللوحة المستوية
masḥ bil-lauḥa al-mustawīya

plane-tabling — رسم مساحي باللوحة المستوية
rasm misāḥī bil-lauḥa al-mustawīya

planimeter — مقياس المسطحات
miqyās al-musaṭṭaḥāt

planimetry — قياس المساحات
qiyās al-masāḥāt

planking — تغطية بالألواح
taġṭīya bil-alwāḥ

planning — تخطيط
takṭīṭ

planning engineer — مهندس تخطيط
muhandis takṭīṭ

planometric projection — إساقط السطح المستوي
isqāṭ as-saṭḥ al-mustawī

planoscope — آلة تسوية
ālat taswiya

plant — معدات
mu'iddāt

plant cost — تكاليف المعدات
takālīf al-mu'iddāt

plant hire — إستئجار المعدات
isti'jār al-mu'iddāt

plant hours — ساعات عمل المعدات
sā'āt 'amal al-mu'iddāt

plant layout — مخطط المعدات
mukaṭṭaṭ al-mu'iddāt

plant mix — خلط دعم التربة
kalṭ da'm at-turba

plant-mixed concrete — خرسانة مخلوطة بالمصنع
karasāna maklūṭa bil-maṣna'

plant programming — برامج عمل المعدات
barāmij 'amal al-mu'iddāt

plant response — تجاوب المعدات
tajāwub al-mu'iddat

plan view drawing — رسم مسقط علوي
rasm masqaṭ 'ulwī

plasterer — مبيِّض . قصّار
mubayyiḍ, qaṣṣār

plastering — تجصيص . تبييض
tajṣīṣ, tabyīḍ

plastic face shield — حجاب وجه بلاستيك
ḥijāb wajh blāstīk

plastic fracture — تصدع مطاوع
taṣaddu' muṭāwi'

plastic hinge — مفصلة بلاستيك
mifṣala blāstīk

plasticity — لدانة
ladāna

plasticity index — دليل اللدانة
dalīl al-ladāna

plasticizer — مُلدن
muladdin

plastic limit — حد اللدانة
ḥadd al-ladāna

plastic modulus — مُعامل اللدانة
mu'āmil al-ladāna

plastic paint — دهان لدن
dihān ladin

plastics — اللدائن
al-ladā'in

plastic welding — لحام بالفولاذ اللدن
liḥām bil-fūlāḏ al-ladin

plastic window glazing — تزجيج نافذة بلاستيك
tazjīj nāfiḏa blāstik

plate — صفيحة . لوح . لوحة
ṣafīḥa, lauḥ, lauḥa

plate bearing test — إختبار تحميل اللوح
iktibār taḥmīl al-lauḥ

plate clutch — قابض لوحي
qābiḍ lauḥī

plated beam — عارضة مصفحة
'āriḍa muṣaffaḥa

plate floor — أرضية بألواح
arḍīya bi-alwāḥ

plate girder — عارضة لوحية
'āriḍa lauḥīya

plate-girder bridge — جسر بعوارض لوحية
jisr bi-'awāriḍ lauḥīya

plate glass — زجاج لوحي
zujāj lauḥī

plate iron — حديد لوحي
ḥadīd lauḥī

plate screws — براغي ألواح
barāġi alwāḥ

plate section — مجموعة لوحية
majmū'a lauḥīya

plate vibrator — هزازة لوحية
hazzāza lauḥīya

platform — منصة . رصيف
minaṣṣa, raṣīf

platform framing — هيكل غربي
haikal ġarbī

P

platform gantry قنطرة مسطحة
qanṭara musaṭṭaḥa
platform grating حاجز المنصة
ḥājiz al-minaṣṣa
platform hoist مرفاع بمنصة
mirfāʿ bi-minaṣṣa
playground ساحة لعب
sāhat laʿib
pliability قابلية الانطواء
qābilīyat al-inṭiwāʾ
pliers زردية
zardīya
Plimsoll mark علامة «پليمسول»
ʿalāmat blimsūl
plinth وطيدة . قاعدة العمود
waṭīda, qāʾidat al-ʿamūd
plot قطعة أرض
qiṭʿat arḍ
plotting تخطيط بياني
takṭīṭ bayānī
plotting scale مقياس التخطيط البياني
miqyās at-takṭīt al-bayānī
plug سداد . قابس
sidād, qābis
plug flow تدفق كتلي
tadaffuq kutlī
plug valve صمام سدادي
ṣimām sidādi
plumb حجر إزاحة
ḥajar izāḥa
plumbago پلمباجو . جرافيت
blumbāgū, grāfīt
plumb bob شاقول
šāqūl
plumber سمكري . سباك
samkarī, sabbāk
plumbing سمكرة
samkara
plumb line خيط الشاقول
kaiṭ aš-šākūl
plunge pool حوض تغطيس
ḥauḍ tagṭīs
plunger pump مضخة بكباس
miḍakka bi-kabbās
plunger valve صمام بكباس غاطس
ṣimām bi-kabbās gāṭis
plywood خشب رقائقي
kašab raqāʾiqī
plywood floor أرضية خشب رقائقي
arḍiyat kašab raqāʾiqi
plywood formwork هيكل من الخشب الرقائقي
haikal min al-kašab ar-raqāʾiqī
pneumatic يعمل بالهواء المضغوط
yaʿmal bil-hawāʾ al-maḍgūṭ
pneumatically-applied mortar رش الملاط بالهواء المضغوط
rašš al-milāṭ bil-hawāʾ al-maḍguṭ
pneumatic caisson قيسون بهواء مضغوط
qaisūn bi-hawāʾ maḍgūṭ
pneumatic controller حاكم بالهواء المضغوط
ḥākim bil-hawāʾ al-maḍgūṭ
pneumatic conveyor ناقلة بالهواء المضغوط
nāqila bil-hawāʾ al-maḍgūṭ
pneumatic drill مثقاب يعمل بالهواء المضغوط
mitqāb yaʿmal bil-hawāʾ al-maḍgūṭ
pneumatic ejector قاذف بالهواء المضغوط
qāḏif bil-hawāʾ al-maḍgūṭ
pneumatic excavator حفارة بالهواء المضغوط
ḥaffāra bil-hawāʾ al-maḍgūṭ
pneumatic feed تغذية بالهواء المضغوط
taġḏiya bil-hawāʾ al-maḍgūṭ
pneumatic float عوامة هوائية
ʿawwāma hawāʾīya
pneumatic hammer مطرقة بالهواء المضغوط
miṭraqa bil-hawāʾ al-maḍgūṭ
pneumatic high-water alarm إنذار بالهواء المضغوط لارتفاع الماء
indār bil-hawāʾ al-maḍgūṭ li-irtifāʾ al-māʾ
pneumatic hoist مرفاع بالهواء المضغوط
mirfāʿ bil-hawāʾ al-maḍgūṭ
pneumatic lift مصعد بالهواء المضغوط
miṣʿad bil-hawāʾ al-maḍgūṭ
pneumatic mortar ملاط يرش بالهواء المضغوط
milāṭ yarušš bil-hawāʾ al-maḍgūṭ
pneumatic pick صاقور بالهواء المضغوط
ṣāqūr bil-hawāʾ al-maḍgūṭ
pneumatic sander مكنة سنفرة بالهواء المضغوط
makinat sanfara bil-hawāʾ al-maḍgūṭ
pneumatic sewer ejector مضخة إزاحية هوائية للمجارير
miḍakka izāḥīya hawāʾīya lil-majārīr
pneumatic shaft sinking حفر البئر بالهواء المضغوط
ḥafr al-biʾr bil-hawāʾ al-maḍgūṭ

pneumatic suction *saḥb bil-hawā' al-maḍġūṭ*	سحب بالهواء المضغوط	polishing *talmī', ṣaql*	تلميع . صقل
pneumatic tool *'iddat ta'mal bil-hawā' al-maḍġūṭ*	عدة تعمل بالهواء المضغوط	polishing agent *'āmil talmī'*	عامل تلميع
pneumatic tyre *iṭār yunfak̠ bil-hawā' al-maḍġūṭ*	إطار ينفخ بالهواء المضغوط	polishing wheel *dūlāb ṣaql*	دولاب صقل
pneumatic-tyred roller *dalfīn bi-iṭārāt hawā'īya*	دلفين باطارات هوائية	pollution *talawwut̠*	تلوث
podger *miftāḥ rabṭ*	مفتاح ربط	polyester *būlīstar*	بوليستر
podzol *turba ḥamḍīya*	تربة حمضية	polyethylene *būlīt̠īlīn*	بوليثيلين
point gauge *miqyās li-mansūb al-mā'*	مقياس لمنسوب الماء	polymer *būlīmar*	بوليمر
point load *ḥiml murakkaz*	حمل مركز	polypropylene *būlībrūblīn*	بوليبروبلين
point-of-sale terminal *jihāz nuqṭat al-bai'*	جهاز نقطة البيع	polystyrene *būlistrīn*	بوليسترين
Poisson's ratio *nisbat būysūn*	نسبة « بويسون »	polysulphide sealant *ġirā' bulīsulfaid*	غراء بوليسلفايد
poker vibrator *hazzāza bi-miḥrāq*	هزازة بمحراك	polytetrafluoroethylene *būlītitraflūrūt̠īlīn*	بوليتترافلوروثيلين
polariscope *mikšāf al-istiqṭāb*	مكشاف الاستقطاب	polythene *būlīt̠īn*	بوليثين
polarized light *ḍau' mustaqṭab*	ضوء مستقطب	polythene membrane *ġišā' būlīt̠īn*	غشاء بوليثين
polarizer *mustaqṭib*	مستقطب	polyurethane *būlīyūrīt̠ān*	بوليوريثان
polar moment of inertia *'azm al-aṭala al-qṭbī*	عزم العطالة القطبي	polyvinyl chloride *klūrīd būlī vīnīl*	كلوريد بوليفنيل
polder *būldar*	پولدر	pond *burka ṭabī'īya*	بركة طبيعية
pole *'amūd, quṭb*	عمود . قطب	pontoon *zauraq musaṭṭaḥ*	زورق مسطح
pole drill *mit̠qāb 'amūdī*	مثقاب عمودي	pontoon bridge *jisr 'ā'im*	جسر عائم
pole guy *šiddādat al-'amūd*	شدادة العمود	pool *burka*	بركة
pole scaffolds *saqālāt 'amūdīya*	سقالات عمودية	population *sukkān*	سكان
police station *markaz aš-šurṭa*	مركز الشرطة	porch *madk̠al kārijī masqūf*	مدخل خارجي مسقوف
poling back *al-ḥafr k̠alf ad-da'm*	الحفر خلف الدعم	pores *t̠uqūb daqīqa*	ثقوب دقيقة
poling boards *alwāḥ ad-da'm*	ألواح الدعم	pore-water pressure *ḍaġṭ al-mā' fiz-zait al-mušabba'*	ضغط الماء في الزيت المشبع
polished-stone value *qīmat al-ḥajar al-maṣqūl*	قيمة الحجر المصقول	porosity *masāmīya*	مسامية

P

port مرفأ . ميناء . منفذ
marfa', mīnā', manfad
portable متنقل
mutanaqqil
portable building مبنى متنقل
mabna mutanaqqil
portable crane مرفاع متنقل
mirfā' mutanaqqil
portable hand saw منشار يدوي متنقل
minšār yadawī mutanaqqil
portable rig وحدة حفر متنقلة
wahdat hafr mutanaqqila
port administration إدارة الميناء
idārat al-mīnā'
portal crane مرفاع قنطري
mirfā' qanṭarī
portal frame هيكل قنطري
haikal qanṭarī
portal tower برج قنطري
burj qanṭarī
port buildings مباني الميناء
mabani al-mīnā'
Portland blast-furnace cement إسمنت بورتلاند مخلوط بخبث الفرن العالي
ismant burtlānd maklūṭ bi-kabt al-furn al-'ālī
Portland cement إسمنت بورتلاند
ismant burtlānd
Portland-cement concrete خرسانة إسمنت بورتلاند
karasānat ismant burtlānd
Portland pozzolana cement إسمنت بزولان بورتلاند
ismant bzūlān burtlānd
port of entry ميناء الدخول
mīnā' ad-dukūl
port planning تخطيط الميناء
takṭīṭ al-mīnā'
port services مرافق الميناء
marāfiq al-mīnā'
port site selection إختيار موقع الميناء
iktiyār mauqi' al-mīnā'
port structures إنشاءات الميناء
inšā'āt al-mīnā'
position sensor جهاز تحسس الموضع
jihāz tahassus al-mauḍi'
post عمود . دعامة
'amūd, di'āma
post-hole auger مثقاب الثقوب الكبيرة
mitqāb at-tuqūb al-kabīra

post office مكتب البريد
maktab al-barīd
post-stressing إجهاد لاحق
ijhād lāhiq
post-tensioning شد لاحق
šadd lāhiq
potable water ماء الشرب
mā' aš-šurb
potable water supply إمدادات ماء الشرب
imdādāt mā' aš-šurb
potential energy طاقة كامنة
ṭāqa kāmina
potomac caisson قيسون خشبي
qaisūn kašabī
pound سحق . دق
sahaqa, daqqa
powder actuated tool عدة تعمل بتفاعل المساحيق
'idda ta'mal bi-tafā'ul al-masāhīq
powder spreader آلة فرش المساحيق
ālat farš al-masāhīq
power قدرة . طاقة
qudra, ṭāqa
power amplifier مضخم الطاقة
mudakkim aṭ-ṭāqa
power barrow عربة آلية
'araba ālīya
power cable كبل نقل الطاقة
kabl naql aṭ-ṭāqa
power consumption إستهلاك الطاقة
istihlāk aṭ-ṭāqa
power earth auger حفارة مركبة على شاحنة
haffāra murakkaba 'ala šāhina
power handling يحرك آليا
yuharrik ālīyan
power house محطة الطاقة
mahaṭṭat aṭ-ṭāqa
power house steelwork تركيبات فولاذية لمحطة الطاقة
tarkībāt fūlādīya li-mahaṭṭat aṭ-ṭāqa
power loss فقد الطاقة
faqd aṭ-ṭāqa
power rammer مدك آلي
midakk ālī
power saw منشار آلي
minšār ālī
power shovel مجرفة آلية
mijrafa ālīya

P

power socket مقبس الطاقة
miqbas aṭ-ṭāqa
power station محطة توليد الطاقة
maḥaṭṭat taulīd aṭ-ṭāqa
power take-off عمود إدارة خارجي
ʿamūd idāra ḵārijī
power transformer محول الطاقة
muḥawwil aṭ-ṭāqa
power transmission نقل الطاقة
naql aṭ-ṭāqa
power wrench مفتاح ربط آلي
miftāḥ rabṭ ālī
pozzolana بزولان
bzūlān
practice drilling تمرين على طرق الحفر
tamrīn ʿala ṭuruq al-ḥafr
Pratt truss جملون «برات»
jamlūn brāt
pre-amplifier مضخم متقدم
muḍakkim mutaqaddim
pre-boring for piles حفر مقدم للركائز
ḥafr muqaddam lil-rakāʾiq
precast صب مسبق
ṣabb musabbaq
precast beam عارضة مسبقة الصب
ʿāriḍa musabbaqat aṣ-ṣabb
precast concrete خرسات مسبقة الصب
ḵarasāna musabbaqat aṣ-ṣabb
precast-concrete liner بطانة خرسانية مسبقة الصب
biṭāna ḵarasānīya musabbaqat aṣ-ṣabb
precast-concrete pile ركيزة خرسانية مسبقة الصب
rakīza ḵarasānīya musabbaqat aṣ-ṣabb
precast floor أرضية مسبقة الصب
arḍīya musabbaqat as-ṣabb
precast on site صب مسبق بالموقع
ṣabb musabbaq bil-mauqiʿ
precast slab بلاطة مسبقة الصب
bilāṭa musabbaqat aṣ-ṣabb
precipitation ترسب
tarassub
precise levelling تسوية دقيقة
taswiya daqīqa
precision دقة
diqqa
pre-coated chippings حصباء مسبقة التغليف
ḥaṣbāʾ musabbaqat at-taḡlīf
pre-consolidated load حمل مسبق الدعم
ḥiml musabbaq ad-daʿm
pre-engineered building مبنى جاهز التركيبات الهندسية
mabnā jāhiz at-tarkībāt al-handasīya
prefabricated slab track سكة بقواعد خرسانية جاهزة
sikka bi-qawāʿid ḵarasānīya jāhiza
pre-formed rope حبل مسبق التشكيل
ḥabl musabbaq at-taškīl
pre-investment report تقرير سابق للاستثمار
taqrīr sābiq lil-istiṯmār
pre-investment studies دراسات سابقة للاستثمار
dirāsāt sābiqa lil-istiṯmār
preliminary layout مخطط تمهيدي
mukaṭṭaṭ tamhīdī
preliminary survey مسح مبدئي
masḥ mabdaʾī
preliminary treatment معالجة مبدئية
muʿālaja mabdaʾīya
preloading تحميل متقدم
taḥmīl mutaqaddim
pre-load tank خزان مسبق الاجهاد
ḵazzān musabbaq al-ijhad
preparation of tender documents إعداد مستندات العطاء
iʿdād mustanadāt al-ʿaṭaʾ
prepared roadbed قاعدة طريق جاهزة
qāʿidat ṭarīq jāhiza
prescribed mix مزيج جاهز
mazīj jāhiz
preservatives for timber مواد حافظة للخشب
mawādd ḥāfiẓa lil-ḵašab
press مضغط
miḍḡaṭ
press conference مؤتمر صحفي
muʾtamar ṣaḥafī
pressure ضغط
ḍaḡṭ
pressure-assisted ram مدك بالضغط
midaqq biḍ-ḍaḡṭ
pressure build-up تزايد الضغط
tazāyud aḍ-ḍaḡṭ
pressure cable كبل ضغطي
kabl ḍaḡṭī
pressure chamber حجرة الضغط
ḥujrat aḍ-ḍaḡṭ
pressure control التحكم بالضغط
at-taḥakkum biḍ-ḍaḡṭ

P

pressure control valve صمام تنظيم الضغط
ṣimām tanẓīm aḍ-ḍaġṭ
pressure creosoting معالجة بالكريوزوت المضغوط
mu'ālaja bil-krīyusūt al-maḍġūṭ
pressure distribution توزيع الضغط
tauzi' aḍ-ḍaġṭ
pressure drilling حفر ضغطي
ḥafr ḍaġṭī
pressure feed tank خزان تغذية الضغط
ḵazzān taġḏīyat aḍ-ḍaġṭ
pressure gauge مقياس الضغط
miqyās aḍ-ḍaġṭ
pressure head علو الضغط
'ulū aḍ-ḍaġṭ
pressure measurement قياس الضغط
qiyās aḍ-ḍaġṭ
pressuremeter مقياس ضغط التربة
miqyās ḍaġṭ at-turba
pressuremeter test إختبار مقياس الضغط
iḵtibār miqyās aḍ-ḍaġṭ
pressure pipe أنبوب ضغطي
unbūb ḍaġṭī
pressure reducer مخفض الضغط
muḵaffiḍ aḍ-ḍaġṭ
pressure-reducing valve صمام تخفيض الضغط
ṣimām taḵfīḍ aḍ-ḍaġṭ
pressure-regulating valve صمام تنظيم الضغط
ṣimām tanẓīm aḍ-ḍaġṭ
pressure regulator منظم الضغط
munaẓẓim aḍ-ḍaġṭ
pressure-release valve صمام تحرير الضغط
ṣimām taḥrīr aḍ-ḍaġṭ
pressure-relief valve صمام تخفيف الضغط
ṣimām taḵfīf aḍ-ḍaġṭ
pressure-retaining valve صمام إحتجاز الضغط
ṣimām iḥtijāz aḍ-ḍaġṭ
pressure switch مفتاح ضغطي
miftāḥ ḍaġṭī
pressure tank خزان ضغطي
ḵazzān ḍaġṭī
pressure valve صمام ضغطي
ṣimām ḍaġṭī
pressure vessel وعاء ضغطي
wi'ā' ḍaġṭi
pressure welding لحام ضغطي
liḥām ḍaġṭī
pressurised water reactor مفاعل ذري يبرد بالماء المضغوط
mufā'il ḏarrī yubarrad bil-mā' al-maḍġūṭ

prestressed concrete خرسانة سابقة الاجهاد
ḵarasāna sābiqat al-ijhād
prestressed concrete bridge جسر خرسانة سابقة الاجهاد
jisr ḵarasāna sābiqat al-ijhād
prestressing إجهاد مسبق
ijhād musabbaq
pre-tensioning شد مسبق
šadd musabbaq
pretesting إختبار مسبق
iḵtibār musabbaq
prevailing winds الرياح السائدة
ar-rīyāḥ as-sā'ida
preventive maintenance الصيانة الوقائية
aṣ-ṣīyāna al-wiqā'īya
Primacord fuse صمامة التفجير
ṣimāmat at-tafjīr
primary breaker كسارة أولية
kassāra awwalīya
primary treatment معالجة أولية
mu'ālaja awwalīya
primary valve صمام إبتدائي
ṣimām ibtidā'ī
primary ventilation تهوية إبتدائية
tahwiya ibtidā'īya
prime mover محرك أساسي
muḥarrik asāsī
primer بطانة دهان
biṭānat dihān
principal contractor مقاول رئيسي
muqāwil ra'īsī
principal stress إجهاد رئيسي
ijhād ra'īsī
printout مطبوعات
maṭbū'āt
priority junction تقاطع الأولوية
taqāṭu' al-auluwīya
prismatic compass بوصلة موشورية
būṣala mūšūrīya
prismatic telescope تلسكوب موشوري
talskūb mūšūrī
probability إحتمالية
iḥtimālīya
probability factor عامل الاحتمالية
'āmil al-iḥtimālīya
probable error خطأ محتمل
ḵaṭa' muḥtamal
probing سبر
sibr

P

process عملية . أسلوب
'amalīya, uslūb
process chart مخطط بياني للعمليات المتعاقبة
mukaṭṭaṭ bayānīlil-'amalīyāt al-muta'āqiba
process control تحكم في العمليات المتعاقبة
taḥakkum fil-'amalīyāt al-muta'āqiba
processing معالجة متعاقبة
mu'ālaja muta'āqiba
process lag تأخر العمليات المتتابعة
ta'akkur al-'amalīyāt al-mutatābi'a
process load حمل العمليات المتعاقبة
ḥiml al-'amalīyāt al-muta'āqiba
process monitoring مراقبة العمليات المتعاقبة
murāqabat al-'amalīyāt al-muta'āqiba
processor construction تركيب البروسسور
tarkīb al-brūsisūr
procurement حصول
ḥusīl
profile قطاع جانبي
qiṭā' jānibī
profile levelling تسوية جانبية
taswiya jānibīya
profile paper ورق مربعات
waraq murabba'āt
profit ربح
ribḥ
profit forecast الأرباح التقديرية
al-arbāḥ at-taqdīrīya
program editing تنقيح البرنامج
tanqīḥ al-barnāmij
program elements عناصر البرنامج
'anāṣir al-barnāmij
program installation تركيب البرنامج
tarkīb al-barnāmij
programming language لغة البرمجة
luġat al-barmaja
programming method طريقة البرمجة
ṭarīqat al-barmaja
programming microprocessor ميكروبروسسور للبرمجة
mīkrūbrūsisūr al-barmaja
progressive collapse تقوض تدريجي
taqawwuḍ tadrījī
progress report تقرير دوري
taqrīr daurī
projected window نافذة نائتة
nāfiḏa nāti'a
project evaluation تقييم المشروع
taqyīm al-mašrū'
project implementation تنفيذ المشروع
tanfīḏ al-mašrū'
projection إسقاط . مسقط
isqāṭ, masqaṭ
projection welding لحام نتوئي
liḥām nutū'ī
project management إدارة المشروع
idārat al-mašrū'
project specification مواصفات المشروع
muwāṣafāt al-mašrū'
promoter معزز المادة
mu'azziz al-mādda
proofing معالجة التصميد
mu'ālajat at-taṣmīd
proof stress إجهاد الصمود
ijhād al-maṣmūd
proof test إختبار الصمود
iktibār aṣ-ṣumūd
prop drawer مسحبة دعائم
masḥabat da'ā'im
propeller fan مروحة داسرة
mirwaḥa dāsira
propeller turbine تربين مروحي
tūrbīn mirwaḥī
property damage liability مسؤولية الأضرار عن الممتلكات
mas'ūlīyat al-aḍrār 'an al-mumtalakāt
proportional control تحكم تناسبي
taḥakkum tanāsubī
proportional drawing رسم تناسبي
rasm tanāsubī
proportional limit حد التناسب
ḥadd at-tanāsub
proportional scale مقياس تناسبي
miqyās tanāsubī
proportioning توزيع نسبي
tauzi' nisbī
proposal إقتراح . عرض
iqtirāḥ, 'arḍ
proposal form نموذج عرض
namuḏaj 'arḍ
protection against wind erosion حماية ضد حت الرياح
ḥimāya ḍidd ḥatt ar-riyāḥ
protective agent عامل وقاية
'āmil wiqāya

P

protective clothing ألبسة واقية
albisa wāqīya
protective equipment معدات واقية
mu'iddāt wāqīya
protective garment ملابس واقية
malābis wāqīya
protective measures اجراءات وقائية
ijrā'āt wiqā'īya
proving ring حلقة إختبارية
ḥalaqa iḵtibārīya
public address system نظام النداء العام
niẓām an-nidā' al-'āmm
public consumption الاستهلاك العام
al-istihlāk al-'āmm
public health engineering هندسة الصحة العامة
handasat aṣ-ṣaḥḥa al-'āmma
public transport وسائل النقل العام
wasā'il an-naql al-'āmm
public utilities المرافق العامة
al-marāfiq al-'āmma
public works الاشغال العامة
al-ašġāl al-'āmma
pull جر. شد
jarr, šadd
pulley بكرة
bakara
pull-lift مرفاع سحب
mirfā' saḥb
pull-out test إختبار التقويم
iḵtibār at-taqwīm
pulverization سحق
saḥq
pulverized coal فحم حجري مسحوق
faḥm hajarī masḥūq
pulverizing mixer خلاط سحق
ḵallāt saḥq
pumice stone حجر الخفاف
ḥajar al-ḵifāf
pump مضخة
miḍaḵḵa
pump-ashore plant محطة ضخ شاطئية
maḥaṭṭat ḍaḵḵ šāṭi'īya
pump capacity طاقة المضخة
ṭāqat al-miḍaḵḵa
pumped concrete خرسانة مضخوخة
ḵarasāna maḍḵūḵa
pumped main خط رئيسي مضخوخ
ḵaṭṭ ra'īsī maḍḵūḵ
pumped storage reservoir صهريج تخزين مضخوخ
sahrīj taḵzīn maḍḵūḵ
pumped supply إمدادات مضخوخة
imdādāt maḍḵūḵa
pumping appliance جهاز الضخ
jihāz aḍ-ḍaḵḵ
pumping cylinder إسطوانة الضخ
isṭiwānat aḍ-ḍaḵḵ
pumping equipment معدات الضخ
mu'iddāt aḍ-ḍaḵḵ
pumping station محطة الضخ
maḥaṭṭat aḍ-ḍaḵḵ
pumping unit وحدة الضخ
waḥdat aḍ-ḍaḵḵ
pump sump حوض الضخ
ḥauḍ aḍ-ḍaḵḵ
punch سنبك
sunbuk
puncheon مثقب حجارة
miṯqab hajāra
punching ثقب. تخريم
ṯuqb, taḵrīm
punching shear قص التخريم
qaṣṣ at-taḵrīm
punner مدك يدوي
midakk yadawī
purlin رافدة أفقية
rāfida ufuqīya
pusher arm ذراع دافع
ḏirā' dāfi'
pusher tractor جرار دفع
jarrār daf'
push shovel مجرفة دفعية
mijrafa daf'īya
putty معجون لتثبيت الزجاج
ma'jūn litaṯbīt al-zujāj
PVC بلاستيك صلب
blāstīk ṣalib
PVC water stop حاجز ماء بلاستيكي
ḥājiz mā' blāstīkī
PVC wrapped pipe أنبوب مغلف بالبلاستيك
unbūb muġallaf bil-blāstīk
pylon برج أسلاك
burj aslāk
pyramid cut قطع هرمي
qaṭ' haramī
pyrometer بيرومتر (مضرم)
bīrūmitar (miḍram)
quadrant ربعي رصف
rub'ī raṣf
quadrantal angle زاوية ربعية
zāwiya rub'īya

quadrantal bearing إتجاه ربعي
ittijāh rubʿī
quadrilateral رباعي الأضلاع
rubāʿī al-aḍlāʿ
quadripods خدة رباعية
ḵidda rubāʿīya
quarry محجر
maḥjar
quarter peg وتد متعامد
watad mutaʿāmid
quartz كوارتز
kwārts
quartzite كوارتزيت
kwārtsīt
quay رصيف تحميل بحري
raṣīf taḥmīl baḥrī
quay wall جدار رصيف التحميل البحري
jidār raṣīf at-taḥmīl al-baḥrī
queen post قائم الجملون
qāʾim al-jumlūn
quenching تبريد سريع
tabrīd sarīʿ
quick-levelling head رأس التسوية السريعة
raʾs at-taswiya as-sarīʿa
quicklime الجير الحي
al-jīr al-ḥayy
quicksand رمل سريع الانهيار
raml sarīʿ al-inhiyār
quick setting cement إسمنت سريع التصلب
ismant sarīʿ at-taṣallub
quick solder لحام سريع
liḥām sarīʿ
racked timbering ألواح خشب مقواة قطرياً
alwāḥ ḵašab muqawwāt quṭriyan
racking course طبقة رصف
ṭabaqat raṣf
rack railway سكة حديدية مسننة
sikka ḥadīdīya musannana
radar رادار
radār
radial gate بوابة نصف قطرية
bawwāba niṣf quṭrīya
radial gate sluice بوابة تحكم نصف قطرية
bawwābat taḥakkum niṣf quṭrīya
radial-sett paving رصف محوري
raṣf miḥwarī
radian زاوية نصف قطرية
zāwīya niṣf quṭrīya
radio راديو. لاسلكي
rādyū lā-silkī
radioactive waste نفايات مشعة
nufāyāt muši‘‘a
radio frequency تردد لاسلكي
taraddud lā-silkī
radio frequency amplifier مضخم ترددات لاسلكية
muḍakkim taraddudāt lā-silkīya
radius نصف القطر
niṣf al-quṭr
radius-and-safe-load indicator مؤشر نصف القطر والحمل المأمون
muʾaššir niṣf al-quṭr wal-ḥiml al-maʾmūn
radius of curvature نصف قطر الانحناء
niṣf quṭr al-inḥināʾ
radius of gyration نصف قطر الدوران
niṣf quṭr ad-dawarān
raft رمث. طوف
ramṯ, ṭauf
rafter رافدة مائلة
rāfida māʾila
raft foundation أساس من الخرسانة المسلحة
asās min al-ḵarasāna al-musallaḥa
rag bolt مسمار أشوك
mismār aš-šūk
rail قضيب سكة حديد
qaḍīb sikka ḥadīd
rail anchor مثبت السكة
muṯabbit as-sikka
rail base قاعدة السكة
qāʿidat as-sikka
rail bender مكنة حني القضبان
makinat ḥanī al-quḍbān
rail chair كرسي تثبيت السكة
kursī taṯbīt
rail-defect detection كشف إنحراف السكة
kašf inḥirāf as-sikka
rail depression هبوط السكة
hubūt as-sikka
rail fastening كرسي مثبت السكة
kursī taṯbīt
rail gauge مقياس الخطوط الحديدية
miqyās al-ḵuṭūṭ al-ḥadīdīya
rail hydro-stressor مكنة إجهاد هيدرولية للسكة
makinat ijhād hīdrūlīya lis-sikka

Q R

railing — درابزين
darābzīn

rail joint — وصلة سكة حديد
waṣlat sikka ḥadīd

rail-joint bar — قضيب وصلة سكة حديد
qaḍīb waṣlat sikka ḥadīd

rail key — وتد لمقعد تثبيت السكة
watad li-maqʿad taṯbīt

rail life — عمر السكك الحديدية
ʿumr as-sikak al-ḥadīdīya

rail quality — جودة السكك الحديدية
jaudat as-sikak al-ḥadīdīya

rail steel — فولاذ القضبان الحديدية
fūlāḏ al-quḍbān al-ḥadīdīya

rail test — اختبار القضبان
ik̲tibār al-quḍbān

rail transportation — النقل بالسكك الحديدية
an-naql bis-sikak al-ḥadīdīya

rail-transportation system — شبكة النقل بالسكك الحديدية
šabakat an-naql bis-sikak al-hadīdīya

railway — سكة حديدية
sikka ḥadīdīya

railway bridge — جسر السكك الحديدية
jisr as-sikak al-ḥadīdīya

railway bridge load — حمل جسر السكك الحديدية
ḥiml jisr as-sikak al-ḥadīdīya

railway curves — تعرجات الخطوط الحديدية
taʿrrujāt al-k̲uṭūṭ al-ḥadīdīya

railway in tunnel — سكة حديد نفقية
sikkat ḥadīd nafaqīya

railway junction — تقاطع خطوط السكك الحديدية
taqātuʿ k̲uṭūt as-sikak al-ḥadīdīya

railway service coach — حافلة صيانة السكك الحديدية
ḥāfilat ṣīyānat as-sikak al-ḥadīdīya

railway signalling — إشارة السكك الحديدية
išārat as-sikak al-ḥadīdīya

railway station — محطة سكة حديدية
maḥaṭṭat sikka ḥadīdīya

railway tunnel — نفق سكة حديدية
nafaq sikka ḥadīdīya

railway under construction — سكة حديدية قيد الانشاء
sikka ḥadīdīya qaid al-inšāʾ

rail web — وترة السكة
watarat as-sikka

rail weight — وزن السكة
wazn as-sikka

rainfall — سقوط المطر
suqūṭ al-maṭar

rain gauge — مقياس كمية المطر
miqyās kammīyat al-maṭar

rainwater — ماء المطر
māʾ al-maṭar

rainwater pipe — ماسورة ماء المطر
māsūrat māʾ al-maṭar

raised kerb — حافة رصيف مرفوعة
ḥāffat raṣīf marfūʿa

raking pile — ركيزة مائلة
rakīza māʾila

raking shore — دعائم مائلة
daʿāʾim māʾila

ram — مدك
midakk

ramp — منحدر
munḥadar

ram pump — مضخة ترددية
miḍak̲k̲a taraddudīya

random access memory — ذاكرة الدخول العشوائي
ḏākirat ad-duk̲ūl al-ʿašwāʾī

random rubble — كسارة عشوائية
kisāra ʿašwāʾīya

random sample — عينة عشوائية
ʿayyina ʿašwāʾīya

range — مدى . مجال
madā, majāl

range pole — شاخص تباعد
šāk̲is tabāʿud

ranging a curve — محاذاة المنحنى
muḥāḏāt al-munḥana

ranging rod — شاخص تباعد
šāk̲is tabāʿud

rapid-hardening cement — إسمنت سريع التصلب
ismant sarīʿ at-taṣallub

rapid-hardening concrete — خرسانة سريعة التصلب
k̲arasāna sarīʿat at-taṣallub

rapids — جنادل
janādil

rapid sand filter — مرشح رملي سريع
muraššiḥ ramlī sarīʿ

rapid sand filtration — ترشيح رملي سريع
taršīḥ ramlī sarīʿ

rapid-transit coach — حافلة النقل السريع
ḥāfilat an-naql as-sarīʿ

rapid-transit lines — خطوط النقل السريع
k̲uṭūṭ an-naql as-sarīʿ

R

rapid-transit system شبكة النقل السريع
šabakat an-naql as-sarī'
rapid-transit tunnel نفق النقل السريع
nafaq an-naql as-sarī'
ratchet spanner مفتاح ربط بسقاطة
miftāḥ rabṭ bis-saqāṭa
rating تقدير
taqdīr
rating flume مسيل مراقبة
musīl murāqaba
rat-tail file مبرد ذيل الفأر
mibrad ḏail al-fa'r
Rawlbolt مسمار ملولب
mismār mulaulab
raw water ماء عسر
mā' 'asir
Raykin fender buffer مخمد المصد «رايكن»
mikmad al-miṣadd
reach أرض منبسطة
'arḍ munbasiṭa
reaction رد فعل
radd fi'l
reaction turbine تربين رد فعلي
tūrbīn radd fi'lī
reactive compensation تعويض مفاعل
ta'wīḍ mufā'il
reactor core قلب المفاعل
qalb al-mufā'il
reactor foundations أساسات المفاعل
asāsāt al-mufā'il
reactor pressure vessel وعاء ضغط المفاعل
wi'ā' ḍaġṭ al-mufā'il
reading قراءة
qirā'a
read only memory ذاكرة قراءة فقط
ḏākirat qirā'a faqaṭ
ready-mix concrete خرسانة جاهزة الخلط
karasāna jāhizat al-kalṭ
realignment تصحيح التراصف
taṣḥīḥ at-tarāṣuf
ream توسيع الثقب
wassa'a aṯ-ṯuqb
reamer موسع ثقوب
muwassi' ṯuqūb
rebound hammer مطرقة إرتدادية
miṭraqa irtidādīya
rebuilding إعادة البناء
i'ādat al-binā'
receiver وعاء إستقبال
wi'ā' istiqbāl
receiving waterways مجاري المصب
majārī al-maṣabb
receiving yard ساحة الاستلام
sāḥat al-istilām
recess تجويف
tajwīf
recessed joint وصلة متداخلة
waṣla mutadākila
reciprocal levelling قياس المناسيب التبادلي
qiyās al-manāsīb at-tabādulī
reciprocating compressor ضاغط ترددي
ḍāġiṭ taraddudī
reciprocating engine محرك ترددي
muḥarrik taraddudī
reciprocating pump مضخة ترددية
miḍakka taraddudīya
recirculating cooling تبريد باعادة الدوران
tabrīd bi-i'ādat ad-dawarān
recirculation إعادة الدوران
i'ādat ad-dawarān
reclamation scheme مشروع إستصلاح الأراضي
mašrū' istiṣlāḥ al-arāḍī
recording gauge مقياس تسجيل
miqyās tasjīl
recovery إسترداد
istirdād
recovery peg وتد إسناد
watad isnād
recreation area منطقة ترفيه
minṭaqat tarfīh
rectangular notch نقر مستطيل
naqr mustaṭīl
rectangular weir سد مستطيل
sadd mustaṭīl
rectification تصفية . تقويم
taṣfīya, taqwīm
reduction تخفيض . تصغير
takfīḍ, taṣġīr
reduction factor عامل الخفض
'āmil al-kafḍ
redwood خشب أحمر
kašab aḥmar
reeving thimble وصلة تسليك
waṣlat taslīk
refectory حجرة الطعام
ḥujrat aṭ-ṭa'ām
reference mark علامة إسناد
'alāmat isnād
reference number رقم الاسناد
raqm al-isnād

R

reference peg وتد إسناد
watad isnād
reference standard معيار الاسناد
ma'yār al-isnād
reflecting level ميزان إستواء عاكس
mīzān istiwā' 'ākis
reflecting studs أوتاد عاكسة
autād 'ākīsa
reflux valve صمام لارجعي
ṣimām lā-raja'ī
refractory concrete خرسانة حرارية
karasāna ḥarārīya
refractory linings بطانات صامدة للحرارة
biṭānāt ṣāmida lil-ḥarāra
refrigerant سوائل التبريد
sawā'il at-tabrīd
refrigerating capacity قدرة التبريد
qudrat at-tabrīd
refrigeration plant وحدة تبريد
waḥdat tabrīd
refrigerator foundations قواعد جهاز التبريد
qawā'id jihāz at-tabrīd
refuge ملجأ . ملاذ
maljā', malāḏ
refusal رفض
rafḍ
refuse collection جمع القمامة
jam' al-qimāma
refuse compression ضغط القمامة
ḍaġṭ al-qimāma
refuse disposal التخلص من القمامة
at-takalluṣ min al-qimāma
regime توازن الحت والترسب
tawāzun al-ḥatt wat-tarassub
registered design تصميم مسجَّل
taṣmīm musajjal
registration تسجيل
tasjīl
regular inspection تفتيش دوري
taftīš daurī
regular maintenance صيانة دورية
ṣīyāna daurīya
regulating course طبقة تسوية
ṭabaqat taswiya
regulations قوانين . أنظمة
qawānīn, anẓima
regulator منظم . معدل
munaẓẓim, mu'addil

reheater مسخن بيني
musakkin bainī
reinforced brickwork بناء بالطوب مسلَّح
binā' biṭ-ṭaub musallaḥ
reinforced concrete خرسانة مسلحة
karasāna musallaḥa
reinforced concrete base قاعدة خرسانية مسلحة
qā'ida karasāna musallaḥa
reinforced concrete facing سطح خرسانة مسلحة
saṭḥ karasāna musallaḥa
reinforcement تقوية . تدعيم
taqwiya, tad'īm
reinforcement hook خطاف التقوية
kaṭṭāf at-taqwiya
reinforcement of a foundation footing تقوية قاعدة الأساس
taqwiyat qā'idat al-asās
relative compaction التراص النسبي
at-tarās an-nisbī
relative density الكثافة النسبية
al-kaṯāfa an-nisbīya
relative density of a sand الكثافة النسبية للرمل
al-kaṯāfa an-nisbīya li-raml
relative humidity الرطوبة النسبية
ar-ruṭūba an-nisbīya
relative settlement الهبوط النسبي
al-hubūt an-nisbī
relaxation تراخ . تخفيف
tarākī, takfīf
relay مرحّل . محرك مؤازر
murāhhil, muḥarrik mu'āzir
release agent مادة تفتيت
māddat taftīt
relief holes ثقوب تصريف
ṯuqūb taṣrīf
relief sewer مجرور تنفيس
majrūr tanfīs
relief well بئر تنفيس
bi'r tanfīs
relieving platform منصة توزيع الحمل
minaṣṣat tauzī' al-ḥiml
remedial work أعمال إصلاحية
a'māl iṣlāḥīya
remote control تحكم ناءٍ
taḥakkum nā'ī
remote control switch مفتاح تحكم ناءٍ
miftāḥ taḥakkum nā'ī
remote parking توقف ناءٍ
tawaqquf nā'ī

R

emoulding إعادة التشكيل
i'ādat at-taškīl
emoval نقل . إنتقال
naql, intiqāl
endering طلية أولى
ṭalya ūla
Reno mattress فرشة « رينو »
faršat rīnū
epair أصلح . تصليح
aṣlaḥa, taṣlīḥ
epetition تكرار
tikrār
eport تقرير
taqrīr
epresentative sample عينة نموذجية
'ayyina namūḏajīya
escue boat قارب إنقاذ
qārib inqāḏ
escue service خدمات الانقاذ
ḵadamāt al-inqāḏ
esection تقاطع . قطع
taqāṭu', qaṭ'
eservoir خزان
ḵazzān
eservoir outlet مخرج الخزان
maḵraj al-ḵazzān
eservoir roof سطح الخزان
saṭḥ al-ḵazzān
esident engineer المهندس المقيم
al-muhandis al-muqīm
esidential accommodation أماكن سكنية
amākin sakanīya
residential wall جدار داخلي
jidār dāḵilī
residual errors أخطاء متخلفة
aḵṭā' mutaḵallifa
residual stress إجهاد متخلف
ijhād mutaḵallif
resilience رجوعية . إرتدادية
rujū'īya, irtidādīya
resistance مقاومة
muqāwama
resistance flash welding لحام ومضي بالمقاومة
liḥām wamḍī bil-muqāwama
resistance seam welding لحام درزي بالمقاومة
liḥām durzī bil-muqāwama
resistance spot welding لحام نقطي بالمقاومة
liḥām nuqṭī bil-muqāwama

resistance welding لحام بالمقاومة
liḥām bil-muqāwama
resisting moment عزم المقاومة
'azm al-muqāwama
resistivity logging تسجيل المقاومة النوعية
tasjīl al-muqāwama al-nau'īya
resistor مقاوم
muqāwim
resistor transistor logic دائرة ترانزستور – مقاوم منطقية
dā'ira trānzistar muqāwim manṭiqīya
resoiling تسوية التربة
taswiyat at-turba
resonance رنين
ranīn
respiratory protective device وسيلة واقية للتنفس
wasīla wāqiya lit-tanaffus
responsibility مسؤولية
mas'ūlīya
restaurant مطعم
maṭ'am
restaurant facilities تسهيلات المطعم
tashīlāt al-maṭ'am
restoration إصلاح
iṣlaḥ
restrained beam عارضة مقيدة الحركة
'āriḍa muqayyadat al-ḥaraka
rest room غرفة الاستراحة
ġurfat al-istirāḥa
retained bank ضفة محتجزة
ḍaffa muḥtajaza
retaining wall جدار ساند
jidār sānid
retarder عائق
'ā'iq
retarding agent عامل معوق
'āmil mu'awwiq
retreading roads تجديد الطرق
tajdīd aṭ-ṭuruq
return sheave بكرة إرجاع
bakarat irjā'
reverse curve منحنى متعاكس
munḥana muta'ākis
reversed filter مرشح معكوس
muraššiḥ ma'kūs
reverse osmosis تناضح عكسي
tanāḍuḥ 'aksī

R

reverse-rotary drilling — حفر دوراني عكوس
ḥafr dawarānī ʿakūs

revet — كسي بالاسمنت
kasī bil-ismant

revetment — جدار إحتجاز . تكسية من الاسمنت
jidār iḥtijāz, taksiya min al-ismant

revision — مراجعة . تنقيح
murājaʿa, tanqīḥ

revolver crane — مرفاع دوار
mirfāʿ dawwār

revolving — دوار . دائر
dawāʾir, dāʾir

revolving doors — أبواب دوارة
abwāb dawwāra

revolving screen — غربال دوار
ġirbāl dawwār

Reynolds number — رقم «رينولدز»
raqm rinūldz

rib — رافدة . ضلع
rāfiḍa, ḍilʿ

ribbed arch — قنطرة مضلعة
qanṭara muḍallaʿa

ribbed construction — إنشاء مضلع
inšāʾ muḍallaʿ

ribbon slab track — سكة حديد بقواعد خرسانية
sikkat ḥadīd bi-qawāʿid karasānīya

rib holes — ثقوب على جانب النفق
ṯuqūb ʿala jānib an-nafaq

rich concrete — خرسانة مستوفرة
karasāna mustaufira

ridge — حرف . نتوء
ḥarf, nutūʾ

ridge covering — تغطية المنحدرين
taġṭiyat al-munḥadarain

riding quality — جودة الركوب
jaudat ar-rukūb

riffle sampler — جامع عينات حلزوني
jāmiʿ ʿayyināt ḥalazunī

rift — صدع . شق
ṣadaʿ, šaqq

rigger — حاجز واق
ḥājiz wāqī

rigging — تجهيزات تركيب
tajhīzāt tarkīb

rigging up — تركيب أجهزة الحفر
tarkīb ajhizat al-ḥafr

rigid — صلب
ṣalib

rigid arch — طرة ثابتة
qanṭara ṯābita

rigid coupling — ارن صلب
taqārun ṣalib

rigid cradle — مل صلب
maḥmal ṣalib

rigid flange — فة صلبة
šaffa ṣaliba

rigid frame — يكل ثابت
haikal ṯābit

rigidity — لابة
ṣalāba

rigid pavement — صيف من قوالب الخرسانة
raṣīf min qawālib al-karasāna

rigid pipe — بوب صلب
unbūb ṣalib

rigid raft — وف صلب
ṭauf ṣalib

rim clutch — ابض حافة
qābiḍ ḥāffa

ring — لقة
ḥalaqa

ring beam — ارضة حلقية
ʿāriḍa ḥalaqīya

ring gate — وابة دائرية
bawwāba dāʾirīya

ring girder — عارضة دائرية
ʿāriḍa dāʾirīya

ring liner — طانة حلقية
biṭāna ḥalaqīya

ring main system — شبكة كهربائية حلقية
šabaka kahrabāʾīya ḥalaqīya

ring support — دعامة حلقية
diʿāma ḥalaqīya

ring tension — نوتر حلقي
tawattur ḥalaqī

riparian — شاطئي
šāṭiʾī

ripper — كسارة
kassāra

ripple — تموج
tamawwuj

rip-rap — دكة حجارة
dakkat hajāra

rise — إرتفاع
irtifāʿ

riser — ماسورة صاعدة
māsūra ṣāʿida

R

ising main مأخذ رئيسي صاعد
ma'kad ra'īsī ṣā'id
ising shaft مهواة صاعدة
mihwāt ṣā'ida
iver نهر
nahr
iver bed قاع النهر
qā' an-nahr
iver intake مدخل نهري
madkal nahrī
iver mouth مصب النهر
maṣabb an-nahr
ivet برشامة
biršāma
iveted connections توصيلات مبرشمة
tauṣīlāt mubaršama
iveting برشمة
baršama
oad طريق
ṭarīq
oad bed قاعدة الخط الحديدي
qā'idat al-kaṭṭ al-ḥadīdī
oad breaker كسارة رصف
kassārat raṣf
oad capacity سعة الطريق
si'at aṭ-ṭarīq
road connection وصلة طريق
waṣlat ṭarīq
road construction plant معدات إنشاء الطرق
mu'iddāt inšā' aṭ-ṭuruq
road drainage صرف الطريق
ṣarf aṭ-ṭarīq
road excavation شق الطريق
šaqq aṭ-ṭuruq
road fencing تسوير الطرق
taswīr aṭ-ṭuruq
road forms هياكل جانبية
hayākil jānibīya
road foundations قواعد الطرق
qawā'id aṭ-ṭuruq
road grader مدرجة طريق
madrajat ṭarīq
road heater مسخن سطح الطريق
musakkin saṭḥ aṭ-ṭarīq
road level مستوى الطريق
mustawa aṭ-ṭarīq
road lighting إضاءة الطرق
iḍā'at aṭ-ṭuruq
road-making plant معدات إنشاء الطرق
mu'iddāt inšā' aṭ-ṭuruq
road markings علامات الطرق
'alāmāt aṭ-ṭuruq
road metal حجارة رصف الطرق
ḥajārat raṣf aṭ-ṭuruq
road network شبكة طرق
šabakat ṭuruq
road panel منطقة مفروشة بالخرسانة
minṭaqa mafrūša bil-karasāna
road pavement رصيف الطريق
raṣīf aṭ-ṭarīq
road roller مدحلة طرق
midḥalat ṭuruq
road signals إشارات الطرق
išārāt aṭ-ṭuruq
road specification مواصفات الطرق
muwāṣafāt aṭ-ṭuruq
road surface طبقة الرصف السطحية
ṭabaqat ar-raṣf as-saṭḥīya
road system نظام الطرق
niẓām aṭ-ṭuruq
robotics إستخدام الآلات
istikdām al-ālāt
rock صخر
ṣakr
rock-boring machine مكنة حفر الصخور
makinat ḥafr aṣ-ṣukūr
rock burst تفجير الصخور
tafjīr aṣ-ṣukūr
rock core قلب صخري
qalb ṣakrī
rock drill ثقابة صخور
ṯaqqābat ṣukūr
rocker bearing محمل الهزازة
maḥmal al-hazzāza
rocker shovel مجرفة قلابة
mijrafa qallāba
rock-fill dam سد ذو حشوة صخرية
sadd ḏū hašwa ṣakrīya
rock mechanics علم ميكانيكا الصخور
'ilm mīkānīka aṣ-ṣukūr
rock-mounded breakwater مصد أمواج ركامي
miṣadd amwaj rukāmī
rock noise ضجيج الصخور
ḍajīj aṣ-ṣukūr
rock rubble كسارة صخرية
kassāra ṣakrīya
rock tunnel نفق صخري
nafaq ṣakrī

R

Rockwell hardness test إختبار الصلادة لـ «روكويل»
iktibār aṣ-ṣalāda li-rukwīl
rocky outcrop نتوء صخري
nutū' sakrī
rod قضيب
qaḍīb
rodding تنظيف بالقضبان
tanẓīf bil-quḍbān
rodent control مكافحة القوارض
mukāfaḥat al-qawāriḍ
rod float عوامة قضيبية
'awwāma qaḍībīya
rod man عامل الشاخص
'āmil aš-šākis
rolled asphalt طبقة الرصف الاسفلتية
ṭabaqat ar-raṣf al-isfaltiya
rolled beam ركيزة دلفينية
rakīza dalfīnīya
rolled-beam bridge جسر ذو ركيزة دلفينية
jisr ḏū rakīza dalfinīya
rolled steel فولاذ مدلفن
fūlāḏ mudalfan
rolled-steel joist رافدة فولاذ مدلفن
rāfida fūlāḏ mudalfan
rolled-steel section قطاع فولاذ مدلفن
qaṭā' fūlāḏ mudalfan
roller مدحلة
midḥala
roller bearings محامل دلفينية
maḥāmil dalfīnīya
roller bit لقمة حفر بمسننات دوارة
luqmat ḥafr bi-musannanāt dawwāra
roller bridge جسر دوار
jisr dawwār
roller cutter آلة قطع دوارة
ālat qaṭ' dawwāra
roller door باب دوار
bāb dawwār
roller gate بوابة دوارة
bawwāba dawwāra
roller mounted مركبة على دلفين
murakkaba 'ala dalfīn
rolling gate بوابة دائرة
bawwāba dā'ira
rolling lift bridge جسر رفع دوار
jisr raf' dawwār
rolling load حمل متنقل
ḥiml mutanaqqil

rolling resistance اومة الدروج
muqāwamat ad-durūj
rolling-up curtain weir . متحرك عمودي
sadd mutaḥarrik 'amūdī
rollway ى الفائض
majrā al-fā'iḍ
ROM كرة قراءة فقط
ḏakirat qirā'a faqaṭ
rood د : مقياس للأراضي
rūd, miqyās lil-'arāḍī
roof boards اح السقف
alwāḥ as-saqf
roof bolt غي السطح
baraḡī as-saṭḥ
roof construction ساء السطح
inšā' as-saṭḥ
roof garden ديقة السطح
ḥadīqat as-saṭḥ
roof girder رضة السطح
'āriḍat as-saṭḥ
roofing asphalt سفلت للسطوح
isfalt lis-suṭūḥ
roof pitch جة إنحدار السطح
darajat inḥidār as-saṭḥ
roof structure يكل السطح
haikal as-saṭḥ
roof truss ملون السطح
jumlūn as-saṭḥ
room slab لاطة الحجرة
bilāṭat al-ḥujra
root ساس
asās
rooter مدشة مقطورة
mukaddiša maqṭūra
Rootes blower فخ دوار
nāfik dawwār
rope بل . كبل
ḥabl, kabl
rope diameter طر الحبل
quṭr al-ḥabl
rope drilling فر كبلي
ḥafr kablī
rope fastenings بتات الحبل
muṯabbitāt al-ḥabl
rope suspension عليق بالحبل
ta'līq bil-ḥabl
rope transmission ل بالحبال
naql bil-ḥibāl
ropeway ريق حبلي
ṭarīq ḥablī

R

rope wheel بكرة محززة
bakara muhazzaza
rotary blower نفاخ دوار
naffāḵ dawwār
rotary booster معزز دوار
mu'azziz dawwār
rotary boring حفر رحوي
ḥafr raḥawī
rotary breaker كسارة رحوية
kassāra raḥawīya
rotary bucket دلو دوار
dalū dawwār
rotary core drilling حفر قلب رحوي
ḥafr qalb raḥawī
rotary drilling حفر رحوي
ḥafr raḥawī
rotary excavator حفارة دوارة
ḥaffāra dawwāra
rotary plug valve صمام سدادي دوار
ṣimām sidadī dawwār
rotary pump مضخة رحوية
miḍakḵa raḥawīya
rotary screen غربال دوار
ġirbāl dawwār
rotary snow plough جرافة ثلج دوارة
jarrāfa ṯalj dawwāra
rotary wall crane مرفاع جداري دوار
mirfā' jidārī dawwār
rotation recorder مسجل الدوران
musajjil ad-dawarān
rotten wood خشب نخر
ḵašab naḵr
rough خشن . وعر
ḵašin, wā'ir
rough-cast ملاط خشن
milāṭ ḵašin
roughness خشونة
ḵušūna
roughness coefficient معامل الخشونة
mu'āmil al-ḵušūna
rough rubble دبش خشن
dabš ḵašin
round دائري
dā'irī
roundabout دائرة مرور
dā'irat murūr
round-headed buttress dam سد خرسانة بدعامات متوازية
sadd ḵarasāna bi-di'āmāt mutawāzīya

round steel فولاذ مبروم
fūlāḏ mabrūm
route selection إختيار الطرق
iḵtiyār aṭ-ṭuruq
route survey مسح الطرق
masḥ aṭ-ṭuruq
r.s.j. جائز من الفولاذ المدلفن
jā'iz min al-fūlāḏ al-mudalfan
rubber flooring أرضية مطاطية
arḍīya maṭṭāṭīya
rubber gloves قفازات مطاطية
qaffāzāt maṭṭāṭīya
rubber-lined pipe ماسورة مبطنة بالمطاط
māsūra mubaṭṭana bil-maṭṭāṭ
rubber rim إطار مطاط
iṭār maṭṭāṭ
rubber-tyred rollers دلافين باطارات مطاطية
dalafīn bi-iṭārāt maṭṭāṭīya
rubber-tyred scraper كاشطة باطارات مطاطية
kāšita bi-iṭarāt maṭṭāṭīya
rubbing حك . إحتكاك
ḥakk, iḥtikāk
rubble دبش
dabš
rubble concrete خرسانة دبشية
ḵarasāna dabšīya
rubble drain أخدود تصريف
uḵdūd taṣrīf
rubble-mound breakwater حائل موجي من الحجارة الضخمة
ḥā'il maujī min al-ḥijara aḍ-ḍaḵma
rugous surface سطح مجعّد
saṭḥ muja''ad
ruling gradient درجة الانحدار الاقصى
darajat al-inḥidār al-aqṣā
rumble strip شريط دوّار
šarīṭ dawwār
runner قدة تحديد
qiddat taḥdīd
running bridge جسر دوّار
jisr dawwār
running costs تكاليف التشغيل
takālīf at-tašġīl
running ground أرض زالقة
arḍ zāliqa
running rail سكة متحركة
sikka mutaḥarrika

R

running time زمن التشغيل
zaman at-tašġīl
run-off تفريغ السائل
tafrīġ as-sā'il
runway مدرج
madraj
runway centreline lighting إضاءة خط المركز للمدرج
iḍā'at kaṭṭ al-markaz lil-madraj
runway clear zones المناطق الخالية على المدرج
al-manāṭiq al-kalīya 'ala al-madraj
runway edge lighting إضاءة جانبي المدرج
iḍā'at jānibayy al-madraj
runway gradient معدل إنحدار المدرج
mu'addal inḥidār al-madraj
runway layout مخطط المدرج
mukaṭṭaṭ al-madraj
runway length طول المدرج
ṭūl al-madraj
runway lighting إضاءة المدرج
iḍā'at al-madraj
runway numbering system نظام ترقيم المدرج
niẓām tarqīm al-madraj
rural roads طرق ريفية
ṭuruq rīfīya
rust صدأ
ṣada'
sacrificial protection وقاية بالأنود الذواب
wiqāya bil-anūd aḏ-ḏawwāb
saddle سرج . حامل
sarj, ḥāmil
safe load حمل مأمون
ḥiml ma'mūn
safe stacking تكديس مأمون
takdīs ma'mūn
safe storage تخزين مأمون
takzīn ma'mūn
safe stress إجهاد مأمون
ijhād ma'mūn
safety barrier حاجز أمان
ḥājiz amān
safety belt حزام أمان
ḥizam amān
safety committee لجنة الأمان
lajnat al-amān
safety factor عامل الأمان
'āmil al-amān
safety fences أسوار أمان
aswār amān

R

safety glass زجاج أمان
zujāj amān
safety glasses نظارات أمان
naẓẓarāt amān
safety officer ضابط أمان
ḍābiṭ amān
safety of men سلامة الرجال
salāmat ar-rijāl
safety of nuclear reactors أمان المفاعلات النووية
amān al-mufā'ilāt an-nawwawīya
safety on site الأمان بالموقع
al-amān bil-mauqi'
safety organisation تنظيم الأمان
tanẓīm al-amān
safety procedures إجراءات الأمان
ijrā'āt al-amān
safety programme برنامج الأمان
barnāmij al-amān
safety relief valve صمام تنفيس للأمان
ṣimām tanfīs lil-amān
safety representative ممثل الأمان
mumaṯṯil al-amān
safety shoes حذاء أمان
ḥiḏā' amān
safety standards مواصفات الأمان
muwāṣafāt al-amān
safety valve صمام أمان
ṣimām amān
sag إرتخى . إرتخاء
irtaka
sag correction تصحيح الارتخاء
taṣḥīḥ al-irtikā'
sagging إرتخاء . هبوط
irtikā', hubūt
sagging moment عزم الإرتخاء
'azm al-irtika'
sag pipe أنبوب عابر للطريق
unbūb 'ābir liṭ-ṭarīq
saltation الحركة الارتدادية للرمل
al-ḥaraka al-irtidādīya lir-raml
salt glaze تزجيج ملحي
tazjīj milḥī
salt-glazed-ware pipe أنبوب صرف مغلف بالملح
unbūb ṣarf muġallaf bil-milḥ
salt spreader آلة رش الملح
ālat rašš al-milḥ
salvage إنقاذ . تخليص
inqāḏ, taklīṣ

salvage tender عطاء إنقاذ
'aṭā' inqāḏ
sample عينة . نموذج
'ayyina, namūḏaj
sampler مختبر العينات
mukṯabar al-'ayyināt
sampling أخذ العينات واختبارها
akḏ al-'ayyināt wa-ikṯibāriha
sampling cock كوب العينات
kūb al-'ayyināt
sampling spoon ملعقة العينات
mil'aqat al-'ayyināt
sand رمل
raml
sand blast سفع رملي
saf' ramlī
sand blasting تنظيف بالسفع الرملي
tanẓīf bis-saf' ar-ramlī
sand catcher مقياس الرمل
miqyās ar-raml
sand drain مصرف رملي
maṣrif ramlī
sander مرملة
mirmala
sand fill ردم رملي
radm ramlī
sand filter مرشح رمل
muraššiḥ raml
sand-grain meter مقياس الرمل
miqyās ar-raml
sanding of wood صنفرة الخشب
ṣanfarat al-kašab
sandpaper surface سطح محبب
saṭḥ muḥabbab
sand patch method طريقة الترميم الرملي
ṭarīqat at-tarmīm ar-ramlī
sand piles ركائز رملية
rakā'iz ramlīya
sand pit منجم رمل
manjam raml
sand pump مضخة رمل
miḍakkat raml
sand pump dredger كراءة بمضخة رمل
karrā'a bi-miḍakkat raml
sandstone حجر رملي
ḥajar ramlī
sand trap مصيدة الرمل
miṣyadat ar-raml
sandwich construction إنشاء مركب
inšā' murakkab
sandwich system نظام الطبقات
niẓām aṭ-ṭabaqāt
sandwick مصرف رملي
maṣrif ramlī
sanitary صحي
ṣiḥḥī
sanitary sewer مجارير صحية
majārīr ṣiḥḥīya
sanitary works أعمال صحية
a'māl ṣiḥḥīya
sapwood خشب رخو
kašab rakū
sash إطار منزلق
iṭār munzaliq
sash window نافذة باطارين منزلقين
nāfiḏa bi-iṭārain munzaliqain
satellite surveying مسح بالقمر الصناعي
masḥ bi-qamr aṣ-ṣinā'ī
satellite terminal محطة الاقمار الصناعية
maḥaṭṭat aqmār aṣ-ṣinā'īya
saturated air هواء مشبع بالرطوبة
hawā' mušabba bir-ruṭūba
saturation line مستوى التشبع
mustawa at-tašabbu'
saucer طبق
ṭabaq
saw منشار
minšār
sawdust نشارة الخشب
nišārat al-kašab
saw files مبارد شحذ المناشير
mabārid šaḥd al-manašīr
scabbing تقشر السطح
takaššur as-saṭḥ
scabbling النحت
an-naḥt
scaffold سقالة
saqāla
scaffolding سقالات
saqāla
scale مقياس
miqyās
scale drawing رسم بمقياس نسبي
rasm bi-miqyās nisbī
scale of a drawing مقياس الرسم
miqyās ar-rasm

S

scale of professional charges جدول الأتعاب المهنية
jadwal al-atʿāb al-mihanīya
scaling إزالة القشور
izālat al-qušūr
scarifier مخدشة
mukaddiša
school مدرسة
madrasa
scissors crossover تحويلة مقصية
taḥwīla miqaṣṣīya
scissors junction تقاطع مقصي
taqātuʿ miqaṣṣī
scleroscope hardness test إختبار الصلادة بالسكلرسكوب
iktibār aṣ-ṣalāda bis-sklīrūskūb
scoop dredger كراءة غرف
karrāʾat ğarf
scotch block كتلة منع الانزلاق
kutlat manʿ al-inzilāq
Scotch derrick مرفاع برج الحفر
mirfaʿ burj al-ḥafr
scour منظف . إنجراف
munaẓẓif, injirāf
scouring sluice حوض تنظيف
ḥauḍ tanẓīf
scour protection الوقاية من التعرية
al-wiqāya min at-taʿrīya
scow صندل
ṣandal
scraper مكشطة
mikšaṭa
scraper bucket قادوس كاشط
qādūs kāšiṭ
scraper excavator حفار كاشط
ḥaffār kāšiṭ
scraper loader حمالة كاشطة
ḥammāla kāšiṭa
screed دليل الثخانة
dalīl aṯ-ṯikāna
screen ستار . غربال
sitār, ğirbāl
screen analysis تحليل بالمنخل
taḥlīl bil-minkal
screened material مواد مغربلة
mawādd muğarbala
screening غربلة
ğarbala
screw برغي
burğī

screw bell ناقوس إلتقاط
nāqūs iltiqāṭ
screw compressor ضاغط لولبي
ḍāğiṭ laulabī
screw conveyor ناقلة لولبية
nāqila laulabīya
screwdriver مفك
mifakk
screw pile ركيزة ملولبة
rakīza mulaulaba
screw shackle شداد ملولب الطرفين
šadād mulaulab aṭ-ṭarafain
screw spike مسمار كبير ملولب
mismār kabīr mulaulab
scrub حك
hakk
SCUBA جهاز تنفس تحت الماء
jiḥāz tanaffus taḥt al-māʾ
S curves منعطفات بشكل S
munʿaṭafāt bi-šakl is
sea-defence works أشغال الدفاع البحري
ašğāl ad-difāʿ al-baḥrī
sealant مادة مانعة للتسرب
mādda maniʿ lil-tasarrub
sea-level correction تصحيح مستوى سطح البحر
taṣḥīḥ mustawa saṭḥ al-baḥr
sealing coat طبقة ختم
ṭabaqat katm
sealing compound مركّب مانع للتسرب
murakkab maniʿ lil-tasarrub
sealing ring حلقة مانعة للتسرب
ḥalaqa maniʿ lil-tasarrub
sealing washer فلكة مانعة للتسرب
falaka maniʿ lil-tasarrub
seam وصلة التئام
waṣlat iltiʿām
seam welding لحام درزي
liḥām durzī
sea outfall مخرج تصريف بحري
makraj taṣrīf baḥrī
seasoned wood خشب مجفف
kašab mujaffaf
sea surveying مسح بحري
masḥ baḥrī
seat مستقر القاعدة
mustaqarr al-qāʿida
seated-valve pump مضخة بصمام مرتكز
miḍakka bi-ṣimām murtakiz
seating تركيز
tarkīz

S

seat of settlement قاعدة إستقرار
qāʿidat istiqrār
seawall جدار بحري
jidār baḥrī
secant modulus مُعامل القاطع
muʿāmil al-qāṭiʿ
secant modulus of elasticity مُعامل اللدونة للقاطع
muʿāmil al-ladūna lil-qāṭiʿ
second ثانية
ṯānīya
secondary beam عارضة ثانوية
ʿāriḍa ṯānawīya
secondary road طريق فرعي
ṭarīq farʿī
secondary sedimentation ترسيب ثانوي
tarsīb ṯānawī
secondary treatment معالجة ثانوية
muʿālaja ṯānawīya
secondary ventilation تهوية ثانوية
tahwiya ṯānawīya
second floor (level 3) الطابق الثاني
aṭ-ṭābiq aṯ-ṯānī
second underlayer طبقة سفلية ثانية
ṭabaqa suflīya ṯānīya
section قسم . قطاع
qism, qiṭāʿ
sectional elevation مسقط رأسي قطاعي
masqaṭ raʾsī qiṭāʿī
section leader رئيس قسم
raʾīs qism
section properties خصائص المقطع
ḵaṣṣāʾiṣ al-maqṭaʿ
section tank خزان قطاعي
ḵazzān qiṭāʿī
sector gate بوابة قطاعية
bawwāba qiṭāʿīya
security fence حاجز أمن
ḥājiz amn
sediment راسب
rāsib
sedimentation ترسب
tarassub
sedimentation tank خزان الترسيب
ḵazzān at-tarsīb
seeding plant غراسة البذور
ġarāsat al-buḏūr
seepage إرتشاح
irtišāḥ
seepage loss فقد بالارتشاح
fuqd bil-irtišāḥ
seepage pit حفرة الارتشاح
ḥufrat al-irtišāḥ
segmental arch قوس قطاعي
qaus qiṭāʿī
segmental gate بوابة قطاعية
bawwāba qiṭāʿīya
segregation عزل
ʿazl
seismic load حمل الزلزال
ḥiml az-zilzāl
seismic prospecting التنقيب بالطريقة الزلزالية
at-tanqīb biṭ-ṭarīqa az-zilzālīya
seismograph سيزموجراف : مرسمة الزلزال
sīzmūġrāf, mirsamat az-zalāzil
seismometer سيزمومتر : مقياس الزلزلة
sīzmūmitar, miqyās az-zalzala
self-aligning level مستوى ذاتي المحاذاة
mustawa ḏātī al-muḥāḏāt
self-cleansing gradient منحدر ذاتي التنظيف
munḥadar ḏātī at-tanẓīf
self-contained متكامل
mutakāmil
self-fill تعبئة ذاتية
taʿbiʾa ḏātīya
self-hopper barges صنادل ذاتية القواديس
ṣanādil ḏātīyat al-qawādīs
self-levelling level مستوى ذاتي الاستواء
mustawa ḏatī al-istiwāʾ
self-locking nut صمولة ذاتية الزنق
ṣamūla ḏatīyat az-zanq
self-operated controller جهاز تحكم ذاتي التشغيل
jihāz taḥakkum ḏātī at-tašġīl
self-purifying capacity سعة التنقية الذاتية
siʿat at-tanqīya aḏ-ḏātīya
self-sustaining winch ونش ذاتي الدعم
winš ḏātī ad-daʿm
self-tensioning توتر ذاتي
tawattur ḏātī
semi-circular arch قنطرة نصف دائرية
qanṭara niṣf dāʾirīya
semi-conductor شبه موصل
šibh muwaṣṣil
semi-diurnal tides مد (أو جزر) شبه نهاري
madd au jazr šibh nahārī

S

semi-natural harbour ميناء شبه طبيعي
mīnā' šibh ṭabi'ī
semi-rigid شبه صلب
šibh ṣalib
semi-rigid cradle حمالة شبه صلبة
ḥammāla šibh ṣaliba
semi-skilled man عامل نصف ماهر
'āmil niṣf māhir
semi-span نصف باع
niṣf bā'
semi-submersible نصف مغمور
niṣf maǧmūr
semi-transverse system نظام نصف مستعرض
niẓām niṣf musta'raḍ
sensible horizon الأفق المرئي
al-ufuq al-mar'ī
sensitivity حساسية
ḥassāsīya
sensitivity ratio نسبة الحساسية
nisbat al-ḥassāsīya
sensors أجهزة إحساس
ajhizat iḥsās
separate system نظام تصريف منفصل المجاري
niẓām taṣrīf munfaṣṣil al-majārī
separator فرازة
farrāza
septic tank خزان تعفين
ḵazzan ta'fīn
serial data transmission إرسال البيانات المتتابعة
irsāl al-bayānāt al-mutatābi'
serial number رقم التسلسل
raqm at-tasalsul
service cable كبل الخدمة
kabl al-ḵidma
service lift مصعد البضائع
miṣ'ad al-baḍā'i'
service mains الموصلات الرئيسية
al-mauṣilāt ar-ra'īsīya
service manual دليل إرشادات الصيانة
dalīl iršādāt aṣ-ṣīyāna
service pipe أنبوب إمداد
unbūb imdād
service reservoir خزان إمداد
kazzān imdād
service riser ماسورة إمداد صاعدة
māsūrat imdād ṣā'ida
service road طريق الخدمة
ṭarīq al-ḵidma
services area منطقة الخدمات
minṭaqat al-ḵadamāt
service stair سلم الخدمة
sullam al-ḵidma
service yard مساحة الخدمة
sāḥat al-ḵidma
servo mechanism آلية مؤازرة
ālīya mu'āzira
set ضبط . تصلب
taṣallub, ḍabṭ
set screw برغي تثبيت
burǧī taṯbīt
set square كُوس
kūs
sett حجر رصف
ḥajar raṣf
setting معايرة
mu'āyara
setting coat طبقة التمليط النهائية
ṭabaqat at-tamlīṭ an-nihā'īya
setting out وضع العلامات
waḍ' al-'alāmāt
setting time of cement زمن التصلب للاسمنت
zaman at-taṣallub lil-ismant
setting up نصب . إقامة
naṣb, iqāma
settlement هبوط (قاعدة البناء)
hubūt qā'idat al-binā'
settlement analysis تحليل الهبوط الاستقراري
taḥlīl al-hubūt al-istiqrārī
settlement crater وحدة الهبوط
waḥdat al-hubūt
settlement joint وصلة إستقرار
waṣlat istiqrār
settlement of humus ترسب الدبال
tarassub ad-dibāl
settling ترسب . ترسيب
tarassub, tarsīb
settling basin حوض الترسيب
ḥauḍ at-tarsīb
set-up تركيب . تركيز
tarkīb, tarkīz
seven سبعة
sab'a
seventeen سبعة عشر
sab'ata 'ašar
seventy سبعون
sab'ūn

S

severity rate *mu'addal al-kuṭūra* معدل الخطورة

sewage *miyāh al-majārīr* مياه المجارير

sewage composition *tarkiz miyāh al-majārīr* تركيز مياه المجارير

sewage disposal *at-takallus min miyāh al-majārīr* التخلص من مياه المجارير

sewage ejector *qāḍif miyāh al-majārīr* قاذف مياه المجارير

sewage farm *mazra'at miyāh al-majārir* مزرعة مياه المجارير

sewage flow *tadaffuq miyāh al-majārīr* تدفق مياه المجارير

sewage gas *ğāz al-majārir* غاز المجارير

sewage pumping *ḍakk miyāh al-majārīr* ضخ مياه المجارير

sewage treatment *mu'ālajat miyāh al-majārīr* معالجة مياه المجارير

sewage treatment works *mahaṭṭat mu'ālajat miyāh al-majāri* محطة معالجة مياه المجاري

sewer *majrūr* مجرور

sewerage *majārī* مجاري

sewerage system *šabakat al-majārī* شبكة مجاري

sewer flushing *rahḍ al-majrūr* رحض المجرور

sewer pill *kurat taslīk al-majārī* كرة تسليك المجاري

sewer tunnel *nafaq al-majārī* نفق المجاري

sextant *sudsīya* سدسية

shackle *šikāl* شكال

shaft *bi'r at-tahwiya* بئر التهوية

shaft construction *inšā' bi' tahwiya* إنشاء بئر التهوية

shaft foundation *qā'idat bi'r at-tahwiya* قاعدة بئر التهوية

shaft plumbing *tafdīn bi'r at-tahwiya* تفدين بئر التهوية

shaft-plumbing wire *silk tafdīn bi'r at-tahwiya* سلك تفدين بئر التهوية

shaft sinking *ḥafr al-mihwāt nuzūlan* حفر المهواة بالغطس

shaft spillway *qanāt taṣrīf bi'r at-tahwiya* قناة تصريف لبئر التهوية

shaft timbering *tad'īm jawānib al-mihwāt bil-akšāb* تدعيم جوانب المهواة بالأخشاب

shaft well *bi'r al-miṣ'ad* بئر المصعد

shaking test *iktibār al-ihtizāzīya* إختبار الاهتزازية

shale *ṭīn ṣafḥī* طين صفحي

shallow foundation *asās ḍaḥl* أساس ضحل

shallow manhole *ḥufrat taftīš ḍaḥla* حفرة تفتيش ضحلة

shallow well *bi'r ḍaḥla* بئر ضحلة

shape of bars *šakl al-quḍbān* شكل القضبان

shaping *taswiyat arḍ aṭ-ṭarīa* تسوية أرض الطريق

shareholder *musāhim* مساهم

sharp-crested weir *sadd ḥādd aḏ-ḏurwa* سد حاد الذروة

shear *qaṭa'a, qaṣṣa* قطع . قص

shear centre *markaz al-qaṣṣ* مركز القص

shear core *qulūb al-qaṣṣ* قلوب القص

shear diagram *rasm bayānī lil-qaṣṣ* رسم بياني للقص

shearing *qaṣṣ* قص

shearing force *quwwat al-qaṣṣ* قوة القص

shearlegs *mirfā' miqaṣṣī* مرفاع مقصي

shear loading *ḥiml al-qaṣṣ* حمل القص

shear modulus *mu'āmil al-qaṣṣ* معامل القص

shear slide *inzilāq al-qaṣṣ* إنزلاق القص

shear strain *juhd al-qaṣṣ* جهد القص

S

shear strength مقاومة القص
muqāwamat al-qaṣṣ
shear stress ضغط القص
ḍağṭ al-qaṣṣ
shear tests إختبارات القص
ikṯibārāt al-qaṣṣ
shear wall جدار مستعرض
jidār musta'raḍ
sheathing غلاف . تغليف
ğilāf, tağlīf
sheave بكرة محززة
bakara muḥazzaza
shed حظيرة
ḥaẓīra
sheepsfoot roller هراس التربة
harrās at-turba
sheeted caisson قيسون مصفح
qaisūn muṣaffaḥ
sheeters ألواح حماية جوانب الخندق
alwāḥ ḥimāyat jawānib al-ḵandaq
sheeting تصفيح
taṣfīḥ
sheet metal لوح معدني
lauḥ ma'danī
sheet pavement رصف ناعم
raṣf nā'im
sheet piles ركائز مستعرضة
rakā'iz musta'raḍa
sheet-pile wall جدار بركيزة مستعرضة
jidār bi-rakīza musta'raḍa
sheet piling ركائز إحتجاز
rakā'iz iḥtijāz
sheets of nylon ألواح نايلون
alwāḥ nīlūn
sheet steel لوح فولاذي
lauḥ fūlāḏī
shelf angle زاوية الرصيف الصخري
zāwīyat ar-raṣīf aṣ-ṣaḵrī
shelf retaining wall جدار ساند للرصيف
jidār sānid li-raṣīf
shell هيكل البناء
haikal al-binā'
shell-and-auger boring حفر بمثقب مجوف
ḥafr bi-miṯqab mujawwaf
shell construction إقامة هيكل البناء
iqāmat haikal al-binā'
Shell-perm process (TM) حقن مستحلب القار
ḥaqn mustaḥlab al-qār

shell pump مضخة نزح
miḍakḵat nazḥ
shelter ملجأ
maljā'
shield حجاب واق
ḥijāb wāqī
shielded-arc welding لحام قوسي محجب
liḥām qausī muḥajjab
shielded block كتلة محجبة
kutla muḥajjaba
shield tunnelling حفر الأنفاق بالتدريع
ḥafr al-anfāq bi-tadrī'
shift أزاح . إزاحة
azāḥa
shift instruction تعليمات إزاحية
ta'līmāt izāḥīya
shim فلكة . رفادة
falaka, rafāda
shin لوح وصل تراكبي
lauḥ waṣl tarākubī
shingle حصباء
ḥaṣbā'
shingle beach شاطئ حصبائي
šāṭi' ḥaṣbā'ī
shin guard واقي لوح الوصل
wāqī lauḥ al-waṣl
ship caisson قيسون عوام
qaisūn 'awwām
shipping-terminal services خدمات محطة الشحن
kadamāt maḥaṭṭaṭ aš-šaḥn
ship servicing facilities مرافق خدمة السفن
marāfiq ḵadamat as-sufun
shoe حذاء
ḥiḏā'
shop weld لحام بالورشة
liḥām bil-wirša
shore شاطئ
šāṭi'
shore protection حماية الشواطئ
ḥimāyat aš-šawāṭi'
shoring دعم . دعائم
da'm, da'ā'im
short bored piles ركائز أساسات قصيرة
rakā'iz asāsāt qaṣīra
short column عمود قصير
'amūd qaṣīr
short-span roof سطح قصير الباع
saṭḥ qaṣīr al-bā'
short ton طن أمريكي
ṭann amrīkī

shot blasting *saf' bil-ḵardaq*	سفع بالخردق
Shotcrete (TM) *milāṭ ramlī*	ملاط رملي
shot firing *saf'*	سفع
shot-loaded *mu'abba' bil-kurayyāt al- ma'danīya*	معبأ بالكريات المعدنية
shot point *nuqṭat at-tafjīr*	نقطة التفجير
shoulder *katif, 'ātiq*	كتف . عاتق
shovel *mijrafa, jārūf*	مجرفة . جاروف
show window *wājihat al-'arḍ*	واجهة العرض
shrinkage *taqalluṣ*	تقلص
shrinkage joint *waṣlat taqalluṣ*	وصلة تقلص
shrinkage limit *ḥadd at-taqalluṣ*	حد التقلص
shut-off valve *ṣimām qaṭ' au īqāf*	صمام قطع أو ايقاف
shuttering *haikal ṣabb al-ḵarasāna*	هيكل صب الخرسانة
side board *lauḥ jānibī*	لوح جانبي
side-channel spillway *barbaḵ bi-majrā jānibī*	بربخ بمجرى جانبي
side entrance *madḵal jānibī*	مدخل جانبي
side-entrance manhole *ḥufrat taftīš bi-madḵal jānibī*	حفرة تفتيش بمدخل جانبي
side heading *'unwān far'ī*	عنوان فرعي
side-jacking test *iḵtibār al-ḥiml bid-daf'*	إختبار الحمل بالدفع
sidelong ground *arḍ munḥadira*	أرض منحدرة
side pond *burkat taḵzīn jānibīya*	بركة تخزين جانبية
sidesway *tamayyul jānibī*	تمايل جانبي
sidetracking *ḥafr mā'il*	حفر مائل
side tree *ā'midat taṯbīt*	أعمدة تثبيت

sidewalk *mamšā jānibī*	ممشى جانبي
siding *alwāḥ al-judrān al-ḵašabīya*	ألواح الجدران الخشبية
siding machine *makinat tahḏīb ḥaffat aṭ- ṭarīq*	مكنة تهذيب حافة الطريق
sight *manẓar*	منظر
sight distance *majāl ar-ru'ya*	مجال الرؤية
sight rail *qaḍīb al-murāqaba*	قضيب المراقبة
signalman *'āmil išāra*	عامل إشارة
sign bit *miṯqāb al-'alāmāt*	مثقاب العلامات
sign board *lāfita*	لافتة
silent pile-driver *midaqq ḵawāzīq hīdrūlī*	مدق خوازيق هيدرولي
silica *silika*	سليكا
silica brick *ṭaub silika*	طوب السليكا
silicaceous limestone *ḥajar jīrī silikūnī*	حجر جيري سليكوني
silicon carbide *karbīd as-silikūn*	كربيد السليكون
silicone *silikūn*	سليكون
sill *'ataba*	عتبة
silo *ṣauma'at ḥubūb*	صومعة حبوب
silt *ṭammī*	طمي
siltation *ṭammī mutarassib*	طمي مترسب
silt box *sundūq tajammu' at-ṭammī*	صندوق تجمع الطمي
silt displacement *izāḥat aṭ-ṭammī*	إزاحة الطمي
silting *tarassub aṭ-ṭammī*	ترسب الطمي
simple beam *'āriḍa basīṭa*	عارضة بسيطة
simple bending *inḥinā' basīṭ*	إنحناء بسيط

S

simple curve منحنى بسيط
munḥana basīṭ
simple framed building مبنى بهيكل كامل
mabna bi-haikal kāmil
simple framework هيكل كامل
haikal kāmil
Simpson's rule قاعدة سمبسون
qā'idat simbsūn
single-acting وحيد الفعل
waḥīd al-fi'l
single-acting hammer مطرقة وحيدة الفعل
miṭraqa waḥīdat al-fi'l
single base قاعدة مفردة
qā'ida mufrada
single bogie عربة نقل واحدة
'arabat naql wāḥida
single-butt construction joint وصلة إنشاء تناكبية
waṣlat inšā' tanāqubīya
single chip microprocessor ميكروبروسسور وحيد الشظية
mīkrūbrūsisūr waḥīd aš-šaẓīya
single-cut file مبرد مفرد القطعية
mibrad mufrad al-qaṭ'īya
single duct مجرى وحيد
majrā waḥīd
single-duct all-air system نظام هواء أحادي المجرى
niẓām hawā' uḥādī al-majrā
single-pass soil stabilizer آلة ترسيخ التربة أحادية المرور
ālat tarsīk at-turba uḥādīyat al-murūr
single screw compressor ضاغط أحادي البرغي
ḍāġiṭ uḥādī al-burġī
single shaft pier ركيزة أحادية العمود
rakīza uḥādīyat al-'amūd
single-sized aggregate ركام موحد الحجم
rukām muwaḥḥad al-ḥajm
single sling حمالة مفردة
ḥammāla mufrada
single source-operand عملية أحادية المصدر
'amalīya uḥādīyat al-maṣdar
single speed motor محرك وحيد السرعة
muḥarrik waḥīd as-sur'a
single-stage compressor ضاغط وحيد المرحلة
ḍāġiṭ waḥīd al-marḥala
single-stage pump مضخة وحيدة المرحلة
miḍakka waḥīdat al-marḥala
single-track railway سكة حديدية مفردة
sikka ḥadīdīya mufrada
single-wall cofferdam سد مفرد الجدار
sadd mufrad al-jidār
sink مرفق بالوعي
marfaq bālū'ī
sinker drill حفارة صخور ضخمة
ḥaffārat ṣukūr ḍakma
sinking pump مضخة حفر آبار المناجم
miḍakkat ḥafr ābār al-manājim
sintered clay صلصال إسمنتي
ṣalṣāl ismantī
sintering تكتّل
takattul
siphon سيفون
sīfūn
siphon spillway قناة تصريف سيفونية
qanāt taṣrīf sīfūnīya
site موقع
mauqi'
site appraisal تخمين الموقع
takmīn al-mauqi'
site engineer مهندس الموقع
muhandis al-mauqi'
site evaluation تقييم الموقع
taqyīm al-mauqi'
site exploration إستكشاف الموقع
istikšāf al-mauqi'
site investigation إستقصاء الموقع
istiqṣā' al-mauqi'
site labour عمال الموقع
'ummāl al-mauqi'
site preparation إعداد الموقع
i'dād al-mauqi'
site selection إختيار الموقع
iktiyār al-mauqi'
site staff موظفي الموقع
muwaẓẓafī al-mauqi'
site survey مسح الموقع
masḥ al-mauqi'
six ستة
sitta
sixteen ستة عشر
sittata'ašar
sixty ستون
sittūn
skeleton مخطط هيكلي
mukaṭṭaṭ haikalī

S

sketch design تصميم تخطيطي
taṣmīm takṭīṭī
skew bridge جسر مائل
jisr mā'il
skidding resistance مقاومة الانزلاق
muqāwamat al-inzilāq
skid resistant surface سطح مقاوم للانزلاق
saṭḥ muqāwim lil-inzilāq
skids خشبان إستواء
kušbān istiwā'
skilled labour عمال مهرة
'ummāl mahara
skilled man عامل ماهر
'āmil māhir
skimmer مكشطة
mikšaṭa
skimmer equipment معدات التسوية
mu'iddāt at-taswiya
skimming تمهيد
tamhīd
skimming tank خزان إستقطار
kazzān istiqṭār
skin friction الاحتكاك السطحي
al-iḥtikāk as-saṭḥī
skip قادوس
qādūs
skirting board إزار الحائط
izār al-ḥā'iṭ
slab بلاطة
bilāṭa
slab band construction إنشاء خط بلاطات
inšā' kaṭṭ bilāṭāt
slab bridge جسر بلاطي
jisr bilāṭī
slab track سكة حديد بعوارض خرسانية
sikkat ḥadīd bi-'awāriḍ karasānīya
slab track installations تركيبات السكة بعوارض خرسانية
tarkībāt as-sikka bi-'awāriḍ karasānīya
slack-line cableway مصعد كبلي بحبل مرتخي
miṣ'ad kablī bi-ḥabl murtakī
slag خبث
kabṯ
slag cements إسمنت الخبث المعدني
ismant al-kabṯ al-ma'dani
slate أردواز
ardwāz
slater أردوازي
ardwāzī
slate roof سقف أردوازي
saqf ardwāzī
sledge hammer مطرقة ثقيلة
miṭraqa ṯaqīla
sleeper راقدة
rāqida
sleeve كم
kumm
slender beam جائز قضيف
jā'iz qaḍīf
slenderness ratio نسبة القضافة
nisbat al-qaḍāfa
slewing إلتفاف
iltifāf
slewing cylinder أسطوانة الالتفاف
isṭiwānat al-iltifāf
sliced blockwork ألواح قاعدة منحدرة
alwāḥ qā'ida munḥadira
slickensides سطوح صخرية ملساء
suṭūḥ sakriya malsā'
slide إنزلاق
inzilāq
slide rail قضيب إنزلاق
qaḍīb inzilāq
slide rule مسطرة قياس منزلقة
misṭarat qiyās munzaliqa
sliding door باب منزلق
bāb munzaliq
sliding forms هياكل منزلقة
hayākil munzaliqa
sliding formwork هيكل مؤقت منزلق
haikal mu'aqqat munzaliq
sliding gate بوابة منزلقة
bawwāba munzaliqa
sliding joint وصلة منزلقة
waṣla munzaliqa
sliding-panel weir سد بألواح إنزلاقية
sadd bi-alwāḥ inzilāqīya
sliding-wedge method طريقة الاسفين المنزلق
ṭarīqat al-isfīn al-munzaliq
sling حبل الرفع
ḥabl ar-raf'
slip إنزلاق
inzilāq
slip dock مرفأ إنزلاقي
mirfā' inzilāqī
slip factor مُعامل الانزلاق
mu'āmil al-inzilāq

S

slip-form تشكل إنزلاقي
tašakkul inzilāqī
slip-form paver آلة رصف إنزلاقية
ālat raṣf inzilāqīya
slip joint وصلة منزلقة
waṣla munzaliqa
slip plane مستوى الانزلاق
mustawa al-inzilaq
slip road طريق فرعي
ṭarīq far'ī
slip surface سطح منزلق
saṭḥ munzaliq
slipway رصيف إنزال
raṣīf inzāl
slope إنحدار
inḥidār
slope correction تصحيح الانحدار
taṣḥīḥ al-inḥidār
slope gauge مقياس الانحدار
miqyās al-inḥidār
slope protection وقاية الانحدار
wiqāyat al-inḥidār
slope staking دعم الانحدار
da'm al-inḥidār
sloping zone منطقة الانحدار
minṭaqat al-inḥidār
slot excavation حفر خُدي
ḥafr ḵuddī
slough سلخ . إنسلخ
salaḵa, insalaḵa
slow sand filter مرشح رملي بطئ
muraššiḥ ramlī baṭī'
sludge حمأة
ḥama'a
sludge age عمر الحمأة
'umr al-ḥama'a
sludge concentration تركيز الحمأة
tarkīz al-ḥama'a
sludge digestion هضم الحمأة
haḍm al-ḥama'a
sludge-digestion plant وحدة هضم الحمأة
waḥdat haḍm al-ḥama'a
sludge-digestion tank خزان هضم الحمأة
ḵazzān haḍm al-ḥama'a
sludge disposal التخلص من الحمأة
at-takallus min al-ḥama'a
sludge drying bed طبقة تجفيف الحمأة
ṭabaqat tajfīf al-ḥama'a
sludge gas غاز الحمأة
ğāz al-ḥama'a

sludge pressing plant وحدة ضغط الحمأة
waḥdat ḍağṭ al-ḥama'a
sludger مضخة الحمأة
miḍaḵḵat al-ḥama'a
sludge recovery إستعادة الحمأة
isti'ādat al-ḥama'a
sludge-thickening equipment معدات تغليظ الحمأة
mu'iddāt tağlīẓ al-ḥama'a
sludge treatment معالجة الحمأة
mu'ālajat al-ḥama'a
slug كتلة معدنية
kutla ma'danīya
sluice بوابة قناة
bawwābat qanāt
sluice dam سد ببوابة هويس
sadd bi-bawwābat hawīs
sluice gates بوابات الهويس
bawwābāt al-hawīs
sluice valve صمام تصريف
ṣimām taṣrīf
slump test إختبار الكزازة
iḵtibar al-kizzāza
slurry ملاط رقيق القوام
milāṭ raqīq al-qimām
slurry explosive متفجر الحمأة
mutafajjir al-ḥama'a
slurry trench خندق حمأة
ḵandaq ḥama'a
small scale integration إندماج على نطاق محدود
indimāj 'ala niṭāq maḥdūd
smithing حدادة
ḥidāda
smoke density metering قياس كثافة الدخان
qiyās kaṯāfat ad-duḵān
smoke detector مكشاف الدخان
mikšāf ad-duḵān
smoke emission إبتعاث الدخان
ibti'āṯ ad-duḵān
smoke plate لوح الدخان
lauḥ ad-duḵān
smoke proof صامد للدخان
ṣāmid lid-duḵān
smoothing iron مدحلة تسوية
madḥalat taswiya
smooth round wire سلك مدوّر منبسط
silk mudawwar munbasiṭ
snatch block بكرة مقطوعة
bakara maqṭu'a

S

snorkel منشاق
minšāq
snow load حمل الثلج
ḥiml aṯ-ṯalj
snow plough جرافة الثلج
jarrāfat aṯ-ṯalj
soakaway حفرة التشرب
ḥufrat at-tašarrub
soakaway system نظام لتصريف الماء بالارتشاح
niẓām lit-taṣrīf al-mā' bil-irtišāḥ
socket تجويف
tajwīf
socket joint قارنة كمية
qārina kummīya
socket outlet مأخذ التيار
ma'ḵaḏ at-tayyār
soda-acid extinguisher مطفأة حامض الصودا
miṭfā' ḥāmiḍ aṣ-ṣūdā
sodium-vapour lamp مصباح بخار الصوديوم
miṣbāḥ buḵār aṣ-ṣūdyūm
soft-ground boring machine مكنة حفر التربة الطرية
makinat ḥafr at-turba aṭ-ṭarīya
software برامج كمبيوتر
barāmij kumbyūtar
soft water ماء يسر
mā' yasir
softwood خشب لين
ḵašab layyin
soil تربة
turba
soil analysis تحليل التربة
taḥlīl at-turba
soil evaluation تقييم التربة
taqyīm at-turba
soil mechanics ميكانيكا التربة
mīkānīka at-turba
soil mixer خلاطة التربة
ḵallāṭat at-turba
soil profile تضاريس التربة
taḍārīs at-turba
soil sample عينة التربة
'ayyinat at-turba
soil sampler فاحص عينات التربة
fāḥiṣ 'ayyināt at-turba
soil scientist عالم التربة
'ālim at-turba
soil shredder آلة تفتيت التربة
ālat taftīt at-turba
soil stabilization تقوية التربة
taqwiyat at-turba
soil survey مسح التربة
masḥ at-turba
solar surveying المسح الشمسي
al-masḥ aš-šamsī
soldering لحام
liḥām
soldering iron كاوية لحام
kāwīyat liḥām
soldier دعامة خشبية عمودية
di'āma ḵašabīya 'amūdīya
soldier beam دعامة خشبية أفقية
di'āma ḵašabīya ufuqīya
soldier waling لوح ربط أفقي
lauḥ rabṭ ufuqī
solenoid ملف لولبي
milaff laulabi
solicited interrupts مقاطعات حث
muqāṭa'āt ḥaṯṯ
solid manganese-steel insert وليجة فولاذ منغنيزي صلبة
walījat fūlāḏ manganīzī ṣaliba
solid map خريطة جيولوجية
ḵarīṭa jīyūlūjīya
solid masonry آجر صلب
ājurr ṣalib
solid-type dock construction إنشاء ذو سطح مصمت
inšā' ḏū saṭḥ muṣmit
solid web وترة مصمتة
watara muṣmita
solid wharf رصيف صلب
raṣīf ṣalib
solution injection حقن المحلول
ḥaqn al-maḥlūl
sonde مسبار رصد
misbār raṣd
sonic bleeder valve صمام تفريغ هزاز
ṣimām tafrīġ hazzāz
sonic pile-driver مدق خوازيق هزاز
midaqq ḵawāzīq hazzāz
sound-absorbing material مادة ماصة للصوت
mādda māṣṣa liṣ-ṣaut
sounding مصوت . سبر
muṣawwit, sibr

S

sounding lead فادن خيط السبر
fādin kait as-sibr
sounding line حبل المسبار
ḥabl al-misbār
sound insulation عزل الصوت
'azl aṣ-ṣaut
sound sensor جهاز الاحساس الصوتي
jihāz al-iḥsās aṣ-ṣautī
source catchment مستجمع المصب
mustajma' al-miṣabb
source program برنامج المنشأ
barnāmij al-manšā'
source text editing تحرير النص الأصلي
taḥrīr an-naṣṣ al-aṣlī
space frame هيكل فراغي
haikal farāğī
space lattice شبكة فراغية
šabaka farāğīya
spacer فاصل
fāṣil
spacing تباعد . مباعدة
tabā'ud, mubā'ada
spacing limits حدود المباعدة
ḥudūd al-mubā'ada
span باع
bā'
spandrel wall جدار القنطرة
jidār al-qanṭara
spare إحتياطي
iḥtiyāṭī
spare parts قطع غيار
qiṭa' ğiyyār
spare wheel دولاب إحتياطي
dūlāb iḥtiyāṭī
spark arresting screen حاجز كبح الشرر
ḥājiz kabh aš-šarrar
special aggregates ركام خاص
rukām kāṣṣ
special conditions ظروف خاصة . شروط خاصة
ẓurūf kāṣṣa, šurūṭ kāṣṣa
specialist إختصاصي
iktiṣāṣī
specialized design services خدمات تصميم إختصاصية
kadamāt tatwīr iktiṣāṣīya
specialized development services خدمات تطوير إختصاصية
kadamāt taṣmīm iktiṣāṣīya
special provision شرط خاص
šarṭ kāṣṣ
special purpose computer كمبيوتر للأغراض الخاصة
kumbyūtar lil-ağrāḍ al-kāṣṣa
special safety equipment معدات أمان خاصة
mu'iddāt amān kāṣṣa
special structural concrete خرسانة إنشائية خاصة
karasāna inšā'īya kāṣṣa
specialty contract عقد إختصاصي
'aqd iktiṣāṣī
specific adhesion إلتصاق نوعي
iltiṣāq nau'ī
specification مواصفات
muwāṣafāt
specification engineer مهندس وضع المواصفات
muhandis waḍ' al-muwāṣafāt
specific gravity الوزن النوعي
al-wazn an-nau'ī
specific retention الاحتجاز النوعي
al-iḥtijāz an-nau'ī
specific speed السرعة النوعية
as-sur'a an-nau'īya
specific surface المساحة السطحية النوعية
al-masāha as-saṭḥīya an-nau'īya
specific yield الانتاج النوعي
al-intāj an-nau'ī
speed factor عامل السرعة
'āmil as-sur'a
spent water ماء مستهلك
mā' mustahlak
spherical plug valve صمام سدادي كروي
ṣimām sidādī kurawī
spigot-and-socket joint وصلة ذكر وأنثى
waṣla ḏakar wa-unṯa
spike مسمار كبير
mismār kabīr
spiling دق الخوازيق
daqq al-kawāzīq
spillway قناة تصريف
qanāt taṣrīf
spillway gate بوابة قناة التصريف
bawwābat qanāt at-taṣrīf
spiral حلزوني . لولبي
ḥalazunī, laulabī
spiral curve منحنى حلزوني
munhana ḥalazūnī
spiral stairway سلم حلزوني
sullam ḥalazūnī

S

splice	جدل
jadl	
splice bar	لوح وصل تراكبي
lauḥ waṣl tarākubī	
split-finger layout	تخطيط متفرع
taḵṭīṭ mutafarri'	
spool	مكب
mikabb	
sports stadium	ملعب رياضي
mal'ab riyāḍī	
spot level	منسوب النقطة
mansūb an-nuqṭa	
spotting	إستطلاع
istiṭlā'	
spot welding	لحام نقطي
liḥām nuqṭī	
spout-delivery pump	مضخة تصريف متدفق
miḍakkat taṣrīf mutadaffiq	
sprag clutch	قابض قدة موقفة
qābiḍ qudda muwaqqafa	
spray bar	قضيب الرش
qaḍīb ar-rašš	
sprayed concrete	خرسانة مرشوشة
ḵarasāna maršūša	
spray lance	أنبوب المرشة اليدوية
unbūb al-miraššaḥ al-yadawīya	
spread	نشر
našr	
spreader	فارشة . ناشرة
fāriša, nāšira	
spreader beam	عارضة ناشرة
'ariḍa nāšira	
spread footing	قاعدة عريضة
qā'ida 'arīḍa	
spreading box	صندوق الفرش
sundūq al-farš	
spread recorder	مقياس الانتشار
miqyās al-intišār	
spring	نابض . نبع
nābiḍ, nab'	
spring catchment	مستجمع النبع
mustajma' an-nab'	
spring closers	طوب سد النبع
ṭaub sadd an-nab'	
springing	بدء التخصر
bad' at-taḵaṣṣur	
spring points	محولات سكة حديد نابضية
muḥawwilāt sikkat ḥadīd nābiḍīya	

spring spike	مسمار نابضي
mismār nābiḍī	
spring tides	مد أو جزر تام
madd au jazr tām	
spring washer	فلكة نابضة
falaka nābiḍa	
sprinkler	مرشة إطفاء
miraššat iṭfā'	
sprinkler irrigation	ري بشبكة مرشات
rayy bi-šabakat miraššāt	
spruce	خشب أبيض
ḵašab abyaḍ	
spud	آلة حفر مستدقة
ālat ḥafr mustadiqqa	
spudding	مستدق الطرف
mustadiqq aṭ-ṭaraf	
spun concrete	خرسانة مغزولة
ḵarasāna maḡzūla	
spun iron	حديد مغزول
ḥadīd maḡzūl	
spur	نتوء
nutū'	
square metre	متر مربع
mitr murabba'	
square thread	سن لولبة مربعة
sinn laulaba murabba'a	
square yard	ياردة مربعة
yārda murabba'a	
squirrel-cage motor	محرك قفص السنجاب
muḥarrik qafaṣ as-sinjāb	
stability	إستقرار
istiqrār	
stabilization basins	أحواض التركيز
aḥwāḍ at-tarkīz	
stabilization of soil	تقوية التربة
taqwiyat at-turba	
stabilized earth	تربة مستقرة
turba mustaqirra	
stabilizer	موازن
muwāzin	
stack	ترصيف . مدخنة
tarṣīf, midḵana	
stack monitor	مراقبة الترصيف
murāqabat at-tarṣīf	
stadia	شاخص تسوية
šāḵiṣ taswiya	
stadia hairs	شعرتا الشبيكة
ša'ratā aš-šabīka	
stadia work	مسح الأبعاد
masḥ al-ab'ād	

S

stadium ملعب مدرج (ستاد)
mal'ab mudarraj, stād
staff شاخص . موظفين
šākiṣ, muwazzafīn
staff development تنمية هيئة الادارة
tanmiyat hayyat al-idāra
staff gauge مقياس منسوب الماء
miqyās mansūb al-mā'
staff man عامل الشاخص
'āmil aš-šākiṣ
staff training تدريب الموظفين
tadrīb al-muwazzafīn
stage مرحلة . سقالة
marḥala, saqāla
staging سقالات البناء
saqālāt al-binā'
stain صبغ
ṣabġ
stainless steel فولاذ لا يصدأ
fūlāḏ lā-yasda'
stair درجة
daraja
staircase سلم . درج
sullam, daraj
stair handrail درابزين الدرج
darābzīn ad-daraj
stair string حبل الدرج
ḥabl ad-daraj
stairway ممر الدرج
mamarr ad-daraj
stair width عرض الدرج
'arḍ ad-daraj
stalk جذع . ساق
juḏ', sāq
stanchion قائم دعم فولاذي
qā'im da'm fūlāḏī
standard عيار
'iyār
Standard Busy Rate معدل الانشغال المعياري
mu'addal al-inšiġāl al-mi'yārī
standard deviation الانحراف المعياري
al-inḥirāf al-mi'yārī
standard diving gear لباس الغوص المعياري
libās al-ġaus al-mi'yārī
standard error خطأ عياري
ḵaṭa' 'iyārī
standard gauge البعد العياري
al-bu'd al-'iyārī
standard gauge plain track سكة حديد ذات بعد عياري
sikkat ḥadīd ḏāt bu'd 'iyārī
standardization توحيد المقاييس
tauḥīd al-maqāyīs
standardization correction تصحيح المعايرة
taṣḥīḥ al-mu'āyara
standardization temperature درجة حرارة المعايرة
darajat ḥarārat al-mu'āyara
standard lift مصعد عياري
miṣ'ad 'iyārī
standard penetration resistance مقاومة الاختراق المعيارية
muqāwamat al-iḵtirāq al-mi'yārīa
standards المواصفات القياسية
al-muwāṣafāt al-qiyāsīya
standard section قطاع عياري
qiṭā' 'iyārī
standards in force المواصفات القياسية المطبقة
al-mūwāṣafāt al-qiyāsīya al-muṭabbaqa
standard specification المواصفات القياسية
al-muwāṣafāt al-qiyāsīya
standard-water level منسوب الماء القياسي
mansūb al-mā' al-qiyāsī
standing pier ركيزة قائمة
rakīza qā'ima
stand pipe ماسورة قائمة
māsūra qā'ima
stank جدار أو سد مسيك
jidār, sadd musayyak
starling ركائز حماية الجسر
rakā'iz ḥimāyat al-jisr
starter بادئ
bādi'
starter bar قضيب حبك الخرسانة
qaḍīb ḥabk al-ḵarasāna
statement تصريح
taṣrīḥ
static axle load حمل المحور الساكن
ḥiml al-miḥwar as-sākin
static equilibrium توازن ساكن
tawāzun sākin
static head علو ساكن
'ulū sākin
static load حمل ساكن
ḥiml sākin

S

static moment *ʿazm sākin*	عزم ساكن
static penetration *iḵtirāq sākin*	إختراق ساكن
statics *ʿilm as-sukūn, istātīka*	علم السكون (استاتيكا)
static suction lift *rafʿ al-maṣṣ as-sākin*	رفع المص الساكن
station *muḥaṭṭa*	محطة
stationary crane *mirfāʿ ṯābit*	مرفاع ثابت
stationary dredger *karrāʾa ṯābita*	كراءة ثابتة
station construction *inšāʾ al-muḥaṭṭa*	إنشاء المحطة
station control room *ġurfat al-murāqaba bil-muḥaṭṭa*	غرفة المراقبة بالمحطة
station platform *raṣīf al-muḥaṭṭa*	رصيف المحطة
station siting *taḥdīd mauqiʿ al-muḥaṭṭa*	تحديد موقع المحطة
statistical uniformity *at-tauḥīd al-iḥṣāʾī*	التوحيد الاحصائي
statistics *ʿilm al-iḥṣāʾ, iḥṣāʾīyāt*	علم الإحصاء . إحصائيات
statute mile *mīl inglīzī*	ميل انكليزي
staunching piece *jidār sīk*	جدار مسيك
staunching rod *qaḍīb maṭṭāṭ sīk*	قضيب مطاط مسيك
stay *diʿāma*	دعامة
stay pile *rakīzat daʿm*	ركيزة دعم
steady flow *tadaffuq muṭṭarid*	تدفق مطرد
steam boiler *ġallāyat bukār*	غلاية بخار
steam curing *inḍāj bil-bukār*	إنضاج بالبخار
steam engine *muḥarrik bukārī*	محرك بخاري
steam generating heavy water reactor *mufāʿil māʾ ṯaqīl muwallid lil-bukār*	مفاعل ماء ثقيل مولد للبخار
steam-powered generator *muwallid yaʿmal bil-bukār*	مولد يعمل بالبخار
steam purging *tanẓīf bil-bukār*	تنظيف بالبخار
steam roller *midḥala bukārīya*	مدحلة بخارية
steam shovel *mijrafa bukārīya*	مجرفة بخارية
steam turbine *tūrbīn bukāri*	تربين بخاري
steel *fūlāḏ*	فولاذ
steel bender *ḥanīyat fūlāḏ*	حانية فولاذ
steel bridge *jisr fūlāḏī*	جسر فولاذي
steel cable *kabl fūlāḏī*	كبل فولاذي
steel connections *tauṣīlāt fūlāḏīya*	توصيلات فولاذية
steel construction *inšāʾ fūlāḏī*	إنشاء فولاذي
steel culvert *majrā fūlāḏī*	مجرى فولاذي
steel-deck surfacing *saṭḥ fūlāḏī*	سطح فولاذي
steel erector *ʿāmil tarkīb al-fūlāḏ*	عامل تركيب الفولاذ
steel filings *birādat al-fūlāḏ*	برادة الفولاذ
steelfixer *muṯabbit al-fūlāḏ*	مثبت الفولاذ
steel formwork *qālab fūlāḏī*	قالب فولاذي
steel frame *iṭār fūlāḏī*	إطار فولاذي
steel framework *haikal fūlāḏī*	هيكل فولاذي
steel grid *šabaka fūlāḏīya*	شبكة فولاذية
steel H-pile *rakīza fūlāḏīya bi-šakl atš*	ركيزة فولاذية بشكل H
steel ladder *sullam fūlāḏī*	سلم فولاذي
steel liner *biṭāna fūlaḏīya*	بطانة فولاذية
steel pile *rakīza fūlāḏīya*	ركيزة فولاذية
steel pipe *unbūb fūlāḏī*	أنبوب فولاذي

S

steel reinforcement bar قضيب تسليح فولاذي
qaḍīb taslīḥ fūlāḏī
steel ring حلقة فولاذية
ḥalaqa fūlāḏīya
steel roof deck سطح فولاذي
saṭḥ fūlāḏī
steel sheet لوح فولاذي
lauḥ fūlāḏī
steel sheet piling ركيزة فولاذية مستعرضة
rakīza fūlāḏīya musta'raḍa
steel sleeper راقدة فولاذية
rāqida fūlāḏīya
steel stanchion قائم دعم فولاذي
qā'im da'm fūlāḏī
steel tower برج فولاذي
burj fūlāḏī
steel wire سلك فولاذي
silk fūlāḏī
steel-wire rope كبل من أسلاك الفولاذ
kabl min aslāk al-fūlāḏ
steelworks slag خبث مصانع الفولاذ
ḳabṯ maṣāni' al-fūlāḏ
steening تبطين البئر
tabṭīn al-bi'r
steeple بُرج
burj
stem جذع
juḏ'
step درجة
daraja
step iron متكأ حديدي
muttaka' ḥadīdī
step ladder سيبة
sība
stepped foundation قاعدة نضدية
qā'ida naḍadīya
stereometric map خريطة مجسمة
ḳarīṭa mujassama
stereoplotter آلة تسجيل بياني
ālat tasjīl bayānī
stereoscope مجسام
mijsām
stevedores' warehouse مستودع متعهد شحن السفن
mustauda' muta'ahhid šaḥn as-sufun
stick welding لحام معدني
liḥām ma'dani
stiff clay صلصال متماسك
ṣalṣāl mutamāsik
stiffened steel plate لوح فولاذي صلب
lauḥ fūlāḏī ṣalib
stiffened suspension bridge جسر معلق صلب
jisr mu'allaq ṣalib
stiffened wall جدار مدعم صلب
jidār mud'am ṣalib
stiffener زاوية تقوية
zāwiyat taqwiya
stiffening beams عوارض تقوية
'awariḍ taqwiya
stiffening girder عارضة تقوية
'āriḍat taqwiya
stiff frame هيكل صلب
haikal ṣalib
stiff-leg derrick مرفاع بقائم صلب
mirfā' bi-qā'im ṣalib
stiffness جسوءة.صلابة
jasu'a ṣalāba
stilling basin حوض تهدئة
ḥauḍ tahdi'a
stilling pool بركة تهدئة
burkat tahdi'a
stilling well بئر تهدئة
bi'r tahdi'a
stimulation تنشيط
tanšīṭ
stinger طوف
ṭauf
stipulation إشتراط
ištirāṭ
stirrup رباط
ribāṭ
stock ladder سلم مدعوم
sullam mad'ūm
STOL ports مرافئ « ستول »
marāfi' stūl
stone حجر
ḥajar
stone-block paving رصف بالبلاط الحجري
raṣf bil-bilāṭ al-ḥajarī
stone drain مصرف حجري
maṣrif ḥajarī
stoneware مصنوعات حجرية
maṣnū'āt ḥajarīya
stopcock محبس
miḥbas
stop logs عوارض السد
'awāriḍ as-sadd

S

stop planks ألواح السد
alwāḥ as-sadd

stop valve صمام قطع
ṣimām qaṭʿ

storage تخزين
takzīn

storage bin صومعة تخزين
ṣaumaʿat takzīn

storage capacity سعة التخزين
siʿat at-takzīn

storage of cargo تخزين البضائع
takzin al-biḍāʾiʿ

storage pond بركة تخزين
burkat takzīn

storage reservoir خزان
kazzān

store مخزن
makzan

store building مبنى المستودع
mabnā al-mustaudaʿ

store compound مجمع مستودعات
mujammaʿ mustaudaʿāt

storekeeper أمين المستودع
amīn al-mustaudaʿ

store yard ساحة المستودع
sāḥat al-mustaudaʿ

storm frequency تواتر العواصف
tawātur al-ʿawāṣif

storm-overflow sewer مجرى تصريف الأمطار
majrā taṣrīf al-amṭār

storm pavement حاجز الأمواج
ḥājiz al-amwāj

storm sewer مجرور ماء المطر
majrūr māʾ al-maṭar

stormwater مياه الأمطار
miyāh al-amṭār

stormwater drain بالوعة مياه الأمطار
bālūʿat miyāh al-amṭār

stormwater drainage صرف مياه الأمطار
ṣarf miyāh al-amṭar

stormwater overflow طفح مياه الأمطار
ṭafḥ miyāh al-amṭar

stormwater tanks خزانات مياه الأمطار
kazzānāt miyāh al-amṭar

straight مستقيم
mustaqīm

straight beams عوارض مستقيمة
ʿawāriḍ mustaqīma

straight-run bitumen قار متخلف
qār mutakallif

straights إستقامات
istiqāmāt

straight stretch امتداد مستقيم
imtidād mustaqīma

strain جهد . توتر
juhd, tawattur

strain ageing تصلد زمني
taṣallud zamanī

strain energy طاقة التوتر
ṭāqat at-tawattur

strainer مصفاة
miṣfāt

strain gauge مقياس الجهد
miqyās al-juhd

strain hardening تصليد إجهادي
taṣlīd ijhādī

strake لوح طولي
lauḥ ṭūlī

strand جديلة . ضفيرة
jadīla, ḍafīra

stranded aluminium ألومنيوم مجدول
aluminyūm majdūl

strap شريط
šarīṭ

stratified flow دفق طباقي
dafq ṭabāqī

stratified soil تربة مترسبة
turba mutarassiba

stratigraphy علم طبقات الأرض
ʿilm ṭabaqāt al-arḍ

stream جدول
jadwal

streamline flow جريان إنسيابي
jarayān insiyābī

street شارع
šāriʿ

street cleansing تنظيف الشوارع
tanẓīf aš-šawāriʿ

street lamp مصباح الشارع
miṣbāḥ aš-šāriʿ

street refuge جزيرة المشاة
jazīrat al-mušāt

street sweeping كنس الشوارع
qans aš-šawāriʿ

strength قوة
quwwa

strength of materials متانة المواد
matānat al-mawādd

stress إجهاد
ijhād

stress analysis تحليل قوى الاجهاد
taḥlīl qiwa al-ijhād
stress and strain الاجهاد والتوتر
al-ijhād wat-tawattur
stress components مكونات الاجهاد
mukawwināt al-ijhād
stress concentration تركيز الاجهاد
tarkīz al-ijhād
stress notation علامة الاجهاد
'alāmat al-ijhād
stress-strain curve منحنى الاجهاد والتوتر
munḥana al-ijhad wat-tawattur
striding level ميزان تسوية راكب
mīzān taswīyat rākib
strike طرق
ṭarq
striker عامل تطريق
'āmil taṭrīq
striking إزالة هيكل الدعم المؤقت
izālat haikal ad-da'm al-mu'aqqat
stringer beam عارضة طولانية
'āriḍa ṭūlānīya
string lining تبطين خيطي
tabṭīn ḵaiṭī
strip قطعة ضيقة
qiṭ'a ḍayyiqa
strip footing قاعدة ضيقة
qā'ida ḍayyiqa
strip foundation أساس ضيق
asās ḍayyiq
stripping نزع الهيكل المؤقت
naz' al-haikal al-mu'aqqat
struck capacity سعة القادوس الصدمية
si'at al-qādūs aṣ-ṣadmīya
structural analysis التحليل الانشائي
at-taḥlīl al-inšā'ī
structural carbon steel فولاذ كربوني للانشاءات
fūlāḏ karbūnī lil-inšā'āt
structural concrete خرسانة إنشاءات
ḵarasānat inšā'āt
structural design تصميم إنشائي
taṣmīm inšā'ī
structural designer مصمم إنشاءات
muṣammim inšā'āt
structural designer-draughtsman مهندس – رسام إنشاءات
muhandis rassām inšā'āt

structural draughtsman رسام إنشاءات
rassām inšā'āt
structural engineer مهندس إنشاءات
muhandis inšā'āt
structural engineering technician فني هندسة إنشاءات
fannī handasat inšā'āt
structural fire resistance مقاومة الانشاءات للحريق
muqāwamat al-inšā'āt lil-ḥarīq
structural lightweight-aggregate concrete خرسانة ركام إنشاءات خفيفة
ḵarasānat rukām inšā'āt ḵafīfa
structural member عضو إنشائي
'uḍū inšā'ī
structural-steel framing هيكل فولاذ إنشاءات
haikal fūlāḏ inšā'āt
structural steelwork فولاذ إنشاءات
fūlāḏ inšā'āt
structural timber أخشاب الانشاءات
aḵšāb al-inšā'āt
structure إنشاء
inšā'
structure excavation أعمال حفر الانشاءات
a'māl ḥafr al-inšā'āt
strut دعامة . عمود
di'āma, 'amūd
stud قائمة خشبية
qā'ima ḵašabīya
studio استوديو
stūdyū
stuffing box صندوق حشو
ṣandūq hašī
stunt end هيكل وصل الانشاءات
haikal waṣl al-inšā'āt
sub-agent وكيل فرعي
wakīl far'ī
sub-base قاعدة تحتية
qā'ida taḥtīya
subcontractor مقاول فرعي
muqāwil far'ī
subcontractor costs تكاليف المقاول الفرعي
takālīf al-muqāwil al-far'ī
subcontractor rates أسعار المقاول الفرعي
as'ār al-muqāwil al-far'ī
subcontract procedure إجراءات التعاقد من الباطن
ijrā'āt at-ta'āqud min al-bāṭin

S

subcritical flow تدفق شبه حرج
tadaffuq šibh ḥarij
sub-grade الأرض الطبيعية
al-arḍ aṭ-ṭabīʿīya
sub-irrigation ري تحتي
rayy taḥtī
sub-main مواسير مطمورة
mawāsīr maṭmūra
submain sewer مجرور تحت المواسير المطمورة
majrūr taḥt al-mawāsīr al-maṭmūra
submerged-arc welding اللحام القوسي المغمور
al-liḥām al-qausī al-maǧmūr
submerged float عامة مغمورة
ʿāma maǧmūra
submerged sleeve valve صمام كمي مغمور
ṣimām kummī maǧmūr
submerged tunnel نفق مغمور
nafaq maǧmūr
submerged weir سد غاطس
sadd ǧaṭis
submersible قابل للغمر
qābil lil-ǧamr
submersible pump مضخة مغمورة
miḍakka maǧmūra
sub-program برنامج فرعي
barnāmij farʿī
sub-program calling تحديد البرنامج الفرعي
taḥdīd al-barnāmij al-farʿī
subsidence إنخساف . إستقرار
inkisāf, istiqrār
subsoil تربة سفلية
turba suflīya
subsoil drain مصرف التربة السفلية
maṣrif at-turba as-suflīya
subsoil stabilization تقوية التربة السفلية
taqwiyat at-turba as-suflīya
subsoil water ماء التربة السفلية
māʾ at-turba as-suflīya
subsonic jet aircraft طائرة دون سرعة الصوت
ṭāʾira dūn surʿat aṣ-ṣaut
substations محطات فرعية
muḥaṭṭāt farʿīya
substructure إنشاء قاعدي
inšāʾ qāʿidī
subsurface drainage صرف الطبقة تحت السطحية
ṣarf aṭ-ṭabaqa taḥt as-saṭḥīya
subsurface erosion تأكل تحت السطح
taʾākul taḥt as-saṭḥ

subsurface float طوف تحت سطح الماء
ṭauf taḥt saṭḥ al-māʾ
subsurface irrigation ري الطبقة تحت السطحية
rayy aṭ-ṭabaqa taḥt as-saṭḥīya
subtense bar قضيب المستقيم المقابل
qaḍīb al-mustaqīm al-muqābil
subway ممشى تحت الأرض
mamšā taḥt al-arḍ
subway ventilation تهوية النفق
tahwiyat an-nafaq
suction إمتصاص
maṣṣ, imtiṣāṣ
suction-cutter dredger كراءة قطع ماصة
karrāʾat qaṭʿ māṣṣa
suction dredger كراءة ماصة
karrāʾa māṣṣa
suction head علو المص
ʿulū al-maṣṣ
suction pad مسند سحب
misnad saḥb
suction pipe ماسورة المص
māsūrat al-maṣṣ
suction tank خزان المص
kazzān al-maṣṣ
suction valve صمام مص
ṣimām maṣṣ
sudden drawdown إنخفاض مفاجئ
inkifāḍ mufājiʾ
sudden outburst إنفجار مفاجئ
infijār mufāji
sue load الحمل المحول
al-ḥiml al-muḥawwal
sulphate-bearing soil تربة حاملة للكبريت
turba ḥāmila lil-kibrīt
sulphate-resisting cement إسمنت صامد للكبريت
ismant ṣāmid lil-kibrīt
sump حفرة المجاري
ḥufrat al-majārī
sumpers ثقوب قطعية
ṯuqūb qaṭʿīya
sunken-tube tunnel نفق أنبوبي مغمور
nafaq unbūbī maǧmūr
supercritical flow تدفق فوق الحرج
tadaffuq fauq al-ḥārij
superelevation إرتفاع إضافي
irtifāʿ iḍāfī
superficial سطحي
saṭḥī

S

superficial compaction دموج سطحي
dumūj saṭḥī
superimposed load الحمل المتراكب
al-ḥiml al-mutarākib
superintendent مشرف عام
mušrif ʿāmm
superload الحمل المضاف
al-ḥiml al-muḍāf
superposition تراكب
tarākub
superstructure إنشاء علوي
inšāʾ ʿalawī
supersulphated cement إسمنت فلزي
ismant fillizī
supervision costs تكاليف الاشراف
takālīf al-išrāf
supplement ملحق
mulḥaq
supplementary agreement عقد تكميلي
ʿaqd takmīlī
supply pipe ماسورة الامداد
māsūrat al-imdād
support دعم
daʿm
support flange شفة دعم
šaffat daʿm
supporting piers ركائز دعم
rakīzat daʿm
supporting ring حلقة دعم
ḥalaqat daʿm
support moment عزم الدعم
ʿazm ad-daʿm
suppressed weir سد قياس
sadd qiyās
surcharge حمل إضافي
ḥiml iḍāfī
surcharged wall جدار محتجز
jidār muḥtajaz
surface access عبور سطحي
ʿubūr saṭḥī
surface-active agent عامل ذو فاعلية سطحية
ʿāmil ḏū fāʿilīya saṭḥīya
surface-aeration system نظام تهوية سطحية
niẓām tahwiya saṭḥīya
surface-course طبقة سطحية
ṭabaqa saṭḥīya
surface detention إحتجاز سطحي
iḥtijāz saṭḥī
surface drainage صرف سطحي
ṣarf saṭḥī
surface dressing طبقة رصف سطحية
ṭabaqat raṣf saṭḥīya
surface float طافية سطحية
ṭāfīya saṭḥīya
surface irrigation ري سطحي
rayy saṭḥī
surface irrigation system نظام ري سطحي
niẓām rayy saṭḥī
surface tension توتر سطحي
tawattur saṭḥī
surface-tension depressant خافض التوتر السطحي
ḵāfiḍ at-tawattur as-saṭḥī
surface texture depth عمق بنية السطح
ʿumq bunyat as-saṭḥ
surface vibrator هزازة سطحية
hazzāza saṭḥīya
surface water مياه سطحية
miyāh saṭḥīya
surface-water drain مصرف المياه السطحية
maṣrif al-miyāh as-saṭḥīya
surface water scheme مشروع المياه السطحية
mašrūʿ al-miyāh as-saṭḥīya
surfacing طبقة السطح
ṭabaqat as-saṭḥ
surfactant خافض للتوتر السطحي
ḵāfiḍ lit-tawattur as-saṭḥī
surge تموُّر
tamawwur
surge pipe ماسورة التموُّر
māsūrat at-tamawwur
surge tank صهريج التموُّر
ṣahrīj at-tamawwur
survey مسح
masḥ
surveying المساحة
al-misāḥa
surveying equipment معدات المساحة
muʿiddāt al-misāḥa
surveying instruments أجهزة المساحة
ajhizat al-misāḥa
survey launch زورق المسح
zauraq al-masḥ
surveyor مسّاح
massāḥ
surveyor's draughtsman رسام مساحة
rassām misāḥa
surveyor's office مكتب المسّاح
maktab al-massāḥ
survey station محطة المساحة
muḥaṭṭat al-misāḥa

S

survey team leader رئيس فريق المساحة
ra'īs farīq al-misāḥa
suspended-frame weir سد بهيكل معلق
sadd bi-haikal mu'allaq
suspended scaffold سقالة معلقة
saqala mu'allaqa
suspended span الباع المعلق
al-bā' al-mu'allaq
suspended structure إنشاء معلق
inšā' mu'allaq
suspender علاقة
'allāqa
suspension bridge جسر معلق
jisr mu'allaq
suspension cable كبل التعليق
kabl at-ta'līq
suspension-cable anchor مثبت كبل التعليق
muṯabbit kabl at-ta'līq
suspension of work توقيف العمل
tauqīf al-'amal
sustained yield الانتاج الدائم
al-intāj ad-dā'im
swamp مستنقع
mustanqa'
swash height ارتفاع الموجة على الشاطئ
irtifā' al-mauja 'ala aš-šaṭṭ
sway تراوح جانبي
tarāwuḥ jānibī
sway rod قضيب مانع للتراوح
qaḍīb mani' lit-tarāwuḥ
swelling pressure ضغط التضخم
ḍağṭ at-taḍakkum
swelling soil تربة قابلة للتمدد
turba qābila lit-tamaddud
swimming pool بركة سباحة
burkat sibāḥa
swing bridge جسر دوار
jisr dawwār
swing door باب دوار
bāb dawwār
swinger قضيب مدبب
qaḍīb mudabbab
swing-jib crane مرفاع بذراع دوار
mirfā' bi-ḏirā' dawwār
swing stage scaffold سقالة بمنصة دوارة
saqāla bi-minaṣṣa dawwāra
switch مفتاح كهربائي . محولة
miftāḥ kahrabā'ī, muḥawwila

switch blade نصل المفتاح السكيني
naṣl al-miftāḥ as-sikkīnī
switchboard لوحة المفاتيح
lauḥat al-mafātīḥ
switchgear مجموعة المفاتيح الكهربائية
majmū'at al-mafātīh al-kahrabā'īya
switch points طرفا خط التحويل
ṭarafa ḵaṭṭ at-taḥwīl
switch side plates ألواح طرفي المحولة
alwāḥ ṭarafayy al-muḥawwila
switch stand قاعدة المحولة
qā'idat al-muḥawwila
switch tie شدادة المحولة
šaddādat al-muhawwila
switch toes مرتكزات المحولة
murtakizāt al-muhawwila
symbol رمز
ramz
symbolic assembler مجمع رمزي
mujammi' ramzī
synchronous serial interface توصيل تتابعي تزامني
tauṣīl tatābu'ī tazāmunī
synthetic resins الراتينجات . راتينج إصطناعي
ar-rātīnjāt, rātinj iṣṭinā'ī
system نظام . شبكة
niẓām, šabaka
systematic errors أخطاء رتيبة
akṭā' ratība
system building بناء نظامي
binā' niẓāmī
table جدول
jadwal
tacheometer مقياس أبعاد
miqyās ab'ād
tacheometric مسح الأبعاد
masḥ al-ab'ād
tack coat طبقة قيرية
ṭabaqa qīrīya
tack rivet برشامة ربط
biršāmat rabṭ
tack weld لحام نقطي
liḥām nuqṭī
tail rope حبل قطر
ḥabl quṭr
taintor gates بوابات نصف قطرية
bawwābāt niṣf qutrīya

T

take-off surfaces سطوح الاقلاع
suṭūḥ al-iqlā'
tally رقعة
ruq'a
tamper مدك. مدقة
midakk, midaqqa
tamping دك
dakk
tamping roller مدحلة
midḥala
tandem roller مدحلة ترادفية
midḥala tarādufīya
tangent distance مسافة المماس
masāfat al-mamās
tangential deviation إنحراف المماس
inḥirāf al-mamās
tangential tension توتر المماس
tawattur al-mamās
tangent modulus مُعامل التماس
mu'āmil iltimās
tangent point نقطة التماس
nuqṭat iltimās
tangent screw لولب مماس
laulab mamās
tank صهريج. خزان
ṣahrīj 'ḵazzān
tanking غشاء إسفلتي مسيك
ḡišā' isfaltī musayyaq
tank sprayer مرشة صهريجية
mirašša ṣahrījīya
tap صنبور. حنفية
ṣunbūr, hanafīya
tape شريط
šarīṭ
tape corrections تصحيحات الشريط
taṣḥīḥāt ăš-šarīṭ
tape recorder مسجل شرائط
musajjil šarā'iṭ
tapered-flange beam عتبة بشفة مستدقة
'ataba bi-šaffa mustadiqqa
tapered washer فلكة مخروطية
falaka makrūṭīya
taper file مبرد مستدق
mibrad mustadiqq
tar قطران. قار
quṭrān, qār
tare الوزن الفارغ
al-wazn al-fāriḡ
tar emulsion مستحلب قطراني
mustaḥlab quṭrānī

target هدف
hadaf
target rod شاخص تسوية
šāḵiṣ taswiya
tarmacadam حصباء مقيّرة
ḥaṣbā' muqayyara
tarmacadam plant معدات تقيير الحصباء
mu'iddāt taqyīr al-ḥaṣbā'
tarpaulin تربولين
tarbūlīn
tar paving رصف قاري
raṣf qārī
tarred road طريق مرصوف بالقار
ṭarīq marṣūf bil-qār
taxiway مدرج
madraj
taxiway guidance signs لافتات دليلية على المدرج
lāfitāt dalīlīya 'ala al-madraj
taxiway lighting إضاءة المدرج
iḍā'at al-madraj
taxiway lights أضواء المدرج
aḍwā' al-madraj
T-beam bridge جسر ذو عتبة بشكل T
jisr ḏū 'ataba bi-šakl tī
teak خشب الساج
ḵašab as-sāj
team tracks سكك جماعية
sikkak jamā'iya
technical bureau مكتب فني
maktab fannī
technical report تقرير فني
taqrīr fannī
technician engineer مهندس فني
muhandis fannī
technology تكنولوجية
tiknulūjīya
tee-beam عتبة بشكل T
'ataba bi-šakl tī
tee-square مسطرة بشكل T
misṭara bi-šakl tī
telecommunication aids معينات الاتصال السلكي واللاسلكي
mu'īnāt al-ittisāl as-silki wal-lā-silkī
telecontrolled power station محطة لتوليد القدرة تدار عن بعد
muḥaṭṭa li-taulīd al-qudra tudār 'an bu'd
telemeter مقياس البعد
miqyās al-bu'd

T

telemetry — علم القياس عن بعد
'ilm al-qiyās 'an bu'd

telephone — هاتف
hātif

telescoping gangplank — معبر وقتي مدرج متداخل
ma'bar waqtī mudarraj mutadākil

television — تلفزيون
talfizyūn

teller — بطاقة
biṭāqa

telltale — دليل المستوى
dalīl al-mustawa

Tellurometer — مقياس مسافات إلكتروني
miqyās masāfāt iliktrūnī

telpher — عربة معلقة
'araba mu'allaqa

temper — لين
layyina

temperature correction — تصحيح القياس للفارق الحراري
taṣḥīḥ al-qiyās lil-fāriq al-harārī

temperature gradient — تدرج الحرارة
tadarruj al-ḥarāra

temperature sensor — جهاز الاحساس بالحرارة
jihāz al-iḥsās bil-ḥarāra

temperature stress — إجهاد حراري
ijhād ḥarārī

tempered glass — زجاج مقسى
zujāj muqassa

tempering — سقي (المعدن)
saqqī al-ma'dan

template — عارضة أفقية
'āriḍa ufuqīya

templet — رافدة أفقية
rāfida ufuqīya

temporary adjustment — ضبط مؤقت
ḍabṭ mu'aqqat

temporary formwork — هيكل مؤقت
haikal mu'aqqat

temporary labour — عمال مؤقتون
'ummāl mu'aqqatūn

temporary staff — موظفون مؤقتون
muwaẓẓafūn mu'aqqatūn

temporary storage — تخزين مؤقت
takzīn mu'aqqat

temporary support — دعم مؤقت
da'm mu'aqqat

temporary toilet — مرحاض مؤقت
mirḥāḍ mu'aqqat

ten — عشرة
'ašara

tenacity — متانة . تماسك
matāna, tamāsuk

tender analysis — تحليل العطاء
taḥlīl al-'aṭā'

tender documents — مستندات العطاء
mustanadāt al-'aṭā'

tendering — التقدم بالعطاء
at-taqaddum bi-'aṭā'

tendon — وتر
watar

tensile force — قوة الشد
quwwat aš-šadd

tensile strength — مقاومة الشد
muqāwamat aš-šadd

tensile test — إختبار الشد
iktibār aš-šadd

tension — توتر . شد
tawattur, šadd

tension carriage — حاضن الشد
ḥāḍin aš-šadd

tension correction — تصحيح الشد
taṣḥīḥ aš-šadd

tensioned beam — جائز مشدود
jā'iz mašdūd

tensioned rope — حبل مشدود
ḥabl mašdūd

tension flange — شفة التوتر
šaffat at-tawattur

tension pile — ركيزة إرساء
rakīzat irsā'

tension sleeve — كم شد
kumm šadd

tension structure — إنشاء منع الاجهاد
inšā' man' al-ijhād

tensor — متوتر
mutawattir

terminal — جناح (مطار)
janāḥ maṭār

terminal area — منطقة الجناح
minṭaqat al-janāḥ

terminal area layout — مخطط الجناح
mukaṭṭaṭ al-janāḥ

terminal buildings — منشآت الجناح
munšā'āt al-janāḥ

terminal station — محطة طرفية
muḥaṭṭa ṭarafīya

terminal valve — صمام طرفي
ṣimām ṭarafī

terminal velocity السرعة الحدية
as-sur'a al-haddīya

terotechnology تصميم وتركيب واختبار
taṣmīm wa-tarkīb wa-iḵtibār

terrace مصطبة
masṭaba

terracing يزود بمصاطب
yuzawwid bi-masāṭib

terrazzo أرضية حجرية
arḍīya ḥajarīya

tertiary treatment المعالجة الثلاثية
al-mu'ālaja aṯ-ṯulāṯīya

test إختبار
iḵtibār

test cube مكعب إختبار الخرسانة
muka''ab iḵtibār al-ḵarasāna

testing machine مكنة إختبار
makinat iḵtibār

test piece قطعة (مشكلة) للاختبار
qiṭa' mušakkala lil-iḵtibār

test pit بئر إختبارية
bi'r iḵtibārīya

tetrahedron شكل رباعي السطوح
šakl rubā'ī as-suṭūḥ

tetrapod مصدم رباعي القوائم
miṣdam rubā'ī al-qawā'im

tetrapod armour unit وحدة مدرعة لمصدم رباعي القوائم
waḥda mudarra' li-miṣdam rubā'ī al-qawā'im

T-head pier ركيزة بشكل T
rakīza bi-šakl tī

theatre مسرح
masraḥ

theft سرقة . إختلاس
sariqa, iḵtilāṣ

theodolite مزواة
muzwāt

theory of errors نظرية الخطأ
naẓarīyat al-ḵaṭā'

thermal حراري
ḥarārī

thermal environment محيط حراري
muḥīṭ ḥarārī

thermal forces قوى حرارية
qiwa ḥarārīya

thermal insulation عزل حراري
'azl ḥarārī

thermal loading تحميل حراري
taḥmīl ḥarārī

thermal property خاصية حرارية
ḵāṣīya ḥarārīya

thermal stress إجهاد حراري
ijhād harārī

thermic boring حفر حراري
ḥafr ḥarārī

thermic lancing شعاع حراري
šu'ā' ḥarārī

thermionic circuit دائرة ثرميونية
dā'ira ṯirmiyūnīya

thermit ثرميت
ṯirmīt

thermit weld لحام بالثرميت
liḥām bi-ṯirmīt

thermocouple مزدوجة حرارية
muzdawija ḥarārīya

thermo-osmosis إنتضاح حراري
intiḍāḥ ḥarārī

thermoplastic لدن بالحرارة
ladin bil-ḥarāra

thermostat ثرموستات
ṯirmūstāt

thick-arch dam سد بقنطرة سميكة
sadd bi-qanṭara samīka

thick slab بلاطة سميكة
bilāṭa samīka

thimble حلقة
ḥalaqa

thin-arch dam سد بقنطرة رقيقة
sadd bi-qanṭara raqīqa

thin section قطاع رقيق
qiṭā' raqīq

thin-shell construction إنشاء هيكلي مفرغ
inša' haikalī mufarraḡ

thin surfacing طبقة رصف قارية
ṭabaqat raṣf qārīya

third floor (level 4) الطابق الثالث
aṭ-ṭābiq aṯ-ṯāliṯ

thirteen ثلاثة عشر
ṯalāṭata 'ašar

thirty ثلاثون
ṯalāṯūn

thixotropy سيلان الهلام
sayyalān al-hulām

thousand ألف
alf

three ثلاثة
ṯalāta

three-amp ثلاثة أمبيرات
ṯalāṯat ambīrāt

three-axle tandem roller دلفين ترادفي ثلاثي المحاور
dalfīn tarādufī ṯulāṯī al-mahāwir
three-hinged arch قنطرة ثلاثية المفاصل
qanṭara ṯulāṯīyat al-mafāṣil
three-lane road طريق ذو ثلاث مسارات
ṭarīq ḏū ṯalāṯ masārāt
three-legged derrick برج ثلاثي القوائم
burj ṯulāṯī al-qawā'im
three-leg sling حمالة ثلاثية الأرجل
ḥammāla ṯulāṯīyat al-arjul
three-point problem المشكلة ثلاثية النقط
muškila ṯulāṯīyat an-nuqaṭ
three-tripod traversing تقاطع ثلاثي القوائم
taqāṭu' ṯulāṯī al-qawā'im
threshold مدخل
madkal
threshold lights أضواء المدخل
aḍwā' al-madkal
throat مجاز ضيق
majāz ḍayyiq
through bridge جسر عبور
'ubūr
through concrete sleeper راقدة خرسانية مستقيمة
rāqida karasānīya mustqīma
through line خط مستقيم
kaṭṭ mustaqīm
thrust دفع . دسر
daf', dasr
tidal dock حوض الجنوح
ḥauḍ al-junūḥ
tidal lag تخلف المد والجزر
takalluf al-madd wal-jazr
tidal outfall مصب مد وجزر
maṣabb madd wa-jazr
tide gauge مقياس المد والجزر
miqyās al-madd wal-jazr
tide investigations بحوث المد والجزر
buḥūṯ al-madd wal-jazr
tie ربط
rabṭ
tie bolt مسمار ربط
mismār rabṭ
tied-back wall جدار مدعم الخلفية
jidār mud'am al-kalfīya
tied retaining wall جدار إحتجاز مدعم القاعدة
jidār iḥtijāz mud'am al-qā'ida
tie in ربطة خفيفة
rabṭa kafīfa
tie line خط تثبيت قطري
kaṭṭ taṯlīṯ quṭrī
tie plate لوح ربط
lauḥ rabṭ
tie rod قضيب تثبيت
qaḍīb taṯbīt
tile آجرة
ājurra
tilt أمال . ميل
amāla, mail
tilting إمالة . ميل
imāla, mail
tilting-drum machine ماكينة ذات برميل قلاب
makina ḏāt barmīl qallāb
tilting gate بوابة قلابة
bawwāba qallāba
tilting level ميزان تسوية قلاب
mīzān taswiya qallāb
tilting mixer خلاطة قلابة
kallāṭa qallāba
timber خشب
kašab
timber arch قنطرة خشبية
qanṭara kašabīya
timber beam عارضة خشبية
'āriḍa kašabīya
timber decking سُطيح خشبي
suṭṭaiḥ kašabī
timbering أعمال خشبية
a'māl kašabīya
timber joist رافدة خشبية
rāfida kašabīya
timber pile ركيزة خشبية
rakīza kašabīya
timber shuttering هيكل خشبي مؤقت
haikal kašabī mu'aqqat
timber sleeper راقدة خشبية
rāqida kašabīya
timber stringers عوارض خشبية
'awāriḍ kašabīya
timber supports دعائم خشبية
da'ā'im kašabīya
time and motion study دراسة الوقت والحركة
dirāsat al-waqt wal-ḥaraka
timekeeper مسجل الوقت
musajjil al-waqt
time of completion زمن الاتمام
zaman al-itmām
timing توقيت
tauqīt

T

tine شعبة . شوكة
šuʿha, šauka
tint لون خفيف
laun kafīf
tip grade خط مرتكز جدار الدعم
katṭ murtakaz jidār ad-daʿm
tipping lorry شاحنة قلابة
šāḥina qallāba
Titan crane مرفاع جبار
mirfāʿ jabbār
TNT ت. ن. ت
tī-in-tī
toe مرتكز جدار الدعم
murtakaz jidār ad-daʿm
toe filter مرشح متدرج الطبقات
muraššiḥ mutadarrij aṭ-ṭabaqāt
toe line مستوى طرف الركيزة
mustawa ṭaraf ar-rakīza
toe wall جدار دعم
jidār daʿm
toggle mechanism آلية مفصلية
ālīya mifṣalīya
toll رسم . ضريبة
rasm, ḍarība
tomb قبر. ضريح
qabr, ḍarīḥ
ton طن
ṭann
tongue and groove joint وصلة حز ولسان
waṣlat ḥazz wa-lisān
tool عدة
ʿidda
toothed cutter قطاعة مسننة
qaṭṭāʿa musannana
top frame هيكل علوي
haikal ʿulwī
topographical surveying مساحة طوبوغرافية
misāḥa tubūğrāfīya
topographical surveyor مسّاح طوبوغرافي
massāḥ tubūğrāfī
topographic map خريطة طوبوغرافية
karīṭa tubūğrāfīya
topographic survey مسح طوبوغرافي
masḥ tubūğrāfī
topographic terms الشروط الطوبوغرافية
aš-šurūṭ at-tubūğrāfīya
topping طبقة تمليط فوقية
ṭabaqat tamlīṭ fauqīya
topsoil التربة الفوقية
at-turba al-fauqīya
topsoil excavation حفر التربة الفوقية
ḥafr at-turba al-fauqīya
top soiling plant معدات فرش التربة السطحية
muʿiddāt farš at-turba as-saṭḥīya
top-storey wall جدار الطابق العلوي
jidār aṭ-ṭābiq al-ʿulwī
torpedo طوربيد
ṭūrbīd
torque عزم اللي
ʿazm al-layy
torque wrench مفتاح ربط
miftāḥ rabṭ
torrent سيل . وابل
sail, wābil
Torshear bolt مسمار ربط إحتكاكي
mismār rabṭ ihtikākī
torsion إلتواء
iltiwāʾ
torsional load حمل اللي
ḥiml al-layy
total flow meter عداد الدفق الكلي
ʿaddād ad-dafq al-kullī
total pressure الضغط الكلي
aḍ-ḍağṭ al-kullī
toughened glass زجاج مقسى
zujāj muqassa
toughness متانة . صلابة
matāna, ṣalāba
toughness index دليل الصلابة
dalīl aṣ-ṣalāba
tough way إتجاه الصخر القاسي
ittijāh aṣ-ṣakr al-qāsī
tower برج
burj
tower building مبنى شاهق الارتفاع
mabnā šāhiq al-irtifāʿ
tower crane مرفاع برجي
mirfāʿ burjī
tower gantry إنشاء قنطري برجي
inšā qanṭarī burjī
towers أبراج
abrāj
tower shell هيكل برجي
haikal burjī
town by-pass طريق حول البلدة
ṭarīq ḥaul al-balda

T

town hall دار البلدية
dār al-baladīya
town planner مهندس تخطيط المدن
muhandis takṭīṭ al-mudun
town planning تخطيط المدن
takṭīṭ al-mudun
tracer مرسمة
mirsama
tracing إستشفاف الرسم
istišfāf ar-rasm
tracing paper ورق إستشفاف
waraq istišfāf
track سكة . طريق
sikka, ṭarīq
track apron مئزر السكة
mi'zar as-sikka
track bolts براغي تثبيت السكة
barāġī taṯbīt as-sikka
track cable كبل تعليق
kabl ta'līq
track construction إنشاء السكك
inša' as-sikak
track drains مصارف السكة
maṣārif as-sikka
tracked air-cushion vehicle مركبة هوائية برية (مجنزرة)
markaba hawā'īya barrīya mujanzara
track gauge المسافة بين خطي السكة
al-masāfa baina kaṭṭayy as-sikka
track gradient إنحدار السكة
inḥidār as-sikka
tracking تتبع
tatabbu'
track-laying tractor جرار مد السكك
jarrār madd as-sikak
track location موقع السكك
mauqi' as-sikak
track lubricator مشحمة السكة
mušaḥḥimat as-sikka
track maintenance صيانة السكك
ṣīyānat as-sikak
track road مسار السكة
masār as-sikka
track safety standards مستويات أمان السكك
mustawayāt amān as-sikak
track spike مسمار (تثبيت) السكة
mismār taṯbīt as-sikka
track stringer عارضة السكة
'āriḍat as-sikka
trackwork plan خطة أعمال السكة
kuṭṭat a'māl as-sikka
traction rope حبل الجر
ḥabl al-jarr
tractive force قوة الجر
quwwat al-jarr
tractive resistance مقاومة الجر
muqāwamat al-jarr
tractor جرار
jarrār
tractor shovel مجراف تحميل جرار
mijrāf taḥmīl jarrār
tradesman تاجر
tājir
traffic حركة المرور
ḥarakat al-murūr
traffic-carrying deck سطح حمل المرور
saṭḥ ḥaml al-murūr
traffic engineering هندسة المرور
handasat al-murūr
traffic lane ممر السير
mamarr as-sair
traffic road طريق مرور
ṭariq murūr
traffic sign إشارة المرور
išārat al-murūr
traffic signals إشارات المرور
išārāt al-murūr
trailer مقطورة
maqṭūra
trailing cable كبل خلفي
kabl kalfī
trailing points محولات خلفية
muḥawwilāt kalfīya
training wall جدار نهري
jidār nahrī
training works أعمال التدريب
a'māl at-tadrīb
train tonnage حمولة القطار بالطن
ḥumūlat al-qiṭār biṭ-ṭann
trammel عاق . قيد
'āqa, qayyada
trammel drain مصرف حقلي مثقب
maṣrif haqlī muṯaqqab
transducer محول وناقل الطاقة
muḥawwil wa-nāqil aṭ-ṭāqa
transfer نقل . تحويل
naql, taḥwīl

T

transfer pipes أنابيب نقل
anābīb naql
transformer محول
muḥawwil
transformer station محطة تحويل
muḥaṭṭat taḥwīl
transient عابر. مؤقت
'ābir, mu'aqqat
transient labour عمل مؤقت
'amal mu'aqqat
transient staff موظفون مؤقتون
muwaẓẓafūn mu'aqqatūn
transistor audio amplifier مضخم سمعي ترانزستوري
muḍakkim sam'ī trānzistūrī
transistor circuit دائرة ترانزستورية
dā'ira trānzistūrīya
transit ترانزيت
tranzīt
transition curve منحنى التحول
munḥana at-taḥawwul
transition length طول الانتقال
ṭūl al-intiqāl
transition zone منطقة تحول
minṭaqat taḥawwul
transit man عامل النقل
'āmil an-naql
transit mixer خلاطة نقل
kallāṭat naql
transit sheds حظائر النقل
ḥaḍā'ir an-naql
transmission gear box جهاز نقل الحركة
jihāz naql al-ḥaraka
transmission length طول النقل
ṭūl an-naql
transmission line خط النقل
kaṭṭ an-naql
transmission shaft جذع نقل الحركة
juḏ' naql al-ḥaraka
transpiration إرتشاح
irtišāḥ
transport نقل. حمل
naql, ḥaml
transportation engineering هندسة النقل
handasat an-naql
transporter bridge جسر نقل
jisr naql
transporter crane مرفاع نقل
mirfā' naql

transverse gradient إنحدار مستعرض
inḥidār musta'raḍ
transverse loading تحميل مستعرض
taḥmīl musta'raḍ
transverse reinforcement تسليح مستعرض
taslīḥ musta'raḍ
transverse system نظام مستعرض
niẓam musta'raḍ
trap مصيدة
miṣyada
trapezoidal profile weir سد شبه منحرف
sadd šibh munḥarif
trap points محولات حجز
muhawwilāt ḥajz
trash rack حاجز القمامة الطافية
ḥājiz al-qimāma aṭ-ṭāfīya
trass صخر بركاني
ṣaḵr burqānī
travelator التنقل
at-tanaqqul
traveller حلقة متحركة
ḥalaqa mutaḥarrika
travelling crane مرفاع متحرك
mirfā' mutaḥarrik
travelling form قالب متحرك
qālab mutaḥarrik
travelling formwork هيكل متحرك
haikal mutaḥarrik
travelling gantry قنطرة متحركة
qanṭara mutaḥarrika
travelling gantry crane مرفاع قنطري متحرك
mirfā' qanṭarī mutaḥarrik
travelling screen غربال متحرك
ġirbāl mutaḥarrik
travel mixer خلاطة متحركة
kallāṭa mutaḥarrika
traverse رافدة مستعرضة
rāfida musta'raḍa
traverse tables جداول إنحراف خطوط العرض
jadāwil inḥirāf kuṭūṭ al-'arḍ
traversing bridge جسر جانبي الحركة
jisr jānibī al-ḥaraka
traversing slipway رصيف إنزال مستعرض
raṣīf inzāl musta'rad
traxcavator مجرفة تحميل
mijrafat taḥmīl
tread موطئ. مداس
muwaṭi', madās
treadway طريق المداس
ṭarīq al-madās

T

treamie قادوس السمتنة تحت الماء
qādūs as-samtana taḥt al-māʾ
treated water ماء معالج
māʾ muʿālaj
treatment plant محطة معالجة
muḥaṭṭat muʿālaja
tree cutter جرار قطع الشجر
jarrār qaṭʿ aš-šajar
trench خندق
ḵandaq
trench box صندوق خندقي
ṣunduq ḵandaqī
trench drain مصرف خندقي
maṣrif ḵandaqī
trencher حفارة خنادق
ḥaffārat ḵanādiq
trench excavator حفارة خنادق
ḥaffārat ḵanādiq
trenching machine حفارة خنادق
ḥaffārat ḵanādiq
trepan مثقب منشاري
miṯqab minšārī
trestle منصب
minṣab
trestle bridge جسر منصبي
jisr minṣabī
trestle pier ركيزة منصبية
rakīza minṣabīya
trial pit بئر إختبارية
biʾr iḵtibārīya
triangle of error مثلث الخطأ
muṯallaṯ al-ḵaṭaʾ
triangular notch فرضة مثلثية
farḍa muṯallaṯīya
triangular profile weir سد جانبي مثلثي
sadd jānibī muṯallaṯī
triangulation التثليث
at-taṯlīṯ
triaxial compression test إختبار الانضغاط المحصور
iḵtibār al-inḍiḡaṭ al-maḥṣūr
tribar قضيب ثلاثي
qaḍīb ṯulāṯī
tribar armour unit وحدة تدريع ثلاثي
waḥdat tadrīʿ ṯulāṯī
tribrach منصب ثلاثي القوائم
manṣab ṯulāṯī al-qawāʾim
tributary خط فرعي
ḵaṭṭ farʿī
trickling filter مرشح أحيائي
muraššiḥ iḥyāʾī
trief concrete إسمنت الفرن العالي
ismant al-furn al-ʿālī
trigonometrical survey مسح مثلثي
masḥ muṯallaṯī
trilateration قياس الأضلاع الثلاثة
qiyās al-aḍlāʿ aṯ-ṯalāṯa
trimming تهذيب
tahḏīb
trinitrotoluene ثالث نتريت التولوين
ṯaliṯ natrīt at-tūlūyīn
trip coil ملف إعتاق
milaff iʿtāq
tripod حامل ثلاثي القوائم
ḥāmil ṯulāṯī al-qawāʾim
trommel غربال أسطواني دوار
ḡirbāl isṭiwānī dawwār
troughing قناة مفتوحة
qanāt maftūḥa
trowel مسطرين
masṭarain
truck شاحنة
šāḥina
truck loading تحميل الشاحنة
taḥmīl aš-šāḥina
truckmixer شاحنة خلاطة
šaḥina ḵallāṭa
truck-mounted crane شاحنة مرفاع
šaḥinat mirfāʿ
true bearing الانحراف الحقيقي
al-inḥirāf al-ḥaqīqī
true section القطاع الحقيقي
al-qiṭāʿ al-ḥaqīqī
true-to-scale print صورة بالحجم الكامل
ṣūra bil-ḥajm al-kāmil
trunk road طريق رئيسي
ṭarīq raʾīsī
trunk sewer مجرور رئيسي
majrūr raʾīsī
trunnion axis محور دوران
miḥwar dawarān
truss جملون
jamlūn
truss bridge جسر جملوني
jisr jamlūnī
trussed arch قوس جملوني
qaus jamlūnī
trussed beam عتبة مسنمة
ʿataba musannana

T

T-shaped pier — ركيزة بشكل T
rakīza bi-šakl tī
T-square — مسطرة بشكل T
misṭara bi-šakl tī
tubbing — بطانة بأطواق فولاذية
biṭāna bi-aṭwāq fūlāḏīya
tube — أنبوب
unbūb
tube railway — سكة داخل نفق
sikka dākil nafaq
tube valve — صمام أنبوبي
ṣimām unbūbī
tubular polythene sleeve — كم بوليثين أنبوبي
kumm būlīṯīn unbūbī
tubular section — قطاع أنبوبي
qiṭā' unbūbī
tubular steel scaffolding — سقالة أنابيب فولاذية
saqālat anābīb fūlāḏīya
tucking board — لوح دعم
lauḥ da'm
tucking frame — إطار تثبيت
iṭār taṯbīt
tug — جر
jarr
tungsten carbide — كربيد التنغستين
karbīd at-tanğastīn
tungsten carbide cutter — قاطعة كربيد التنغستين
qāṭi'at karbīd at-tanğastīn
tunnel — نفق
nafaq
tunnel construction — إنشاء الأنفاق
inšā' al-anfāq
tunnel drainage — صرف النفق
ṣarf an-nafaq
tunnel lighting — إضاءة النفق
iḍā'at an-nafaq
tunnelling speed — سرعة حفر النفق
sur'at ḥafr an-nafaq
tunnel lining — بطانة النفق
biṭānat an-nafaq
tunnel mole — حفارة أنفاق
ḥaffārat anfāq
tunnel vault — عقد نفقي
'aqd nafaqī
tunnel waterproofing — تصميد النفق للماء
taṣmīd an-nafaq lil-mā'
tup — مطرقة ساقطة
miṭraqa sāqiṭa
turbidity current — تيار التكدر
tayyār at-taqaddur
turbine — تربين
tūrbīn
turbine house — محطة التربين
muḥaṭṭat at-tūrbīn
turbine-house crane loading — تحميل مرفاع محطة التربين
taḥmīl mirfā' muḥaṭṭat at-tūrbīn
turbine housing — تثبيت التربين
tabyīt at-tūrbīn
turbo-drill — حفارة تربينية
ḥaffāra tūrbīnīya
turbo-generator — مولد تربيني
muwallid tūrbīnī
turbulence — إضطراب
iḍṭirāb
turbulent flow — دفق دوامي
dafaq dawāmī
turfing — تخضير
takḍīr
turn bridge — جسر دوار
jisr dawwār
turned bolt — برغي مخروط
burğī makrūṭ
turning basin — حوض دوار
ḥauḍ dawwār
turning point — نقطة الانعطاف
nuqṭat al-in'iṭāf
turnout — تحويلة . تفرع
tahwīla, tafarru'
turntable — قرص دوار
qurṣ dawwār
turntable ladder — سلم منصة دوارة
sullam 'ala minaṣṣa dawwāra
turpentine — تربنتين
tarbantīn
twelve — إثنا عشر
iṯna'šar
twenty — عشرون
'išrūn
twenty-eight — ثمانية وعشرون
ṯamānya wa-'išrūn
twenty-five — خمسة وعشرون
kamsa wa-'išrūn
twenty-four — أربعة وعشرون
arba'a wa-'išrūn
twenty-nine — تسعة وعشرون
tis'a wa-'išrūn

T

twenty-one واحد وعشرون
wāḥid wa-ʿišrūn
twenty-seven سبعة وعشرون
sabʿa wa-ʿišrūn
twenty-six ستة وعشرون
sitta wa-ʿišrūn
twenty-three ثلاثة وعشرون
ṯalāṯa wa-ʿišrūn
twenty-two إثنان وعشرون
iṯnān wa-ʿišrūn
twin-cable ropeway طريق حبلي مزدوج الكبل
ṭarīq ḥablī muzdawaj al-kabl
twist فتل . لوى
fatala, lawa
twisted square bar قضيب مربع ملتوي
qaḍīb murabbaʿ multawī
two إثنان
iṯnān
two-block concrete sleeper راقدة خرسانية مزدوجة الكتلة
rāqida ḵarasānīya muzdawajat al-kutla
two-dimensional stress إجهاد ذو بعدين
ijhād ḏū buʿdain
two-hinged arch قنطرة ثنائية المفصل
qanṭara ṯunāʾīyat al-mifṣal
two-leg sling حمالة ثنائية القائم
ḥammāla ṯunāʾīyat al-qāʾim
two-part sealant مادة مانعة للتسرب ثنائية
mādda mániʿa lit-tasarrub ṯunāʾīya
two-speed motor محرك ذو سرعتين
muḥarrik ḏū surʿatain
two-stage compression الضغط على مرحلتين
aḍ-ḍaġṭ ʿala marḥalatain
two-step control مضبط ذو خطوتين
miḍbaṭ ḏū ḵuṭwatain
two-way grid شبكة مزدوجة الاتجاه
šabaka muzdawajat al-ittijāh
two-way slab construction إنشاء بلاط مزدوج الاتجاه
inšāʾ bilāṭ muzdawaj al-ittijāh
ultimate bearing capacity سعة التحميل القصوى
siʿat at-taḥmīl al-quṣwa
ultimate bearing pressure ضغط التحميل الأقصى
ḍaġṭ at-taḥmīl al-aqṣā
ultimate compressive strength مقاومة الانضغاط القصوى
muqāwamat al-inḍiġāṭ al-quṣwa
ultimate development التطوير النهائي
at-taṭwīr an-nihāʾī
ultimate strength المقاومة النهائية
al-muqāwamat an-nihāʾīya
ultimate tensile strength مقاومة الشد القصوى
muqāwamat aš-šadd al-quṣwa
ultra-filtration ترشيح فائق الدقة
taršiḥ fāʾiq ad-diqqa
ultra-high-early-strength cement إسمنت فائق المتانة وسريعها
ismant fāʾiq al-matāna wa-sarīʿuha
ultrasonic flow meter عداد دفق بالتموجات فوق السمعية
ʿaddād dafq bit-tamawwujāt fauq as-samʿiya
ultrasonic pulse attenuation توهين النبضات فوق السمعية
tauhīn an-nabaḍāt fauq as-samʿīya
ultrasonic testing إختبار بالتموجات فوق السمعية
iḵtibār bit-tamawwujāt fauq as-samʿiya
unbacked vinyl flooring أرضية ڤينيل بدون ظهر
arḍīya fīnīl bidūn ẓahr
unconfined compression test إختبار الانضغاط غير المحصور
iḵtibār al-inḍiġāt ġair al-maḥṣur
unconfined water ماء غير محتجز
māʾ ġair muḥtajjaz
undercoat طلية سفلية
ṭalya suflīya
underflow حركة الماء في التربة السفلية
ḥarakat al-māʾ fit-turba as-suflīya
underground power station محطة توليد تحت الأرض
muḥaṭṭat taulīd taḥt al-arḍ
underground railway سكة حديد تحت الأرض
sikkat ḥadīd taḥt al-arḍ
underground scheme مشروع تحت الأرض
mašruʿ taḥt al-arḍ
underground water مياه جوفية
miyāh jaufīya

U

underpass ممر سفلي
mamarr suflī
underpin دعم الأساس
da'm al-asās
underpinning تدعيم الأساس
tad'īm al-asās
underplanting نقص المعدات المكنية
naqṣ al-mu'iddāt al-makinīya
under-reaming توسيع تجويف القاعدة
tausi' tajwīf al-qā'ida
undersize صغير الحجم
saǧīr al-ḥajm
underwater tunnel نفق تحت الماء
nafaq taḥt al-mā'
undisturbed sample عينة بكر
'ayyina bikr
undrained shear test إختبار القص غير المصفى
iḵtibār al-qaṣṣ ǧair al-muṣaffa
unequal angle زاوية غير متساوية الضلعين
zāwīya ǧair mutasāwīyat aḍ-ḍil'ain
uniform load حمل منتظم التوزيع
ḥiml muntaẓẓam at-tauzi'
uniform sand رمل منتظم الحبيبات
raml muntaẓẓam al-ḥubaibāt
unimproved road طريق غير محسن
ṭarīq ǧair muḥassan
unit hydrograph رسم مائي وحدي
rasm mā'ī waḥdī
unit of measurement وحدة القياس
waḥdat al-qiyās
unit-price contract عقد بسعر شامل
'aqd bi-si'r šāmil
unit strain وحدة الانفعال
waḥdat al-infi'āl
unit stress إجهاد وحدي
ijhād waḥdī
unit terminals أجهزة الكمبيوتر
ajhizat al-kumbjūtar
unit weight وحدة الوزن
waḥdat al-wazn
universal beam ركيزة عامة
rakīza 'āmma
universal joint وصلة عامة الحركة
waṣla 'āmmat al-ḥaraka
universal motor محرك عام
muḥarrik 'āmm
university building مقر الجامعة
maqarr al-jāmi'a
unlined ditches خنادق غير مبطنة
ḵanādiq ǧair mubaṭṭana
unpaved surface سطح غير مرصوف
saṭḥ ǧair marṣūf
unplasticised PVC بلاستيك صلب
blāstīk ṣalib
unreinforced concrete خرسانة غير مسلحة
ḵarasāna ǧair musallaḥa
unsolicited interrupts مقاطعات غير حاثة
muqāṭa'āt ǧair ḥāṯṯa
unstable غير ثابت
ǧair ṯābit
unstable frame هيكل ضعيف
haikal ḍa'īf
untensioned beam عارضة غير متوترة
'āriḍa ǧair mutawattira
uplift رفع . إرتفاع
raf', irtifā'
upper bar قضيب علوي
qaḍīb 'ulwī
upper floor طابق علوي
ṭābiq 'ulwī
upper reservoir خزان علوي
ḵazzān 'ulwī
upper surface سطح علوي
saṭḥ 'ulwī
upper transit نقل علوي
naql 'ulwī
upsetting الفلطحة بالطرق
al-falṭaḥa biṭ-ṭuruq
uranium يورانيوم
yūrānyūm
urban motorway طريق سريع حضري
ṭarīq sari'a ḥaḍarī
utility buildings منشآت المرافق
munša'āt al-marāfiq
vacuum concrete خرسانة خوائية
ḵarasāna ḵawā'īya
vacuum lifting رفع خوائي
raf' ḵawā'ī
vacuum mat حصيرة معدنية خوائية
ḥaṣīra ma'danīya ḵawā'īya
vacuum method of testing sands طريقة التفريغ الهوائي لاختبار الرمل
tarīqat at-tafrīǧ al-hawā'ī li-iḵtibār ar-raml
vacuum pad مسند سحب
misnad saḥb

U

vacuum pump مضخة تفريغ
miḍakkat tafrīğ
vacuum separation فصل بالتفريغ الهوائي
faṣl bit-tafrīğ al-hawā'ī
vacuum sweeper مكنسة كهربائية
miknasa kahrabā'iya
vadose water ماء الڤادوز
mā' al-fādūz
vadose zone منطقة الڤادوز
minṭaqat al-fādūz
valve صمام
ṣimām
valve chamber غرفة الصمام
ğurfat aṣ-ṣimām
valve tower برج الصمام
burj aṣ-ṣimām
vanadium ڤاناديوم
fanādyūm
vane pump مضخة ذات أرياش
miḍakka ḏāt aryāš
vane shear test إختبار القص المروحي
ik̲tibār al-qaṣṣ al-mirwaḥī
vane test إختبار المروحة
ik̲tibār al-mirwaḥa
vapour barrier حاجز بخاري
ḥājiz buk̲ārī
vapour pressure ضغط البخار
ḍağṭ al-buk̲ār
variable area flow meter مقياس التدفق لمساحة متغيرة
miqyās at-tadaffuq li-masāḥa mutağayyara
variable displacement pump مضخة إزاحة متنوعة
miḍak̲k̲at izaḥa mutanawwi'
variable speed couplings قارنات متغيرة السرعة
qārināt mutağayyirat as-sur'a
variance تباين
tabāyun
variation تغير
tağayyur
varnish ورنيش
warnīš
VDU وحدة العرض البصري
wiḥdat al-'arḍ al-baṣarī
vegetation نَبْت
nabt
vehicle عربة
'araba
vehicular live load الحمل المتحرك للمركبة
al-ḥiml al-mutaḥarrik lil-markaba
vel مقياس الحبيبات المحمولة جواً
miqyās al-ḥubaibāt al-maḥmūla jawwan
velocities in pipes معدل السرعة في الأنابيب
mu'addal as-sur'a fil-anābīb
velocity head الطاقة السرعية
aṭ-ṭāqa as-sur'īya
velocity of approach سرعة الاقتراب
sur'at al-iqtirāb
velocity of retreat سرعة الانسحاب
sur'at al-insiḥāb
velocity-profile method طريقة منحنى السرعة
ṭarīqat munḥana as-sur'a
velocity ratio نسبة السرعة
nisbat as-sur'a
vena contracta تخصر النافورة
tak̲aṣṣur an-nāfūra
veneer قشرة خشبية
qišra k̲ašabīya
veneered wall جدار ملبس
jidār mulabbas
ventilating air flow pattern نمط تدفق هواء التهوية
namaṭ tadaffuq hawā' at-tahwiya
ventilating shaft بئر التهوية
bi'r at-tahwiya
ventilation تهوية
tahwiya
ventilation building مبنى التهوية
mabnā at-tahwiya
ventilation of sewers تهوية المجارير
tahwiyat al-majārīr
ventilation plant وحدة تهوية
waḥdat tahwiya
ventilation power requirements متطلبات طاقة التهوية
mutaṭallabāt ṭāqat at-tahwiya
ventilation rates معدلات التهوية
mu'addalāt at-tahwiya
ventilation requirement متطلبات التهوية
mutaṭallabāt at-tahwiya
ventilator مهواة
mihwāt
vent pipe أنبوب تنفيس
unbūb tanfīsī

V

vent shaft عمود التهوية
ʿamūd at-tahwiya
Venturi flume قناة «ڤنتوري»
qanāt fantūrī
Venturi meter عداد «ڤنتوري»
ʿaddād fantūrī
Venturi scrubber كاشطة «ڤنتورية»
kāšiṭa fantūrīya
Venturi tube أنبوب «ڤنتوري»
unbūb fantūrī
verandah شرفة مسقوفة
šurfa masqūfa
verge حافة الطريق
ḥāffat aṭ-ṭarīq
verge trimmer مكنة تهذيب حواف الطريق
makinat tahḏīb ḥawāf aṭ-ṭariq
vertex ذروة
ḏurwa
vertical alignment تراصف رأسي
tarāṣuf raʾsī
vertical biparting doors أبواب ثنائية عمودية
abwāb ṯunāʾīya ʿamūdīya
vertical circle دائرة رأسية
dāʾira raʾsīya
vertical control تحكم رأسي
taḥakkum raʾsī
vertical curve منعطف رأسي
munʿaṭaf raʾsī
vertical-lagging cofferdam سد مبطن عمودياً
sadd mubaṭṭan ʿamūdīyan
vertical lift gate بوابة ترتفع عمودياً
bawwāba tartafiʿ ʿamūdīyan
vertical line خط رأسي
ḵaṭṭ raʾsī
vertical load حمل شاقولي
ḥiml šaqūlī
vertical photograph صورة عمودية
ṣūra ʿamūdīya
vertical profile قطاع جانبي رأسي
qiṭāʿ jānibī raʾsī
vertical roller sluice gate بوابة هويس دوارة رأسية
bawwāba hawīs dawwāra raʾsīya
vertical sand drain مصرف رملي عمودي
maṣrif ramlī ʿamūdī
vertical sliding doors أبواب منزلقة رأسياً
abwāb munzaliqa raʾsīyan
vertical-wall breakwaters حواجز أمواج بجدار رأسي
hawājiz amwāj bi-jidar raʾsī

VHF receiver جهاز إستقبال ذو تردد عال جداً
jihāz istiqbāl ḏū taraddud ʿālī jiddan
VHF transmitter أجهزة إرسال ذات تردد عال جداً
ajhizat irsāl dāt taraddud ʿāli jiddan
viaduct جسر متعدد القناطر
jisr mutaʿaddid al-qanātir
vial قارورة
qārūra
vibrated concrete خرسانة مدمجة بالاهتزاز
ḵarasāna mudmaja bil-ihtizāz
vibrating machinery معدات هزازة
muʿiddāt hazzāza
vibrating pile-driver مدق ركائز هزاز
midaqq rakāʾiz hazzāz
vibrating plate لوح هزاز
lauḥ hazzāz
vibrating poker محراك هزاز
miḥrāk hazzāz
vibrating roller محدلة إهتزازية
midḥala ihtizāzīya
vibrating tamper مدك هزاز
midakk hazzāz
vibration إهتزاز
ihtizāz
vibration of foundations هز الأساسات
hazz al-asāsāt
vibrator هزازة
hazzāza
vibratory driver محرك هزاز
muḥarrik hazzāz
Vibroflot تعويم هزاز
taʿwīm hazzāz
Vicat needle إبرة «ڤيكات»
ibrat fīkāt
view point وجهة نظر
wijhat naẓar
vignole rail سكة منبسطة القاع
sikka munbasiṭat al-qāʿ
villa ڤيلّا
fīlla
vinyl-asbestos tile بلاط الڤينيل والاسبستوس
bilāṭ al-fīnīl wal-asbistūs
virtual memory الذاكرة الافتراضية
aḏ-ḏākira al-iftirāḍīya
viscometer مقياس اللزوجة
miqyās al-luzūja

V

viscosity لزوجة
luzūja
viscous flow إنسياب لزج
insiyāb lazij
visibility distance مسافة الرؤية
masāfat ar-ru'ya
Visual Approach Slope Indicators مؤشرات إنحدار الاقتراب البصري
mu 'ašširāt inḥidār al-iqtirāb al-baṣarī
visual flight rules قواعد الطيران البصري
qawā'id at-tayarān al-baṣarī
V-notch ثلم مثلثي
ṯulm muṯallaṯī
voided slab بلاطة مجوفة
bilāṭa mujawwafa
void former كتلة مطاطية بأسفل البلاطة الخرسانية
kutla maṭṭāṭīya bi-asfal al-bilāṭa al-karasānīya
voids الفراغ بين حبات الرمل
al-farāġ baina ḥabbat ar-raml
voids ratio نسبة الفراغ
nisbat al-farāġ
volt فولت
fūlt
voltage transformer محول الفلطية
muḥawwil al-fūlṭīya
volumetric efficiency الكفاية الحجمية
al-kifāya al-ḥajmīya
volute حلزوني
ḥalazūnī
vortex دوامة
dawwāma
vortex shedding السفع الدوامي
as-saf' ad-dawāma
waffle تجاويف
tajāwif
waffle floor أرضية مجوفة
arḍīya mujawwafa
wages أجور
ujūr
wagon drill عربة الحفر
'arabat al-ḥafr
wagon retarder معوق العربة
mu'awwiq al-'araba
waiting room غرفة الانتظار
ġurfat al-intiẓār
wake إستيقظ
istaiqaẓa
waling لوح ربط أفقي
lauḥ rabṭ ufuqī
walking dragline حفارة متحركة
ḥaffāra mutaḥarrika
walking ganger ناظر عمال متجول
nāẓir 'ummāl mutajawwil
walkway ممشى
mamša
wall جدار
jidār
wallboard lining بطانة الألواح الجدارية
biṭānat al-alwāḥ al-jidāriya
wall concrete pier ركيزة خرسانية جدارية
rakīza karasānīya jidārīya
wall friction إحتكاك جداري
iḥtikāk jidārī
wall paint دهان الجدران
dihān al-judrān
wallpaper ورق الجدران
waraq al-judrān
walnut جوز
jauz
warehouse مستودع
mustauda'
warning device نبيطة إنذار
nabīṭat inḏār
warning signal إشارة إنذار
išārat inḏār
warning system جهاز إنذار
jihāz inḏār
warping joint وصلة إعوجاج
waṣlat i'wijāj
wash boring حفر إجترافي
ḥafr ijtirāfī
washer فلكة . غسالة
falaka, ġassāla
washing facilities تسهيلات الغسيل
tashīlāt al-ġasīl
washland أرض الغسل
arḍ al-ġasl
washout إجتراف
ijtirāf
washout valve صمام تصريف
ṣimām taṣrīf
wastage إهدار
ihdār
waste هدْر
hadr
wastewater ماء فائض
mā' fā'iḍ

W

wastewater channel قناة الماء الفائض
qanāt al-māʾ al-fāʾiḍ

wastewater disposal التخلص من الماء المهدور
at-takalluṣ min al-māʾ al-mahdūr

wastewater treatment معالجة الماء المهدور
muʿālajat al-māʾ al-mahdūr

wasteway مجرى الصرف
majrā aṣ-ṣarf

water authority مصلحة المياه
maṣlaḥat al-miyāh

waterbar رواسب مائية
rawāsib māʾia

water-bearing ground أرض حاوية للماء
arḍ ḥāwīya lil-māʾ

water-bearing strata طبقات حاوية للماء
ṭabaqāt ḥāwīya lil-māʾ

waterbound macadam طريق مرصوف وكابح للماء
ṭarīq marṣūf wa-qābiḥ lil-māʾ

water box صندوق الماء
ṣundūq al-māʾ

water/cement ratio نسبة الماء إلى الاسمنت
nisbat al-māʾ ila al-ismant

water cistern صهريج ماء
ṣahrīj māʾ

water closet دورة المياه
daurat al-miyāh

water content المحتوى المائي
al-muḥtawa al-māʾī

watercourse مجرى مائي
majrā māʾī

water extinguisher مطفأة مائية
miṭfāʾa māʾīya

waterfall شلال
šallāl

water-filled structure إنشاء مملوء بالماء
inšāʾ mamlūʾ bil-māʾ

water gauge مقياس مستوى الماء
miqyās mustawa al-māʾ

water inlet مدخل الماء
madkal al-māʾ

water jet نافورة مائية
nāfūra māʾīya

water level منسوب الماء
mansūb al-māʾ

waterlogged land أرض مشبعة بالماء
arḍ mušabbaʿa bil-māʾ

water meter عداد الماء
ʿaddād al-māʾ

water of capillarity ماء الجاذبية الشعرية
māʾ al-jāḏibīya aš-šaʿrīya

water outlet مخرج الماء
makraj al-māʾ

water paint دهان مائي
dihān māʾī

water pollution تلوث الماء
talawwuṯ al-māʾ

water pollution control مكافحة تلوث الماء
mukāfaḥat talawwuṯ al-māʾ

waterproof صامد للماء
ṣāmid lil-māʾ

waterproof clothing ملابس صامدة للماء
malābis ṣāmida lil-māʾ

waterproofed concrete خرسانة صامدة للماء
karasāna ṣāmida lil-māʾ

waterproofing التصميد للماء
at-taṣmīd lil-māʾ

waterproofing agent مادة صامدة للماء
mādda ṣamida lil-māʾ

waterproofing basement الجزء القاعدي الصامد للماء
al-juzʾ al-qāʿidī aṣ-ṣāmid lil-māʾ

waterproof membrane غشاء صامد للماء
ġišā ṣāmid lil-māʾ

water purification تنقية الماء
tanqiyat al-māʾ

water reducer الملدن
al-muladdin

water-repellent صاد للماء
ṣādd lil-māʾ

water requirement متطلبات الماء
mutaṭallabāt al-māʾ

water-resistant concrete خرسانة مقاومة للماء
karasāna muqāwima lil-māʾ

water resource مصدر الماء
maṣdar al-māʾ

water ring حلقة الماء
ḥalaqat al-māʾ

watershed حوض الصرف
ḥauḍ aṣ-ṣarf

water supply إمدادات الماء
imdādāt al-māʾ

water supply system شبكة إمدادات الماء
šabakat imdādāt al-māʾ

water table سطح الماء الباطني
saṭḥ al-māʾ al-bāṭinī

water tank خزان ماء
kazzān māʾ

water test إختبار هيدرولي للصرف
iktibār hīdrūlī liṣ-ṣarf
watertight cesspool حفرة مجاري سدودة للماء
ḥufrat majārī sadūda lil-mā'
watertight concrete خرسانة مسيكة
karasāna masīka
water tower برج الماء
burj al-mā'
water treatment معالجة الماء
mu'ālajat al-mā'
water-tube boiler مرجل بأنابيب مائية
mirjal bi-anābīb mā'īya
water turbine تربين مائي
tūrbīn mā'ī
water-well casing قميص بئر الماء
qamīṣ bi'r al-mā'
water wheel ناعورة
nā'ūra
waterworks محطة المياه
muḥaṭṭat al-miyāh
watt واط
waṭ
wave height إرتفاع الموجات
irtifā' al-maujāt
wave pressure ضغط الموجات
ḍaġṭ al-maujāt
waves موجات
maujāt
wc wastewater الماء المهدور من المرحاض
al-mā' al-mahdūr min al-mirhāḍ
wearing course طبقة الرصف السطحية
ṭabaqat ar-raṣf al-saṭḥīya
weathercocking دوارة الريح
dawwārat ar-rīḥ
weather-resistant صامد للعوامل الجوية
ṣāmid lil-'awāmil al-jawwīya
weaving section قسم النسيج
qism an-nasīj
web وترة
watara
wedge cut قطع إسفيني
qaṭ' isfīnī
wedge theory نظرية الاسفين
naẓarīyat al-isfīn
wedge-wire screen حجاب من أسلاك إسفينية
ḥijāb min aslāk isfīnīya
weephole ثقب تصريف إرتشاحي
tuqb taṣrīf irtišāḥī
weigh batcher وحدة مزج مكونات الخرسانة بالوزن
waḥdat mazj mukawwināt al-karasāna bil-wazn
weighted average متوسط موزون
mutawassiṭ mauzūn
weighting تثقيل
tatqīl
weight of rail وزن السكة
wazn as-sikka
weir سد صغير
sadd ṣaġīr
weir head علو الماء في السد
'ulū al-mā' fis-sadd
Weisbach triangle مثلث «وايزباخ»
mutallat waizbāk
weld لحم
laḥm . liḥām
weldability قابلية اللحام
qābilīyat al-liḥām
welded ملحوم
malḥūm
welded connections توصيلات ملحومة
tauṣīlāt malḥūma
welded frame هيكل ملحوم
haikal malḥūm
welded steel plate لوح فولاذي ملحوم
lauḥ fūlāḏī malḥūm
welded track سكة ملحومة
sikka malḥūma
welding لحام
liḥām
well بئر
bi'r
well borer آلة حفر الآبار
ālat ḥafr al-ābār
well casing قميص البئر
qamīṣ al-bi'r
well curbing ألواح البئر المحتجزة
alwāḥ al-bi'r al-muḥtajaza
well drain مصرف بئري
maṣrif bi'rī
well foundation أساس بئري
asās bi'rī
well hole حفرة البئر
ḥufrat al-bi'r
well logging تسجيل القياسات البئرية
tasjīl al-qiyāsāt al-bi'rīya
wellpoint أنبوب البئر
unbūb al-bi'r

W

well sinker حفارة آبار
ḥaffārat ābār
westing نحو الغرب
naḥwa al-ġarb
wet analysis تحليل سائلي
taḥlīl sā'ilī
wet cube strength إختبار محتوى الرطوبة
ikṯibār muḥtawa ar-ruṭūba
wet dock حوض مائي
ḥauḍ mā'ī
wet drilling حفر مرطب
ḥafr murattab
wet galvanizing غلفنة بالزنك المذاب
ġalfana biz-zink al-muḏāb
wet mix خليط زائد الرطوبة
ḵalīṭ zā'id ar-ruṭūba
wet-sand process عملية الرمل الرطب
'amalīyat ar-raml ar-raṭib
wetted perimeter المحيط المبتل
al-muḥīṭ al-mubtal
wet welding اللحام تحت سطح الماء
al-liḥām taḥt saṭḥ al-mā'
wet well حوض محطة الضخ
ḥauḍ muḥaṭṭat aḍ-ḍakk
wharf رصيف المرفأ
raṣīf al-marfa'
wharf wall جدار رصيف المرفأ
jidār raṣīf al-marfa'
wheelabrating سفع بخردق الفولاذ
saf' bi-ḵardaq al-fūlāḏ
wheelbarrow عربة يد
'arabat yadd
wheel drop pit دولاب إنزل بالمنجم
dūlāb inzāl bil-manjam
wheeled tractor جرار على عجلات
jarrār 'ala 'ajalāt
wheel excavator حفارة بعجلات
ḥaffāra bi-'ajalāt
wheel gauge مقياس العجلات
miqyās al-'ajalāt
wheel grinder جلاخة ذات دواليب
jallāḵa ḏāt dawālīb
wheel load حمل الدولاب
ḥiml ad-dūlāb
wheel mounted مركب على عجلات
murakkab 'ala 'ajalāt
wheel scraper مكشطة ذات دواليب
mikšaṭa ḏāt dawālīb
wheel size مقاس الدولاب
maqās ad-dūlāb

whisker مادة بلورية صلبة
mādda ballūrīya ṣaliba
white cement إسمنت أبيض
ismant abyaḍ
white metal معدن أبيض
ma'dan abyaḍ
whole-circle bearing الاتجاه الزاوي في دائرة
al-ittijāh az-zāwī fid-dā'ira
whole-tide cofferdam سد المد والجزر
sadd al-madd wal-jazr
wicket بويب
buwaib
wicket dam سد ببويبات
sadd bi-buwaibāt
wide-gauge railway سكة حديدية عريضة
sikkat ḥadīd 'arīḍa
width of landing strip عرض مسار الحط
'arḍ masār al-ḥaṭṭ
width of runway عرض المدرج
'arḍ al-madraj
wilting coefficient مُعامل الذبول
mu'āmil aḏ-ḏubūl
winch ونش
winš
wind beam شكال تقوية ضد الرياح
šikāl taqwiya ḍid ar-riyāḥ
wind force قوة الريح
quwwat ar-rīḥ
wind indicator مؤشر إتجاه الريح
mu'aššir ittijāh ar-rīḥ
winding stairs درج حلزوني
daraj ḥalazūnī
wind load حمل الرياح
ḥiml ar-riyāḥ
wind loading تحميل الرياح
taḥmīl ar-riyāḥ
windmill طاحونة هوائية
ṭāḥūna hawā'īya
window نافذة
nāfiḏa
window seat مقعد النافذة
maq'ad an-nāfiḏa
window sill عتبة النافذة
'atabat an-nāfiḏa
wind portal مدخل مقاوم للريح
madḵal muqāwim lir-rīḥ
wind pressure ضغط الريح
ḍaġṭ ar-rīḥ
wind resistance مقاومة الريح
muqāwamat ar-rīḥ

W

wind rose دائرة الرياح
dā'irat ar-riyāḥ
windrow ركمُ الرياح
rukm ar-riyāḥ
wind shear قص الريح
qaṣṣ ar-rīḥ
windshield حاجب الريح
ḥājib ar-rīḥ
windshield design تصميم حاجب الريح
taṣmīm ḥājib ar-rīḥ
wind tunnel نفق هوائي
nafaq hawā'ī
windy عاصف
'āṣif
wing dam سد مجنح
sadd mujannaḥ
wing of a building جناح المبنى
janāḥ al-mabnā
wing wall جدار جانبي
jidār jānibī
wired glass زجاج مقوى بالأسلاك
zujāj muqawwa bil-aslāk
wire drawing سحب الأسلاك
saḥb al-aslāk
wire gauge مقياس أسلاك
miqyās aslāk
wire guidance system نظام إرشاد سلكي
niẓām iršād silkī
wire lath شبكة سلكية
šabaka silkīya
wire-line core barrel كبل أسطوانة الحفر
kabl isṭiwānat al-ḥafr
wire-mesh reinforcement تسليح بشبكة سلكية
taslīḥ bi-šabaka silkīya
wire rope حبل سلكي
ḥabl silkī
wobble-wheel roller مدحلة بدواليب متراوحة
midḥala bi-dawālīb mutarāwiha
wood خشب
ḵašab
wood-block paving رصف بقوالب خشبية
raṣf bi-qawālib ḵasabīya
wood column عمود خشب
'amūd ḵašab
wood concrete خرسانة الخشب
ḵarasānat al-ḵašab
wooden fence سور خشبي
sūr ḵašabī
wood-fender pile ركيزة حاجز خشب
rakīzat ḥājiz ḵašab
wood floor أرضية خشب
arḍīyat ḵašab
wood framing هيكل خشبي
haikal ḵašabī
wood-pile clusters عناقيد ركام الحطب
'anāqīd rukām al-ḥaṭab
wood scaffolding سقالة خشب
saqālat ḵašab
wood-stave pipe أنبوب من الشرائح الخشبية
unbūb min aš-šarā'iḥ al-ḵašabīya
wood truss جملون خشب
jamlūn ḵašab
wood-wool concrete خرسانة صوف الخشب
ḵarasānat ṣūf al-ḵašab
word processor جهاز تنضيد الحروف
jihāz tanḍīd al-ḥurūf
work عمل
'amal
workability إمكانية التشغيل
imkānīyat at-tašğīl
workboat زورق عمل
zauraq 'amal
workers' entrance مدخل العمال
madḵal al-'ummāl
working chamber غرفة العمل
ğurfat al-'amal
working drawing رسم تشغيلي
rasm tašğīlī
working shaft مهواة التشغيل
mihwāt at-tašğīl
working shift نوبة العمل
naubat al-'amal
working stress إجهاد عملي
ijhād 'amalī
workmanship specification مواصفات المصنعية
muwāṣafāt al-maṣna'īya
workmen's compensation insurance تأمين تعويضات العمال
ta'mīn ta'wīḍāt al-'ummāl
workshop ورشة
wirša
worm conveyor ناقلة بترس دودي
nāqila bi-turs dūdī
wracking forces قوى التحطيم
qiwa at-taḥṭīm

W

wrench مفتاح ربط
miftāḥ rabṭ
wrought aluminium alloy سبيكة ألومنيوم مشكلة
sabīka aluminyūm mušakkala
wrought iron الحديد المطاوع
al-ḥadīd al-muṭāwi'
wythe جدار حاجز في المدخنة
jidār ḥājiz fil-midḵana
Xerox copy (TM) نسخة جافة
nusḵa jāffa
yard ياردة . حظيرة
yarda, ḥaḍīra
yield مطاوعة
muṭāwa'a
yield point نقطة الخضوع
nuqṭat al-ḵudū'
yoke مقرن
miqran
Young's modulus مُعامل ينج «لقياس المرونة»
mu'āmil yunḡ li-qiyās al-murūna
zeolite زيوليت
ziyūlīt
zero bit لقمة صفر
luqmat ṣifr
zero point نقطة الصفر
nuqṭat aṣ-ṣifr
zigzag bond ترابط متعرج
tarābuṭ muta'arrij
zinc زنك
zink
zone منطقة
minṭaqa
zoned construction إنشاء مقسم الى مناطق
inšā' muqassam ila manāṭiq
zoned dam سد مقسم الى مناطق
sadd muqassam ila manāṭiq
zone of aeration منطقة التهوية
minṭaqat at-tahwiya
zone of saturation منطقة التشبع
minṭaqat at-tašabbu'

محتويات المعجم

تمهيد

تم تجميع المصطلحات العلمية والفنية والهندسية التي يضمها هذا المعجم ومتطلبات مهندسي الموقع نصب العينين. إنه معجم عملي في المقام الأول ويتحاشى التعريفات الايضاحية، لأن من يستعملون مصطلحا يعرفون معناه.

وتمتد التغطية الفنية للمعجم لتشمل جميع فروع الهندسة المدنية، بما في ذلك هندسة انشاء المطارات، والجسور، والبناء، والهدم، وأحواض السفن والموانئ، والأنفاق، والمرافق العامة والمياه. وقد أخذنا في الاعتبار جميع الأنشطة المرتبطة بأعمال الهندسة المدنية - مثل أعمال التصميم وإدارة العقود وتركيب المواد والمساحة وهندسة البيئة - وذلك أثناء انتقاء قائمة المصطلحات. كما وضع في الاعتبار كذلك الأساليب الالكترونية أحدث الاساليب الالكترونية - خاصة في تقنية الكمبيوتر - لكن مع التدقيق في الاختيار.

إنه أول معجم من نوعه لمصطلحات الهندسة المدنية ولذا فهو يعتبر سلاحا ضروريا للمهندسين المدنيين، سواء الذين يعملون بالموقع أو بمكاتب المركز الرئيسي. ونظراً لتغطيته الشاملة للموضوع، فسوف يجده جميع المرتبطين - بصورة أو بأخرى - بمجال الهندسة المدنية، بما في ذلك العاملون في المجالات الهندسية الأخرى، معجماً قيماً. كما سيجده المعلمون وطلاب الهندسة بالمعاهد والكليات الفنية والجامعات والمترجمون وجميع المضطلعين بمسؤولية توفير المصطلحات العربية المعادلة للمصطلحات الانكليزية مرجعاً لا غنى عنه.

لقد تم انتقاء المصطلحات بعناية بغرض الاحتفاظ بالمعجم في حجم مناسب، وفي عديد من الحالات تم إدراج المصطلح كجزء من عبارة أو جملة يتكرر استعمالها بكثرة. هذا وقد تم حذف بعض المصطلحات عن عمد لأنها شائعة الاستعمال ومدرجة بالمعاجم العامة، وإذا سهونا عن بعضها الآخر فنحن نرحب بالانتقادات البناءة التي من شأنها رفع مستوى الطبعات التالية.

عربي - انكليزي

ARABIC - ENGLISH

ابتعاث الدخان	smoke emission
إبرة	needle
إبرة الميل لعامل المنجم	miner's dip needle
إبرة ڤيكات	Vicat needle
إبطال العقد	annulment of contract
أبواب بمصاريع كونسرتينية	concertina shutter doors
أبواب تنطوي	folding doors
أبواب ثنائية عمودية	vertical biparting doors
أبواب دوارة	revolving doors
أبواب منزلقة أفقياً	horizontal sliding doors
أبواب منزلقة رأسياً	vertical sliding doors
اتجاه الألياف في الخشب	grain
اتجاه الصخر القاسي	tough way
اتجاه ربعي	quadrantal bearing
الاتجاه الزاوي المغنطيسي	magnetic bearing
الاتجاه الزاوي في دائرة	whole-circle bearing
اتجاه خط المسح	course
اتجاه نحو الشرق	easting
اتجاهات زاوية	bearings
الاتحاد الأمريكي للمهندسين المدنيين	ASCE
اتزان	balancing
اتصال أفقي	bed joint
الاتصالات بالتردد العالي	HF communications
اتفاق تسديد التكاليف	cost-reimbursable agreement
اتفاقية	agreement
إتمام	finishing
أثر السير	body track
أثر دائم	permanent set
اثنا عشر	twelve
اثنان	two
اثنان وعشرون	twenty-two
اجتراف	washout
اجتياز	passage
إجراءات الأمان	safety procedures

إجراءات التركيب	erection procedure
إجراءات التعاقد	contracting procedure
إجراءات التعاقد من الباطن	subcontract procedure
إجراءات وقائية	protective measures
آجر أسفلتي	asphalt tile
آجر صلب	solid masonry
آجر مفرغ	hollow clay blocks
آجرة	tile
آجرة مفرغة	hollow tile
أجزاء الكترونية	electronic components
إجهاد	stress
إجهاد الانحناء	bending stress
إجهاد الانكسار	breaking stress
إجهاد التحديب	buckling stress
إجهاد التحميل	bearing stress
إجهاد الترابط	bond stress
إجهاد الصمود	proof stress
إجهاد تضاغطي مستقيم	direct stress
إجهاد حراري	thermal stress, temperature stress
إجهاد حلقي	hoop stress
إجهاد ذو بعدين	two-dimensional stress
إجهاد رئيسي	principal stress
الاجهاد العادي	normal stress
إجهاد عملي	working stress
الاجهاد الفعال	effective stress
إجهاد لاحق	post-stressing
إجهاد مأمون	safe stress
إجهاد متخلف	residual stress
إجهاد محوري	axial stress
إجهاد مسبق	prestressing
الاجهاد المسموح	admissible stress, allowable stress
إجهاد موحد	combined stresses
الاجهاد والتوتر	stress and strain
أجهزة إحساس	sensors

أجهزة الرصد	meteorological equipment
أجهزة الكمبيوتر	unit terminals
أجهزة المساحة	surveying instruments
أجور	wages
أحادي الكبل	monocable
احتجاز سطحي	surface detention
الاحتجاز النوعي	specific retention
احتكاك	friction
احتكاك الأنابيب	pipe friction
احتكاك جداري	wall friction
احتكاك داخلي	internal friction
الاحتكاك السطحي	skin friction
احتكاك سطحي سالب	negative skin friction
احتمالية	probability
احتياطي	spare
الاحتياطي المركزي	central reserve
إحداثيات	coordinates
الاحداثيات الرأسية	ordinates
الاحداثي الأوسط	mid-ordinate
الاحداثي السيني	abscissa
أحد عشر	eleven
احديداب	camber
احصاءات الحوادث	accident statistics
أحواض التركيز	stabilization basins
اختبار	test
اختبار «آيزود»	Izod test
اختبار الاختراق	penetration test
اختبار الاختراق المخروطي	cone penetration test
اختبار الأداء	performance testing
اختبار الانحناء	bend test
اختبار الانضغاط المحصور	triaxial compression test
اختبار الاهتزازية	shaking test
اختبار التحميل	bearing test
اختبار التصادم بالقضيب المحزز	notched bar test
اختبار التفتت	crushing test
اختبار القضبان	rail test
اختبار التقويم	pull-out test
اختبار الحمل	load test
اختبار الحمل بالدفع	side-jacking test
اختبار «شاربي» للصدم	Charpy test
اختبار الشد	tensile test
اختبار الصدم	impact test
اختبار الصلادة البرينيلية	Brinell hardness test
اختبار الصلادة بالسكلرسكوب	scleroscope hardness test
اختبار الصلادة لـ «روكويل»	Rockwell hardness test
اختبار الصمود	proof test
اختبار الضبط	control test
اختبار القص الصندوقي	box shear test
اختبار القص غير المصفى	undrained shear test
اختبار القص المروحي	vane shear test
اختبار الكزازة	slump test
اختبار الكلال	fatigue test
اختبار المروحة	vane test
اختبار الانضغاط غير المحصور	unconfined compression test
اختبار امتصاص السطح الأولي	initial surface absorption test
اختبار أوتوماتي	automatic testing
اختبار بالتموجات فوق السمعية	ultrasonic testing
اختبار تحميل اللوح	plate bearing test
اختبار غير هدام	non-destructive testing, NDT
اختبار محتوى الرطوبة	wet cube strength
اختبار مسبق	pretesting
اختبار مُعامل التمزق	beam test
اختبار مُعامل الدموج	compacting factor test
اختبار مقاومة الخرسانة	cube test
اختبار مقياس الضغط	pressuremeter test
اختبار هيدرولي للصرف	water test

أ

اختبارات التربة في الموقع	in-situ soil tests
اختبارات القص	shear tests
اختبارات فرط التحميل	overload tests
اختراق	penetration
اختراق ساكن	static penetration
اختصاصي	specialist
اختلاس	theft
اختيار الطريق	route selection
اختيار الموقع	site selection
اختيار موقع الميناء	port site selection
أخدود	gullet
أخدود تصريف	rubble drain
أخذ العينات العشوائي	grab sampling
أخذ العينات جزئياً	fractional sampling
أخذ العينات واختبارها	sampling
أخشاب الإنشاءات	structural timber
أخصائي	expert
أخشاب دعم	crib
أخصائي تجميع الانشاءات	constructional fitter
أخطاء رتيبة	systematic errors
أخطاء متراكمة	cumulative errors
أخطاء متخلفة	residual errors
إخفاق الاشعال	misfires
إخفاق المسعى	dead lock
إخلال بالعقد	default of contract
أداء	performance
أداة استخراج العينات	bottom sampler
أداة التحذير بالحواجز	obstruction warning device
أداة الرياح المعترضة	cross-wind component
أداة ثلم	indenter
إدارة الذاكرة	memory management
إدارة المشروع	project management
إدارة المواد	materials management
إدارة الميناء	port administration
أدوات البناء	bricklayer's tools
أدوات الحفر بالدق	percussion tools
أدوات النتوء	nose devices
أدوات تثبيت	fastenings
أدوات تجويف	coring tools
أدوات عاطلة	defective tools
أدوات كهربائية	electric tools
الأرباح التقديرية	profit forecast
أربعة	four
أربعة عشر	fourteen
أربعة وعشرون	twenty-four
أربعون	forty
أربنت : وحدة قياس فرنسية	arpent
ارتخاء	sagging
ارتخى	sag
ارتدادية	resilience
ارتشاح	seepage
ارتفاع	altitude, rise
الارتفاع الإجمالي	overall height
ارتفاع إضافي	superelevation
ارتفاع الجدار	height of wall
ارتفاع السقف	headroom
ارتفاع المنحنى	elevation of curve
ارتفاع الموجات	wave height
ارتفاع الموجة على الشاطئ	swash height
الارتفاع الحرج	critical height
ارتفاع شعري	capillary rise
أردواز	slate
أردوازي	slater
إرساء السفينة بالمرفأ	docking of ships
إرساء بعيد عن الشاطئ	off-shore mooring
إرسال البيانات المتتابعة	serial data transmission
أرض التدفق	flowing ground
أرض الغسل	washland
أرض حاوية للماء	water-bearing ground
أرض دائمة التجمد	permafrost
أرض زالقة	running ground

أرض شعرية	capillary fringe
الأرض الطبيعية	natural ground, sub-grade
أرض فاصلة بين طريقين	central reservation
أرض قاحلة	aquifuge
أرض مشبعة بالماء	waterlogged land
أرض منبسطة	reach
أرض منحدرة	sidelong ground
أرضية	floor
أرضية بدون عوارض	beamless floor
أرضية ألواح مستعمدة	orthotropic plate floor
أرضية بألواح	plate floor
أرضية حجرية	terrazzo
أرضية خرسانية	concrete floor
أرضية خشب	wood floor
أرضية خشب رقائقي	plywood floor
أرضية خشبية مزخرفة	parquet floor
أرضية ذات ركيزة وبلاطة	beam and slab floor
أرضية شبكية	open-grid flooring
أرضية صب الخرسانة	deck
أرضية ڤينيل بدون ظهر	unbacked vinyl flooring
أرضية لا إنزلاقية	non-slip floor
أرضية متشابكة	diagrid floor
أرضية مجوّفة	waffle floor
أرضية مزخرفة	inlaid parquet
أرضية مسبقة الصب	precast floor
أرضية مطاطية	rubber flooring
أرضية مكشوفة	open floor
أرضية من آجر مفرغ	hollow-tile floor
أرضية من القوالب المفرغة	hollow-block floor
أزاح	shift
إزاحة الترسبات بالماء المتدفق	hydraulicking
إزاحة الطمي	silt displacement
إزار الحائط	skirting board
أزاز	buzzer
إزالة	clearing
إزالة الغبار	dusting
إزالة القشور	scaling
إزالة الملوحة	desalination
إزالة هيكل الدعم المؤقت	striking
إزميل	chisel
إزميل قطع على البارد	cold chisel
أساس	footing, foundation
أساس بئري	well foundation
أساس حصيري	mat foundation
أساس خرساني	concrete foundation
أساس ضحل	shallow foundation
أساس ضيق	strip foundation
أساس طبيعي	natural foundation
أساس عائم	buoyant foundation
أساس على ركائز	piled foundation
أساس كابولي	cantilever foundation
أساس متوازن	balanced foundation
أساس مدّرج	benched foundation
أساس مشترك	combined footing
أساس من الخرسانة المسلحة	raft foundation
أساسات	footings
أساسات الركائز	pile foundation
أساسات المفاعل	reactor foundation
أساسات عائمة	floating foundation
أساسات عميقة	deep foundations
أساسات مبنى الغلايات	boiler-house foundations
أسبستوس	asbestos
استئجار المعدات	plant hire
استبدال فلز القاعدة بآخر	base exchange
استجذاب	adsorption
استخدام الآلات	robotics
استرداد	recovery
استشفاف الرسم	tracing
استصلاح الأرض	land reclamation
استطاعة السلم المتحرك	escalator capacity

استطاعة المحمل	bearing capacity
استطالة	elongation
استطلاع	spotting
استعادة الحمأة	sludge recovery
استعاضة الماء الباطني	beach replenishment
استعمال الأرض	land use
استقامات	straights
استقرار	stability, subsidence
استقرار الركيزة	pile settlement
استقصاء الموقع	site investigation
استكشاف	exploration
استكشاف الموقع	site exploration
استكمال	interpolation
استكمل بالاستقراء	extrapolate
استمرار	continuity
استنزاف	bleeding
استهلاك الطاقة	power consumption
استهلاك الوقود	fuel consumption
استهلاك بلدي	municipal consumption
استهلاك صناعي	industrial consumption
الاستهلاك العام	public consumption
الاستهلاك المحلي	domestic consumption
استوديو	studio
استيقظ	wake
أسطوانة	cylinder, drum
أسطوانة الالتفاف	slewing cylinder
أسطوانة الحفر الموجة	oriented-core barrel
أسطوانة الضخ	pumping cylinder
أسطوانة تفاضلية	differential cylinder
أسطوانة رفع ذراع التطويل	boom-lifting cylinder
أسطوانة مفصلية	hinged cylinder
أسعار المقاول الفرعي	subcontractor rates
أسفلت	asphalt
أسفلت الرصف	paving asphalt
أسفلت رصف دقيق	fine cold asphalt
أسفلت سائل	liquid asphalt
أسفلت صمغي	mastic asphalt
أسفلت طبيعي	natural asphalt
أسفلت للسطوح	roofing asphalt
أسفلت مانع للتسرب	asphalt tanking
أسفلت مضغوط	compressed asphalt
إسقاط	projection
إسقاط أكسونومتري	axonometric projection
إسقاط السطح المستوي	planometric projection
إسقاط سمتي	azimuthal projection
إسقاط مائل	oblique projection, cabinet projection
إسقاط متعامد	orthographic projection
إسقاط متناظر	isometric projection
أسلوب التصنيف	file structure
أسلوب باروكي	baroque style
إسمنت	cement
إسمنت أبيض	white cement
إسمنت الخبث	clinker concrete
إسمنت الخبث المعدني	slag cements
إسمنت المغنسيت	oxychloride cement
إسمنت الفرن العالي	trief concrete, blast-furnace cement
أسمنت ألوميني	aluminous cement
أسمنت بزولان بورتلاند	Portland pozzolana cement
إسمنت بطيء الشك	oil-well cement
إسمنت بنسبة ألومينا مرتفعة	high-alumina cement
إسمنت بورتلاند	Portland cement
إسمنت بورتلاند (للبناء)	masonry cement
إسمنت بورتلاند عادي	ordinary Portland cement
إسمنت بورتلاند مخلوط بخبث الفرن العالي	Portland blast-furnace cement
إسمنت تمددي	expanding cement
إسمنت جبسي	gypsum cement
إسمنت حشو جاف	dry pack

إسمنت سائل	cement grout
إسمنت سريع التصلب	high-early-strength cement, rapid-hardening cement, quick setting cement
إسمنت صامد للكبريت	sulphate-resisting cement
إسمنت صامد للنار	fireproof cement
إسمنت طبيعي	natural cement
إسمنت فائق المتانة وسريعها	ultra-high-early-strength cement
إسمنت فلزي	supersulphated cement
إسمنت للأغراض العامة	general-purpose cement
إسمنت مائي	hydraulic cement
إسمنت مصلد	hardened cement
إسمنت مضاد للماء	hydrophobic cement
إسمنت معادن	metallurgical cement
إسمنت منخفض الحرارة	low heat cement
أسوار أمان	safety fences
أسوار خشبية مؤقتة	hoardings
أسود	black
أسيتات السليولوز	cellulose acetate
إشارات الطرق	road signals
إشارات المرور	traffic signals
إشارات مستمرة بحجرة القيادة	continuous cab signals
إشارة السكك الحديدية	railway signalling
إشارة المرور	traffic sign
إشارة إنذار	warning signal
إشارة خطر	danger sign
إشتراط	stipulation
إشعال	firing
أشعة «بيكريل»	becquerel
أشعة تحت الحمراء	infra-red beam
أشعة ليزر	laser
الأشغال البلدية	municipal works
أشغال الدفاع البحري	sea-defence works
الأشغال العامة	public works
الاشارة الهادية الجائزة المطلقة	absolute permissive block signalling
الاشارة الهادية المتداخلة	overlap block signalling
اصطناعي	artificial
أصل	origin
إصلاح	restoration
إصلاح جزئي	partial overhaul
أصلح	repair
إضاءة	lighting
إضاءة الاقتراب	approach lighting
إضاءة المدرج	runway lighting
إضاءة المدرج في المطار	taxiway lighting
إضاءة الزينة	festoon lighting
إضاءة الصاري العالي	high mast lighting
إضاءة الطرق	road lighting
إضاءة المطار	airfield lighting
إضاءة النفق	tunnel lighting
إضاءة جانبي المدرج	runway edge lighting
إضاءة خط المركز للمدرج	runway centreline lighting
إضاءة غير مباشرة	indirect lighting
إضاءة فلورية	fluorescent light
إضاءة كهربائية	electric lighting
إضاءة مباشرة	direct lighting
اضطراب	turbulence
اضطراب جوي	air turbulence
أضواء الحواجز	obstruction lights
أضواء خط المركز الخضراء	green centreline lights
أضواء المدرج	taxiway lights
أضواء المدخل	threshold lights
أضواء جانبية زرقاء	blue edge lights
أضواء مدرج قوية الاضاءة	high-intensity runway lights
إطار تثبيت	tucking frame
إطار فولاذي	steel frame
إطار كامل	perfect frame

إطار متوسط	middling frame
إطار مطاط	rubber rim
إطار منزلق	sash
إطار ينفخ بالهواء المضغوط	pneumatic tyre
الأطراف المعنية	parties concerned
إطفاء	blow out
إعادة البناء	rebuilding
اعادة التشكيل	remoulding
إعادة الدوران	recirculation
إعادة شحن اصطناعي	artificial recharge
إعادة ضبط أوتوماتي	automatic reset
اعتراض	interception
اعتماد الخطط	approval of plans
إعداد الحافة	edge preparation
إعداد الموقع	site preparation
إعداد مستندات العطاء	preparation of tender documents
إعصار	cyclone
إعلان عن العطاءات	advertisement for bids
أعلى طبقات التربة	A-horizon
أعمال أساسات الركائز	pile foundation work
أعمال إصلاحية	remedial work
أعمال حفر الانشاءات	structure excavation
أعمال التدريب	training works
أعمال الدهان	paint work
أعمال الهندسة الكهربائية	electrical engineering work
الأعمال الترابية	earthwork
أعمال ترابية متوازنة	balanced earthworks
أعمال خشبية	timbering
أعمال صحية	sanitary works
أعمال صيانة السكة الحديدية	on-track maintenance work
أعمال فكرية	headworks
أعمال هندسية ميكانيكية	mechanical engineering work
أعمدة تثبيت	side tree
أعمدة دعم الغلاية	boiler support columns
أعمدة دعم خشبية	cribwork
أغصان مقطوعة	brushwood
أغطية الرصيف	pavement overlays
افتراض «برنولي»	Bernoulli's assumption
افتراض «ناڤيير»	Navier's hypothesis
إفريز	curb
إفريز السطح المائل	eaves
إفريز ذو كتيفات	bracketed cornice
إفريز مائل	coping
أفق	horizon
أفق اصطناعي	artificial horizon
الأفق الجغرافي	apparent horizon
الأفق المرئي	sensible horizon
إقامة	setting up
إقامة قالب نضوي	horsing-up
إقامة هيكل البناء	shell construction
إقتراح	proposal
أقواس مزدوجة	braces
أكال	corrosive
أكريليك	acrylics
أكسدة	oxidation
أكسونومتري	axonometric
أكسيد أسود	black oxide
أكسيد الألومنيوم	aluminium oxide
أكسيد المعادن	metal oxide
التئام ذاتي	autogenous healing
آلات مكنية	machine tools
آلة	machine
آلة الدك على السكة	on-track tamping machine
آلة إنهاء يدوي	hand finisher
آلة بدون مسننات	gearless machine
آلة تدار بمسننات	geared machine
آلة ترسيخ التربة أحادية المرور	single-pass soil stabilizer

آلة تسجيل النقد	cash register
آلة تسجيل بياني	stereoplotter
آلة تسوية	planoscope
آلة تفتيت التربة	soil shredder
آلة تنظيف الحصى عن السكة	on-track ballast cleaner
آلة حفر الآبار	well borer
آلة حفر مستدقة	spud
آلة رش الملح	salt spreader
آلة رصف إنزلاقية	slip-form paver
آلة شق القناة	channeller
آلة فرش الحصباء	gritter
آلة فرش المساحيق	powder spreader
آلة فرش المواد السائبة	bulk spreader
آلة قطع أسياخ الفولاذ	cropper
آلة قطع دوارة	roller cutter
آلة قياس أطوال الكترونية	distomat
آلة قياس الأطوال الالكترونية	geodimeter
آليات هيدرولية	hydraulic machinery
آلية مفصلية	toggle mechanism
آلية مؤازرة	servo mechanism
التصاق	adhesion
التصاق نوعي	specific adhesion
التفاف	slewing
التقاط ثابت	fixed pick
التواء	torsion
ألف	thousand
إلكترود (قطب)	electrode
إلكترود لحام عميق الصهر	deep-penetration electrodes
إلكترونيات صناعية	industrial electronics
إلكتروليت	electrolyte
ألماس أسود	black diamond
ألماس صناعي	industrial diamond
ألماسة	diamond
ألومينا	alumina
ألومنيوم	aluminium
ألياف زجاجية	glass fibre
ألومنيوم مجدول	stranded aluminium
ألياف زفتية	pitch fibre
أماكن سكنية	residential accommodation
أمال	tilt, cant
إمالة	tilting
أمان المفاعلات النووية	safety of nuclear reactors
الأمان بالموقع	safety on site
أمبير	ampere
امتداد مستقيم	straight stretch
امتصاص	absorption, suction
إمداد بالجاذبية	gravity supply
إمداد مُقاس	metered supply
إمدادات عامة	general provision
إمدادات الماء	water supply
إمدادات ماء الشرب	potable water supply
إمدادات مضخوخة	pumped supply
أمر التحديد المطلق	absolute addressing
أمر عنونة موسع	extended addressing
أمر غير مباشر لايجاد المعلومات	indirect addressing
إمكانية التشغيل	workability
إمكانية التصميم	design capacity
أمين المستودع	storekeeper
أنابيب	piping
أنابيب ثقوب الرفع	pitching ferrules
أنابيب معلقة	overhead pipes
أنابيب نقل	transfer pipes
إنارة	lighting
أنبوب	pipe, tube
أنبوب البئر	wellpoint
أنبوب التصريف	delivery pipe
أنبوب المرشة اليدوية	spray lance
أنبوب الوصل	connecting pipe
أنبوب ألومنيوم	aluminium pipe

أنبوب إمداد	service pipe
أنبوب بنصف جلبة	half-socket pipe
أنبوب تصريف	draft tube
أنبوب تصريف الخرسانة	concrete delivery pipe
أنبوب تنفيس	vent pipe
أنبوب دخول الزيت	oil inlet pipe
أنبوب رئيسي للمياه	header
أنبوب صرف مغلف بالملح	salt-glazed-ware pipe
أنبوب صلب	rigid pipe
أنبوب الضغط	pressure pipe
أنبوب عابر للطريق	sag pipe
أنبوب فخار	earthenware pipe
أنبوب فخار مزجج	glazed earthenware pipe
أنبوب ڤنتوري	Venturi tube
أنبوب فولاذي	steel pipe
أنبوب متفرع	branch pipe
أنبوب مدرع	armoured pipe
أنبوب مركب	compound pipe, composite pipe
أنبوب مرن	flexible pipe
أنبوب مغلف بالبلاستيك	PVC wrapped pipe
أنبوب مغلف بزجاج ليفي	fibreglass-wrapped pipe
أنبوب ملحوم تناكبياً	butt-welded tube
أنبوب من الأسبستوس	asbestos pipe
أنبوب من الشرائح الخشبية	wood-stave pipe
أنبوب من الفولاذ المموج	corrugated steel pipe
أنبوب من حديد الزهر	cast iron pipe
أنبوب منفوخ	ductube
أنبوب هيدرولي	hydraulic pipe
أنبوب مجرى مسيك	back-inlet gulley
الإنتاج الدائم	sustained yield
الإنتاج النوعي	specific yield
إنتاجية عالية	high productivity
إنتاجية منخفضة	low productivity
انتشار	diffusion

انتضاح حراري	thermo-osmosis
انتقال الحرارة	heat-transfer
انحدار	slope
الانحدار الأقصى	maximum cant
انحدار السكة	track gradient
الانحدار الحدي	limiting gradient
انحدار طبيعي	natural slope
انحدار طولي	longitudinal gradient
انحدار مستعرض	transverse gradient
انحراف	deflection, deviation
انحراف الدعائم	deflection of supports
انحراف الشعاع	beam deflection
انحراف المماس	tangential deviation
الانحراف الحقيقي	true bearing
الانحراف المعياري	standard deviation
انحناء	bending, curvature
انحناء بسيط	simple bending
انحناء معاكس	contraflexure
انخساف	subsidence
انخفاض مفاجئ	sudden drawdown
اندماج	flocculation
اندماج على نطاق محدود	small scale integration
انذار	alarm
انذار بالحريق	fire alarm
انذار بالهواء المضغوط لإرتفاع الماء	pneumatic high-water alarm
انزال على الشاطئ	beaching
انزلاق	slide, slip
انزلاق القص	shear slide
انزلاق بطيء	detritus slide
انزلاق مائي	aquaplaning
انسداد هوائي	air embolism
انسياب لزج	viscous flow
إنشاء	structure, construction
إنشاء الأنفاق	tunnel construction

إنشاء الركائز	pier construction
إنشاء السطح	roof construction
إنشاء السقف	ceiling construction
إنشاء السكك	track construction
إنشاء المحطة	station construction
إنشاء الهيكل	frame construction
إنشاء بئر التهوية	shaft construction
إنشاء بلاط مزدوج الاتجاه	two-way slab construction
إنشاء خرساني	concrete construction
إنشاء خرساني مقوى بالزجاج	glass-concrete construction
إنشاء خط بلاطات	slab band construction
إنشاء خلوي	cellular construction
إنشاء ذو بلاطة مسطحة	flat-slab construction
إنشاء ذو جائز وعارضة	beam-and-girder construction
إنشاء ذو سطح مصمت	solid-type dock construction
إنشاء ذو سطح مضلع	folded-plate construction
إنشاء ذو عقد وقناطر	arch and catenary construction
إنشاء ذو لوحة مستوية	flat-plate construction
إنشاء ذو هيكل	framed structure
إنشاء عائم	flotation structure
إنشاء علوي	superstructure
إنشاء فولاذي	steel construction
إنشاء قاعدي	substructure
إنشاء قنطري برجي	tower gantry
إنشاء لإرساء بحري	marine mooring structure
إنشاء مدعم بركائز	piling
إنشاء مركب	sandwich construction, composite construction
إنشاء مضلع	ribbed construction
إنشاء معلق	suspended structure
إنشاء مفتوح	open-type construction
إنشاء مقسم إلى مناطق	zoned construction
إنشاء مملوء بالماء	water-filled structure
إنشاء من البلاط المفرغ	hollow-slab construction
إنشاء من السطوح والأعمدة	mushroom construction
إنشاء منخفض التكاليف	low-cost construction
إنشاء منع الاجهاد	tension structure
إنشاء هيدرولي	hydraulic structure
إنشاء هيكل الرفع	hoistway construction
إنشاء هيكلي مفرغ	thin-shell construction
إنشاء المباني	building construction
إنشاءات الميناء	port structures
إنضاج الاسمنت بالترطيب	curing, maturing
إنضاج بالبخار	steam curing
إنضاج مُعجَّل	accelerated curing
إنضغاط	compression
انطلاق	blow out
أنظمة	regulations
أنظمة إمداد الكهرباء	electricity supply regulations
انعطاف	cant
إنفار	invar
إنفجار مفاجئ	sudden outburst
إنقاذ	salvage
إنقطاع لانخفاض الماء	low water fuel cut-off
إنقلاب	overturning
إنهاء	finishing
إنهاء جيد	making good
أنود	anode
أنيون	anion
اهتزاز	vibration
اهتزاز خارجي	external vibration
اهتزاز داخلي	internal vibration
اهتزاز عميق	deep vibration

اهتزاز ميكانيكي	mechanical vibration
اهدار	wastage
أوقف	park
أول أكسيد الكربون	carbon monoxide
أوم	ohm
ايثان	ethane
ايقاف العمل	decommissioning
إيلاج	insert
أيون	ion
باب	door
باب أوتوماتي	automatic door
باب بعوارض خشبية	batten door
باب بمصراع مفصلي	hinged leaf gate
باب جراج (كراج)	garage door
الباب الخارجي	outer door
باب دوار	swing door, roller door
باب رئيسي	main door
باب صامد للنار	fireproof door
باب للطوارئ	emergency door
باب مزدوج	double door
باب مفصلي	hinged door
باب منزلق	sliding door
بادئ	starter
بئر	well
بئر ارتوازية	artesian well
بئر الامتصاص	absorbing well
بئر التصريف	disposal well, drain well
بئر تنفيس	relief well
بئر تنفيس الهواء	air-relief shaft
بئر تهدئة	stilling well
بئر حبشي	Abyssinian well
بئر التهوية	ventilating shaft
بئر الحقن	injection well
بئر ضحل	shallow well
بئر عميق	deep well
بئر محفور	drilled shaft

بئر المصعد	lift well, shaft well
بئر معكوس	inverted well
بئر مفتوح القاع	open-bottom shaft
بئر الوصول	access shaft
بارامترات	parameters
بارايت	barytes
بارود أسود	black powder
بارومتر: مقياس الضغط الجوي	barometer
بارومتر معدني	aneroid barometer
بازالت	basalt
باع	span
الباع الأوسط	midspan
الباع الداخلي	interior span
باع صافٍ	clear span
باع طرفي	end span
باع طويل	long span
الباع الفعال	effective span
الباع المعلق	hung span, suspended span
باكليت	Bakelite (TM)
بالوعة مجاري	cesspit
بالوعة مجاري مسربة	leaching cesspool
بالوعة مياه الأمطار	stormwater drain
بانتوغراف	pantograph
برج الصمام	valve tower
برج الماء	water tower
بحث جيوفيزيائي	geophysical investigation
بحث جيولوجية المياه	hydrogeological investigation
بحوث المد والجزر	tide investigations
بحيرة	lake
بحيرة اصطناعية	artificial lake
بخار زئبقي مع هاليد	mercury vapour with halides

بدء التخصر	springing
برادة الفولاذ	steel filings
براغي ألواح	plate screws
براغي تثبيت السكة	track bolts
براغي زنق احتكاكي عالية المتانة	high-strength friction-grip bolts
برافين	paraffin
برامج كمبيوتر	software
بربخ : قناة ضبط جريان الماء	penstock
بربخ	culvert
بربخ بمجرى جانبي	side-channel spillway
بربخ سحب	inlet culvert
بربخ صندوقي	box culvert
برج	steeple, tower
برج أسلاك	pylon
برج الارساء	anchor tower
برج الحفر	derrick
برج السحب	intake tower
برج المرفاع	king tower, crane tower
برج بئر النفط	oil-well derrick
برج تبريد	cooling tower
برج تبريد بالسحب الطبيعي	natural-draught cooling tower
برج ثلاثي القوائم	three-legged derrick
برج حفر آبار الزيت	oil-well derrick
برج ضيق القاعدة	narrow-base tower
برج عريض القاعدة	broad-base tower
برج فولاذي	steel tower
برج قنطري	portal tower
برج متعدد الطوابق	multi-storey tower
برج مدعم بالحبال	guy derrick
برج مراقبة	control tower
برج مصعد حبلي مقوى	luffing cableway mast
برج منصوب	standing derrick
برج هيكلي مقوى بشكالات مائلة	diagonally braced frame tower

پرسپكس	Perspex (TM)
برشامة	rivet
برشامة ربط	tack rivet
برشمة	riveting
برغي	screw
برغي السطح	roof bolt
برغي الضبط الدقيق	fine-adjustment screw
برغي بشكل ذيل السمكة	fishtail bolt
برغي بيان الحمل	load-indicating bolt
برغي تثبيت	set screw
برغي تسوية	levelling screw
برغي سدادة	gland bolt
برغي قمط	clamping screw
برغي مخروط	turned bolt
برغي معايرة	adjusting screw
بركة	pool
بركة الأكسدة	oxidation pond
بركة تبريد	cooling pond
بركة تخزين	storage pond
بركة تخزين جانبية	side pond
بركة تهدئة	stilling pool
بركة سباحة	swimming pool
بركة طبيعية	pond
برمجة رمز الآلة	machine code programming
برمجة العمل	labour programming
برمجة عمل المعدات	plant programming
برمجة المواد	materials programming
برميل	drum
برنامج الأمان	safety programme
برنامج المنشأ	source program
برنامج تجميع الكمبيوتر	macro assembler
برنامج تسليم المواد	materials scheduling
برنامج ثنائي	binary program
برنامج رئيسي	main program
برنامج فرعي	sub-program

ب

برنامج كمبيوتر	bootstrapping
برنامج نتائج الكمبيوتر	object program
بروز	extrusion
بروسسور متعدد المسجلات	multi-register processor
بروسسور مركزي	central processor
البرونز	bronze
برونز المدافع	gunmetal
بزولان	pozzolana
بستان	orchard
بستنة التجميل الهندسي	landscape gardening
بضائع عامة	general cargo
بطارية	battery
بطاقة	teller
بطانة	lining
بطانات الأنابيب	pipe liners
بطانة السقف	lining of the ceiling
بطانة الألواح الجدارية	wallboard lining
بطانة النفق	tunnel lining
بطانة بأطواق فولاذية	tubbing
بطانة تقوية	packer
بطانة حلقية	ring liner
بطانة خرسانية مسبقة الصب	precast-concrete liner
بطانة داخلية	internal lining
بطانة دهان	primer
بطانة صامدة للحرارة	refractory linings
بطانة عازلة	insulating lining
بطانة فولاذية	steel liner
بُعد	dimension
بُعد أفقي مائل	oblique offset
البُعد العياري	standard gauge
بغير مقياس الرسم	not to scale
بكارة بأعلى هيكل الركيزة	cat-head sheave
بكارة بسلسلة	chain block
بكارة تفاضلية	differential pulley
بكارة رافعة	lifting block
بكتريا	bacteria
بكتريا عضوية الشكل	Escherichia coli
بكرات تعليق باب مخزن الحبوب	barn-door hanger
بكرة	pulley
بكرة إرجاع	return sheave
بكرة التثبيت	anchor block
بكرة لرفع الأثقال	block
بكرة متحركة	fall block
بكرة محززة	rope wheel, sheave
بكرة مقطوعة	snatch block
بلاتة : صمغ لزج جيد العزل	balata
بلاستيك صلب	unplasticised PVC
بلاستيك	PVC
بلاط الأرضيات	floor tiles
بلاط الڤينيل والأسبستوس	vinyl-asbestos tile
بلاط خزفي	ceramic tiles
بلاط رخامي	marbled tiles
بلاط زجاجي	glass tiles
بلاط سطح إسمنتي	cement roof tiles
بلاط سنامي	hip tiles
بلاط مزجج	glazed tiles
بلاطة	slab, tile
بلاطة الأرضية	floor slab
بلاطة الحجرة	room slab
بلاطة خرسانة الحشوة	filler joist floor
بلاطة خرسانية	concrete slab
بلاطة سطح صلصالية	clay roofing tile
بلاطة سميكة	thick slab
بلاطة فلين	cork tile
بلاطة مجوفة	voided slab
بلاطة مسبقة الصب	precast slab
بلاطة مسطحة	flat slab
پلمباجو : غرافيت	plumbago
بلوط	oak
بنّاء	bricklayer
بناء الطوب	brickwork

ب

بناء بالطوب مسلح	reinforced brickwork
بناء بالطوب مقاوم للحوامض	acid-resisting brickwork
بناء بحجارة مختلفة	coursed rubble
بناء ذو إطار صندوقي	box frame construction
بناء من الحجر المنحوت	ashlar masonry
بناء من الكتل	blockwork
بناء نظامي	system building
بناية سكنية	block of dwellings, housing block
بناية شقق	block of flats
بناية متعددة الطوابق	multi-storey building
بناية مكاتب	office building
بنتاريثريت نترانيترات	PETN
بنزين	petrol
بنية	fabric
بهو التفتيش	inspection gallery
بهو الدخول	entrance hall
بوابات الدخول	entrance gates
بوابات عبور	crossing gates
بوابات نصف قطرية	taintor gates
بوابات الهويس	sluice gates
بوابة	gate
بوابة إسطوانية	cylinder gate
بوابة الارساء	anchor gate
بوابة السد الرئيسية	head gate
بوابة المصعد	lift gate
بوابة الهويس	lock gate
بوابة بمحبس دعم	bear trap gate
بوابة تحكم ضخمة	caterpillar gate
بوابة تحكم عمودية	penning gate
بوابة تحكم نصف قطرية	radial gate sluice
بوابة ترتفع عمودياً	vertical lift gate
بوابة دائرة	rolling gate
بوابة دائرية	ring gate
بوابة دوارة	roller gate
بوابة سد ساقطة	falling weir sluice
بوابة قطاعية	segmental gate, sector gate
بوابة قلاّبة	tilting gate, bascule gate, flap gate
بوابة قناة	sluice
بوابة قناة التصريف	spillway gate, crest gate
بوابة قناة دائرية	drum gate
بوابة منزلقة	sliding gate
بوابة نصف قطرية	radial gate
بوابة هويس دوارة رأسية	vertical roller sluice gate
بوابة هويس مائلة	inclined lift gate
بورت	bort
بوصة	inch
بوصلة	compass
بوصلة الميل المغنطيسي	dip needle
بوصلة مغنطيسية	magnetic compass
بوصلة موشورية	prismatic compass
بوكسيت	bauxite
بولدر : منطقة منخفضة مستصلحة من البحر	polder
بولدوزر	bulldozer
بوليبروپلين	polypropylene
بوليترا فلوروثيلين	polytetrafluoroethylene
بوليثيلين	polyethylene
بوليثين	polythene
بوليستر	polyester
بوليسترين	polystyrene
بوليمر	polymer
بوليمرات هيدروكسيلية	hydroxylated polymers
بوليوريثان	polyurethane
بويب	wicket
بيئة داخلية	internal environment
بيانات المياه الجوفية	hydrological data
بيانات تتابعية رقمية	digital sequential data
بيانات نسبية	analogue data

ت

بيت الشباب	hostel
بيرومتر	pyrometer
بيسترو	BISTRO
تابع	follower
تأثير الارساء	berthing impact
تأثير البيئة	environmental impact
تأثير التحزيز	notch effect
تأثير كهرضوئي	photoelectric effect
تأثيرات الاشعاع	effects of radiation
تاج	crown
تاج العمود	capital
تاجر	tradesman
تأخر العمليات المتتابعة	process lag
تاريخ إعادة الشحن	date of recharging
تاريخ الاتمام	completion date
تاريخ التسليم	handover date
تأريض المعدات	earthing of equipment
تأكسد	oxidation
تآكل	corrosion
تآكل إلكتروليتي	electrolytic corrosion
تآكل تحت السطح	subsurface erosion
تاكومتر ذو قراءة مباشرة	Beaman stadia arc
تأمين المسؤولية العامة لحماية المالك	owner's protective public liability
تأمين تعويضات العمال	workmen's compensation insurance
تأمين ضد المسؤولية القانونية لحماية المقاول	contractor's protective liability insurance
تبادل	interchange
التبادل الأيوني	ion exchange
تباعد	spacing
تباين	variance
تبديد الطاقة	energy dissipation
تبريد اصطناعي	artificial cooling
تبريد باعادة الدوران	recirculating cooling
تبريد سريع	quenching
تبريد طبيعي	natural cooling
تبريد مباشر	direct cooling
تبطين	lining
تبطين البئر	steening
تبطين المداخن	flue lining
تبطين بالزفت	pitching
تبطين بالخرسانة	concrete lining
تبطين بالقار	bitumen lining
تبطين خيطي	string lining
التبليط بكتل مستطيلة	block pavement
تبليط مزدوج	lift-slab construction
تبييت التربين	turbine housing
تبييض	bleaching
تتبع	tracking
التثبيت الأولي	fist fastening
مثبت بالمسامير	pinned
تثبيت غير مباشر للسكة	indirect rail fastening
تثبيت مباشر ببرغي كبير	coach screw direct fastening
تثقيل	weighting
التثليث	triangulation
تجاذب موضعي	local attraction
تجاوب المعدات	plant response
تجاوز التحميل	overloading
تجاوز الحمل المسموح	permissible overload
تجاويف	waffle
تجديد الطرق	retreading roads
تجربة	experiment
تجزع	crazing
تجزيع كالرخام	marbling
تجصيص	plastering
تجصيص بالجبس	gypsum plaster
تجفيف	drainage
تجفيف الأرض	land drainage
تجفيف الحقل	field drainage
تجفيف بالهواء الساخن	hot flow drying

تحليخ	grinding
تجمد	freezing
تجميع	assembling
تجميع البيانات	collection of data
تجهيزات	fittings
تجهيزات تركيب	rigging
تجهيزات قمة الصاري	masthead gear
تجويف	cavitation, bore
تجيّر السطح المدهون	chalking
تحات	erosion
تحديب	buckle
تحديد البرنامج الفرعي	sub-program calling
تحديد ايجاد المعلومات المسجلة	index addressing
تحديد موقع المحطة	station siting
تحديد نقاط المنحنى	curve ranging
تحرك الرطوبة	moisture movement
تحركات الطائرة	aircraft movements
تحرير النص الأصلي	source text editing
تحرير مكتبة خرج الكمبيوتر	object library editing
تحكم	control
تحكم أفقي	horizontal control
تحكم أوتوماتي	automatic control
تحكم أوتوماتي في العمليات	automatic process control
تحكم أوتوماتي في القطارات	automatic train control
تحكم بالضغط	pressure control
تحكم تناسبي	proportional control
تحكم رأسي	vertical control
التحكم في الرقم الهيدروجيني	pH control
التحكم في الرهج	haze control
التحكم في الضوضاء	noise control
التحكم في العمليات المتعاقبة	process control
التحكم في المرسى	deadman control
التحكم في المروحة	fan control
التحكم في تأرجح ذراع المرفاع	jib swing control
تحكم ناءٍ	remote control
تحلية الماء بالبثق متعددة المراحل	multistage flash distillation
تحليل	analysis
تحليل بالمنخل	screen analysis
تحليل سائلي	wet analysis
تحليل غازي	gas analysis
تحليل قوى الاجهاد	stress analysis
تحليل كهربائي	electrolysis
تحليل للشبكة	network analysis
تحليل الأبعاد	dimensional analysis
تحليل آلي	mechanical analysis
تحليل انشائي	structural analysis
تحليل التربة	soil analysis
تحليل العطاء	tender analysis
تحليل المتطلبات	analysis of requirements
تحليل منافع التكاليف	cost-benefit analysis
تحليل النماذج	model analysis
تحليل الهبوط الاستقراري	settlement analysis
تحميل احتكاكي	frictional loading
تحميل الرياح	wind loading
تحميل الشاحنة	truck loading
تحميل حراري	thermal loading
تحميل دينامي	dynamic loading
تحميل لا تمركزي	eccentric loading
تحميل متحد المركز	concentrically loaded
تحميل متقدم	preloading
تحميل مرفاع محطة التربين	turbine-house crane loading
تحميل مرن	elastomeric bearing
تحميل مستعرض	transverse loading
تحميل مفرط	overburden
تحول النباتات إلى سماد عضوي	composting
تحويل	transfer
تحويل الأرقام إلى نسب	digital-to-analogue conversion
تحويل المعلومات	down-line loading

تحويل النسب إلى أرقام	analogue-to-digital conversion
تحويلات توصيل	connecting transitions
تحويلة	turnout
تحويلة مقصية	scissors crossover
تخريم	punching
تخريم مثل خلية النحل	honeycombing
تخزين	storage
تخزين البضائع	storage of cargo
تخزين البيانات	data storage
تخزين الوقود	fuel storage
تخزين سائب	bulk storage
تخزين سنوي	annual storage
تخزين مأمون	safe storage
تخزين مؤقت	temporary storage
تخصُّر	necking
تخصّر النافورة	vena contracta
تخضير	turfing
تخطيط	planning
تخطيط المدن	town planning
تخطيط المرفأ	harbour planning
تخطيط الميناء	port planning
تخطيط بياني	plotting
تخطيط متفرع	split-finger layout
تخفيض	reduction
تخفيف الضغط	decompression
التخلص من الحمأة	sludge disposal
التخلص من القمامة	refuse disposal
التخلص من الماء المهدر	wastewater disposal
التخلص من النفاية النشطة	active waste disposal
التخلص من مياه المجارير	sewage disposal
التخلف (المغنطيسي)	hysteresis
تخلف المد والجزر	tidal lag
تخلل	percolation
تخمين	estimation
تخمين الموقع	site appraisal
تداعي الأساس	foundation failure
تدرج الحرارة	temperature gradient
تدريب الموظفين	staff training, employee training
تدريب ملاحظ العمال	foreman training
تدريبي	instruction type
تدعيم الأساس	underpinning
تدعيم الأساس بركائز	pile underpinning
تدعيم النفق الأولي	forepoling
تدعيم جوانب المهواة بالأخشاب	shaft timbering
التدعيم الشبكي للخرسانة	mesh reinforcement
التدفئة بالهواء الساخن	hot-air heating
تدفئة غير مباشرة	indirect heating
تدفئة مركزية	central heating
تدفق الماء الباطني	base flow, interflow
تدفق شبه حرج	subcritical flow
تدفق فوق الحرج	supercritical flow
تدفق مطرد	steady flow
تدفق كتلي	plug flow
تدفق مياه المجاري	sewage flow
تدقيق	audit, check
تدقيق الخطط	checking of plans
تدميج عميق	deep compaction
تدوين	booking
ترابط	bond
ترابط بنائي	masonry bond
ترابط متعرج	zigzag bond
ترابط ميكانيكي	mechanical bond
تراخ	relaxation
تراصف رأسي	vertical alignment
التراص النسبي	relative compaction
تراكب	lap, superposition
ترانزستور ثنائي القطب	bipolar transistor
ترانزستور ضوئي	phototransistor
ترانزيت	transit

ت

ترانزيت الممر الآلي	Automated Guideway Transit
تراوح جانبي	sway
تربة	soil
تربة ثقيلة	heavy soil
تربة حاملة للكبريت	sulphate-bearing soils
تربة حبيبية	loose ground
تربة حمضية	podzol
تربة خشنة الحبيبات	coarse-grained soil
تربة دقيقة الحبيبات	fine-grained soil
تربة رملية طينية	loam
تربة سفلية	subsoil
تربة سيئة	bad soil
تربة طينية	mud soil
تربة عديمة التماسك	non-cohesive soil
تربة فوقية	topsoil
تربة قابلة للتمدد	expansive soil
تربة قابلة للانتفاخ	swelling soil
تربة مترسبة	stratified soil
تربة متماسكة	cohesive soil
تربة مستقرة	stabilized earth
التربة المضغوطة تحت الأساس	bulb of pressure
تربولين	tarpaulin
تربنتين	turpentine
تربين	turbine
تربين بخاري	steam turbine
تربين دفعي	impulse turbine
تربين دفعي للتصادم	Pelton wheel
تربين الدفق المختلط	mixed-flow turbine
تربين رد فعلي	reaction turbine
تربين غازي كهربائي	gas-turbine electric
تربين غازي هيدرولي	gas-turbine hydraulic
تربين مائي	water turbine
تربين مروحي	propeller turbine
تربين هيدرولي	hydraulic turbine
ترتيب الأقدمية	rder of precedence
تردد طبيعي	atural frequency
تردد لاسلكي	adio frequency
ترسب	recipitation, edimentation, settling
ترسب الحصى	etritus settlement
ترسب الدبال	ettlement of humus
ترسب الطمي	ilting
ترسب معاق	indered settling
ترسبات محمولة بالماء المتدفق	ydraulic fill
ترسيب ثانوي	econdary edimentation
ترسيب كيماوي	hemical precipitation
ترسيب مشترك	oprecipitation
ترشيح	iltration, percolation, nfiltration
ترشيح رملي سريع	apid sand filtration
ترشيح رملي متقطع	ntermittent sand filter
ترشيح فائق الدقة	ıltra-filtration
ترشيح مزدوج	louble filtration
ترشيح موجّه	controlled filtration
ترصيف	stack
تركيب	fitting, assembling, installation, set-up
تركيب أجهزة الحفر	rigging up
تركيبات كهربائية	electrical installation
تركيب البرنامج	program installation
تركيب البروسسور	processor construction
تركيب الزجاج	glazing
تركيب جائز في الجدار	needling
تركيب كيماوي	chemical composition
تركيب ميكانيكي	mechanical construction
تركيبات	fittings
تركيبات السكة بعوارض خرسانية	slab track installations
تركيبات فولاذية لمحطة الطاقة	power house steelwork

تركيبات فولاذية مغلفة بالخرسانة	concrete-encased steelwork
تركيبات للوقود	fuel installation
تركيز	concentrating, seating, set-up
تركيز الاجهاد	stress concentration
تركيز الحمأة	sludge concentration
تركيز مياه المجاري	sewage composition
ترميم	overhaul
ترويق	clarification
تزايد الضغط	pressure build-up
تزايد ذاتي	auto-incrementing
تزجيج ملحي	salt glaze
تزجيج نافذة بلاستيك	plastic window glazing
تسجيل	registration
تسجيل الصوت الموشوري	optical sound recording
تسجيل القياسات البئرية	well logging
تسجيل المقاومة النوعية	resistivity logging
تسجيل أوتوماتي للبيانات	automatic data logging
تسخين العازل	dielectric heating
تسخين الفولاذ حتى الزرقة	blueing
تسرب	leakage
تسرب للخارج	exfiltration
تسطيح مرن	flexible surfacing
تسعة	nine
تسعة عشر	nineteen
تسعة وعشرون	twenty-nine
تسعون	ninety
تسليح	reinforcement
التسليح بالأسياخ	bar reinforcement
تسليح بشبكة سلكية	wire-mesh reinforcement
تسليح بنية الخرسانة	fabric reinforcement
تسليح طولي	longitudinal reinforcement
تسليح مستعرض	transverse reinforcement

تسليح مضاد للتصدع	anti-crack reinforcement
تسليك بالقضبان	rodding
تسليم	delivery
التسمم بالرصاص	lead poisoning
تسهيلات إرساء	facilities for mooring
تسهيلات التخزين	bunkering facilities
تسهيلات الغسيل	washing facilities
تسهيلات المطعم	restaurant facilities
تسوية	align, levelling
تسوية أرض الطريق	shaping
تسوية التربة	resoiling
تسوية القطاع العرضي	cross-section levelling
تسوية بالتثليث الأولي	first order levelling
تسوية تفاضلية	differential levelling
تسوية جانبية	profile levelling
تسوية دقيقة	precise levelling
تسوية عادية	ordinary lay
تسوير الطرق	road fencing
تشابك	interlock
تشابك الركام	aggregate interlock
تشابه حركي	kinematic similarity
تشتيت	dispersion
تشبيع بالأوزون	ozonizing
تشرب	impregnation
تشغيل الميكروبروسسور	microprocessor operation
تشغيل آلي	automation
تشغيل أوتوماتي للقطارات	automatic train operation
تشكل إنزلاقي	slip-form
تشكيل بالحرارة والتطريق	forging
تشكيل بالطرق	cogging
تشكيل على الساخن	hot working
تشوّه	deformation
تشوّه مرن	elastic strain

التشييد بالحجارة المنحوتة	block-in-course
تصادم	impact
تصحيح الارتخاء	sag correction
تصحيح الانحدار	slope correction
تصحيح التراصف	realignment
تصحيح الشد	tension correction
تصحيح القياس للفارق الحراري	temperature correction
تصحيح المعايرة	standardization correction
تصحيح بمستوى سطح البحر	sea-level correction
تصحيحات الشريط	tape corrections
تصدع الخرسانة	cracking in concrete
تصدع مطاوع	plastic fracture
تصدع هيدرولي	hydrofracture
تصريح	statement
تصريف	discharge
تصريف (البخار)	blow off
تصريف بالجاذبية	gravitation drainage
تصريف سفلي	blow down
تصريف سيئ	bad drainage
تصفية	rectification
تصفيح	sheeting
تصفيح بالألومنيوم	Alclad (TM)
تصفيح مشكل بالبثق	integrally-stiffened plating
تصلد	hardening
تصلد الخرسانة النهائي	final set
التصلد بمرور الزمن	ageing
تصلد زمني	strain ageing
تصليح	repair
تصليد اجهادي	strain hardening
تصليد السطح	hard facing
تصليد الفولاذ بالحرارة	blank nitriding
تصليد ببخار ذو ضغط عالٍ	high-pressure steam-curing
تصليد المعادن	emper
التصليد المغلف	ase-hardening
تصميد النفق للماء	unnel waterproofing
التصميد للماء	vaterproofing
تصميد للماء بالفراغات الهوائية	avity tanking
تصميم	lesign
تصميم التوصيل	unction design
تصميم الحاجز الترابي	embankment design
تصميم الحالة الحدية	imit state design
تصميم الخلط	mix design
تصميم الطرقات	design lanes
تصميم العارضة	girder design
تصميم القوة الجانبية	lateral-force design
تصميم المرفأ	dock design
تصميم المرونة	elastic design
تصميم المطار	airport design
تصميم المفاهيم	concept design
تصميم إنشائي	structural design
تصميم تخطيطي	sketch design
تصميم حاجب الريح	windshield design
تصميم قنطري	arch design
تصميم مفصل	detailed design
تصميم مجرى الغازات	flue design
تصميم مسجَّل	registered design
تصميم نهائي	final design
تصميم وتركيب واختبار	terotechnology
تصنيع	fabrication
تصنيف	grading
تصنيف التربة	classification of soils
تصنيف تربة المطار	airfield soil classification
تصوير بالأشعة دون الحمراء	infra-red photography
تصوير مساحي	photogrammetry
تضاد	anticlastic
تضاريس التربة	soil profile
تضاريس رملية متنقلة	drifting sand terrain
تطهير	disinfection

ت

تطهير بالكباش	grab dredging
تطويل المدة	extension of time
التطوير النهائي	ultimate development
تعبئة	filling, packing
تعبئة الحفر	filling-up
تعبئة بالخرسانة	concrete infill
تعبئة ذاتية	self-fill
تعدٍ	encroachment
تعديل	amendment
تعديل العقد	contract revision
تعدين	mining
تعرجات الخطوط الحديدية	railway curves
تعريض للاشعاع	irradiation
تعزيز حلزوني	helical reinforcement
تعزيل الصخر	mucking
تعشيق اللسان في النقر	boxing
تعقيم بالبخار المحمي وبالضغط	autoclaving
تعليق بالحبل	rope suspension
تعليق بالسلاسل	chain suspension
تعليق سلسلي	catenary suspension
تعليم الحافة	edge marking
تعليم الحساب	arithmetic instruction
تعليم منطق الحساب البوليني	Boolean logic instruction
تعليمات ازاحية	shift instruction
تعليمات الحركة	move instruction
تعليمات جزئية	branch instruction
تعليمات جزئية مشروطة	conditional branch instruction
تعلية	overtopping
تعهد بتنفيذ العقد	contract bond
تعويض بصري	optical compensation
تعويض مفاعل	reactive compensation
تعويضات نهائية	liquidated damages
تعويم	flotation
تعويم هزاز	Vibroflot

تغذية	feed
تغذية بالهواء المضغوط	pneumatic feed
تغطية الركيزة	pile planking
تغطية الوحل	mudcapping
تغطية المنحدرين	ridge covering
تغطية بالألواح	planking
تغطية بالألواح الخشبية	close timbering
تغطية بالخرائط	cartographic cover
تغطية بالقار	bitumen sheathing
تغطيس الركائز دون دق	jetting
تغليف	cladding, caulking, sheathing, lagging
تغليف الأنابيب	pipe casing
تغليف الجدران	curtain walling
تغليف دائم	permanent shuttering
تغليف كهربائي	electroplating
تغير	variation
تغير الحمل	load change
التغير السنوي	annual variation
التغير المغنطيسي	magnetic variation
تغيير التسلسل	change order
تغيير سريع للسكك الحديدية	lumping
تغيير معدل الانحدار	change of gradient
تفاصيل الانشاءات	construction detail
تفاعل حراري ألوميني	alumino-thermic reaction
تفاعل نووي	nuclear reaction
تفاوت كبير	high tolerance
تفاوت متوسط	medium tolerance
تفاوت منخفض	low tolerance
تفتيت الأرض في المنجم	breaking ground
تفتيت التربة بالنار	fire setting
تفتيش دوري	regular inspection
التفتيش على المراجل	boiler inspection
تفجير	detonation, chambering
تفجير الصخور	rock burst

ت

تفجير عميق	deep blasting
تفدين المهواة بالمزواة	instrumental shaft plumbing
تفدين بئر التهوية	shaft plumbing
تفرع	turnout
تفريغ البيانات	dump
تفريغ السائل	run-off
تفريغ خاضع للتحكم	controlled tipping
تفكيك : تفكك	disassembly
تقارن	coupling
تقارن صلب	rigid coupling
تقاطع	intersection, resection, crossing
تقاطع الأولوية	priority junction
تقاطع الزاوية الحادة	acute angle crossing
تقاطع بشكل معين	diamond interchange
تقاطع ثلاثي القوائم	three-tripod traversing
تقاطع خطوط السكك الحديدية	railway junction
تقاطع خطوط حديدية	crossing frog
تقاطع طرق	crossroads
تقاطع طريق بسكة	level crossing
تقاطع معين الشكل	diamond crossing
تقاطع مقصي	scissors junction
تقاطع منفرج الزاوية	obtuse angle crossing
تقاطعات متصلة	connected frogs
تقدم	headway
التقدم بالعطاء	tendering
تقدير	rating, estimation
تقدير التكاليف	cost estimate
تقدير حركة المرور	estimation of traffic
تقرير	report
تقرير دوري	progress report
تقرير سابق للاستثمار	pre-investment report
تقرير فني	technical report
تقشر السطح	scabbing
تقصير	leaching
تقلص	hrinkage
تقلص بالمساحة	ontraction in area
تقلص طرفي .	nd contraction
تقنية الرافع	ift technology
تقوس	urvature, camber
تقوس سالب	egative camber
تقوض تدريجي	rogressive collapse
تقوض متعمد	eliberate collapse
تقوية التربة	oil stabilization, tabilization of soil
تقوية التربة السفلية	ubsoil stabilization
تقوية بالشكل	racing
تقوية عكسية بالشكالات	ounter bracing
تقوية قاعدة الأساس	einforcement of a oundation footing
تقوير	ounter bore
تقويم	ectification
التقويم الفلكي والبحري الأمريكي	American Ephemeris and Nautical Almanac
تقويم بحري	Nautical Almanac
تقييم التربة	oil evaluation
تقييم العطاءات	valuation of bids
تقييم المشروع	roject evaluation
تقييم الموقع	ite evaluation
تكاليف	osts
تكاليف الاشراف	upervision costs
تكاليف التشغيل	perating costs, running osts
التكاليف الرأسمالية للانشاءات	capital construction cost
تكاليف الصيانة	maintenance costs
تكاليف المعدات	plant cost
تكاليف المقاول الفرعي	subcontractor costs
تكاليف أولية منخفضة	low initial cost
تكاليف صيانة منخفضة	low maintenance cost
تكامل واسع النطاق	large scale integration

تكبد علوي	upper transit
تكبير	enlargement
تكتل	sintering, coagulation
تكثيف : تكاثف	condensation
تكديس مأمون	safe stacking
تكرار	repetition
تكسية من الاسمنت	revetment
تكسير	cracking
تكلفة العمل	labour cost
تكلفة المواد	materials cost
التكنولوجيا	technology
تكوين جيولوجي منخفض النفاذية	aquiclude
تكوين صخري منخفض الانفاذية	aquitard
تلبيس	facing
تلبيس بالرخام	marble facing
تلبيس من الحجر المنحوت	ashlar facing
تلدين	annealing
تلسكوب لمسح الأبعاد	anallatic telescope
تلسكوب موشوري	prismatic telescope
تلف الخرسانة	concrete deterioration
تلف عرضي	accidental damage
تلفزيون	television
تلميع	polishing
تلوث	pollution
تلوث الماء	water pollution
التلوثات العالقة بالهواء	airborne contaminants
تماسك	consistency, tenacity
تمايل جانبي	sidesway
التمايل المسموح	lurching allowance
تمرين على طرق الحفر	practice drilling
تمهيد	skimming
تموج	ripple
تموُّر	surge
تمويل	financing
تميؤ	hydration
تميع	liquefaction
التناضح	osmosis
تناضح عكسي	reverse osmosis
التناضح الكهربائي	electro-osmosis
تناظر	analogy
تناقص ذاتي	auto-decrementing
تنبؤ	forecast
تنبؤ السيولة النقدية	cash flow forecast
ت . ن . ت .	TNT
تنشيط	stimulation
تنظيف	cleaning
تنظيف الأخدود	gully cleansing
تنظيف الشوارع	street cleansing
تنظيف المزراب	gutter cleansing
تنظيف بالبخار	steam purging
تنظيف بالسفع الرملي	sand blasting
تنظيف جيد للموقع	good site housekeeping
تنظيم الأمان	safety organisation
تنظيم التدريب	instruction organisation
تنظيم الدفق	flow regulation
تنظيم المسار الحرج	critical path scheduling
تنفيذ	execution
تنفيذ العقد	execution of contract
تنفيذ المشروع	project implementation
تنقر	pitting
التنقل	travelator
تنقيب	excavation
التنقيب بالطريقة الزلزالية	seismic prospecting
تنقية	clarification
تنقية الماء	water purification
تنقيح	revision
تنقيح البرنامج	program editing
تنمية هيئة الادارة	staff development
تنويت	octal
تنوير	bloom

ت

تهذيب	trimming
تهشير الرسم الهندسي	hatching
تهوية	ventilation
تهوية ابتدائية	primary ventilation
تهوية المجارير	ventilation of sewers
تهوية النفق	subway ventilation
تهوية بالكباس	piston-type ventilation
تهوية ثانوية	secondary ventilation
تهوية طبيعية	natural ventilation
تهوية قسرية	forced ventilation
تواتر التفتيش	frequency of inspection
تواتر العواصف	storm frequency
توازن	balance, equilibrium
توازن الحت والترسب	regime
توازن ساكن	static equilibrium
توتر	strain, tension
توتر التحديب	buckling strain
توتر المماس	tangential tension
توتر حلقي	ring tension
توتر ذاتي	self-tensioning
توتر سطحي	surface tension
التوحيد الاحصائي	statistical uniformity
توحيد المقاييس	standardization
توربين «كابلان»	Kaplan turbine
توربين «فرانسز»	Francis turbine
توزيع	distribution
توزيع اعتراضي	baffle-type distribution
توزيع التردد	frequency distribution
توزيع الحمل	distribution of loading
توزيع الضغط	pressure distribution
توزيع العزم	moment distribution
توزيع حجم الجسيمات	particle-size distribution
توزيع نسبي	proportioning
توسيع تجويف القاعدة	under-reaming
توشيج	interlocking
توصيل	linking
توصيل تتابعي تزامني	synchronous serial interface
توصيل متتابع لا متزامن	asynchronous serial interface
توصيلات الموقد	burner interlocks
توصيلات بالمسامير الملولبة	bolt connections
توصيلات فولاذية	steel connections
توصيلات مبرشمة	riveted connections
توصيلات ملحومة	welded connections
توصيلات منزلية	house connections
توقف ناءٍ	remote parking
توقيت	timing
توقيف العمل	suspension of work
توهين النبضات فوق السمعية	ultrasonic pulse attenuation
تيار التكدر	turbidity current
تيار متدفق	influent stream
ثابت تماماً	fully-fixed
ثابت جمعي	additive constant
ثالث نتريت التولوين	trinitrotoluene
ثانية	second
ثخانة الرصف	pavement thickness
ثرموستات	thermostat
ثرميت	thermit
ثقابة صخور	rock drill
ثقابة عمودية	pillar drill
ثقب	bore, punching, drill
ثقب تصريف ارتشاحي	weephole
ثقب الحفر	borehole
ثقب تزييت	nipple
ثقب خلوصي	clearance hole
ثقب لحشوة النسف	lifter hole
ثقب مشكل بقالب	cored hole
ثقل الفادن البصري	optical plummet
ثقوب تصريف	relief holes
ثقوب على جانب النفق	rib holes

وب قطعية	sumpers
وب دقيقة	pores
لاثة	three
لاثة أمبيرات	three-amp
لاثة عشر	thirteen
لاثة وعشرون	twenty-three
لاثون	thirty
ث الوسط	middle third
م مثلثي	V-notch
لمة	notch
انية	eight
انية عشر	eighteen
انية وعشرون	twenty-eight
انون	eighty
اني العنصر	binary
اني القطب	BIPOLAR
ائز قصيف	slender beam
ائز مشدود	tensioned beam
ائز من الفولاذ المدلفن	r.s.j.
لجاذبية	gravity
لجاذبية الشعرية	capillarity
جاروف	shovel
جاروف أمامي	face shovel
جامع عينات حلزوني	riffle sampler
جامعة معترضة	cross-assembler
جانب دعامة الجسر	cutwater
جبس	gypsum
جدار	wall
جدار احتجاز	counterfort wall, revetment
جدار احتجاز ثقالي	gravity retaining wall
جدار احتجاز مدعم القاعدة	tied retaining wall
جدار ارتكاز	abutment wall
جدار الارساء	anchor wall
جدار أساسي	basement wall
جدار الحاجز الترابي	embankment wall

جدار الطابق العلوي	top-storey wall
جدار القنطرة	spandrel wall
جدار أو سد مسيك	stank
جدار بحري	seawall
جدار بركيزة مستعرضة	sheet-pile wall
جدار بسمك نصف طوبة	half-brick wall
جدار تحميل بالثقل	gravity-quay wall
جدار ثقالي	gravity wall
جدار جانبي	wing wall
جدار حاجز	parapet wall
جدار حاجز أو فاصل	partition wall
جدار حاجز في المدخنة	wythe
جدار حامل	loadbearing wall
جدار خارجي	external wall
جدار خرسانة عادية	plain concrete wall
جدار خرساني	concrete wall
جدار داخلي	residential wall
جدار داعم	bearing wall
جدار داعم لصد التراب	chemise
جدار دعم	head wall, toe wall
جدار رصيف التحميل البحري	quay wall
جدار رصيف المرفأ	wharf wall
جدار رقي	diaphragm wall
جدار ساتر	bagwork
جدار ساند	camp sheathing
جدار ساند للرصيف	shelf retaining wall
جدار ساند من الخرسانة	concrete retaining wall
جدار طرف الخندق	facing wall
جدار على ركائز مصبوبة في الموقع	contiguous bored pile wall
جدار غير حامل	non-load-bearing wall
جدار غير داعم	non-bearing wall
جدار فجوي	cavity wall
جدار كابولي	cantilever wall
جدار مائل	pitched work
جدار مانع لتسرب الماء	core wall, cut-off wall

ج

ج

جدار مجوف	hollow wall
جدار محتجز	surcharged wall
جدار مخرّم	honeycomb wall
جدار مدعم	buttressed wall
جدار مدعم الخلفية	tied-back wall
جدار مدعم صلب	stiffened wall
جدار مرسى	breasting dolphin
جدار مرن	flexible wall
جدار مساند	retaining wall
جدار مستعرض	shear wall
جدار مسيك	staunching piece
جدار مشترك	common wall, party wall
جدار مكسو	faced wall
جدار ملبس	veneered wall
جدار نهري	training wall
جداول انحراف خطوط العرض	traverse tables
جدران احتجاز ثابتة	fixed retaining walls
جدل	splice
جدول	table, creek, brook, stream
جدول الأتعاب المهنية	scale of professional charges
جدول الانحناء	bending schedule
جدول جاف	dry brook
جدول فرعي للنفايات	effluent stream
جدول مواعيد الاختبار	commissioning schedule
الجدوى الاقتصادية	economic viability
جديلة	strand
جديلة « لانج »	Lang lay
جذاذة	chippings
جذر	root
جذع	stem, stalk
جذع سحب مضلع	kelly
جذع نقل الحركة	transmission shaft
جر	drag, pull, tug

الجر بالقوة الكهربائية	lectric traction
جراثيم	acteria
جراثيم عضوية	Bacillus coli
جراج	arage
جرار	ractor
جرار دفع	usher tractor
جرار قطع الشجر	ree cutter
جرار على عجلات	vheeled tractor
جرار مد السكك	rack-laying tractor
جرار نصف مزنجر	alf-track tractor
جرافة	dredger
جرافة الثلج	now plough
جرافة بقواديس دوارة	ucket-ladder dredger
جرافة ثلج دوارة	otary snow plough
جرافة لتسوية الطرق	ngledozer
جرة (اسطوانة) غاز	LPG cylinder
جرف ساحلي	ittoral drift
جريان إنسيابي	streamline flow
جزء	member
جزء السفينة الظاهرة من الماء	freeboard
الجزء القاعدي الصامد للماء	waterproofing basement
الجزء الناتئ من الدرجة	nose of a step
جزر اصطناعية	artificial islands
جزيرة المشاة	street refuge
جسر	bridge
جسر الرافعة المتنقلة	gantry
جسر السكك الحديدية	railway bridge
جسر بجائز صندوقي	box girder bridge
جسر بحري	causeway
جسر بسطح مستعمد	orthotropic-deck bridge
جسر بعوارض لوحية	plate-girder bridge
جسر بعوارض مركبة	composite-girder bridge
جسر بلاطي	slab bridge
جسر « بيلي »	Bailey bridge
جسر ترابي	earth bank
جسر ثابت	fixed bridge

جسر جانبي الحركة	traversing bridge
جسر جملوني (محدب)	truss bridge
جسر خرسانة	concrete bridge
جسر خرسانة سابقة الاجهاد	prestressed concrete bridge
جسر دوار	turn bridge, swing bridge, roller bridge, running bridge
جسر ذو ركيزة دلفينية	rolled-beam bridge
جسر ذو عتبة بشكل T	T-beam bridge
جسر رفع دوار	rolling lift bridge
جسر سطحي	deck bridge
جسر طريق عام	highway bridge
جسر عائم	pontoon bridge
جسر عبور	through bridge
جسر على ركائز	pile bridge
جسر فولاذي	steel bridge
جسر قلاّب	bascule bridge
جسر قلاب متوازن	balance bridge
جسر قنطري	arch bridge
جسر مائل	skew bridge
جسر متحرك	drawbridge, movable bridge
جسر متعدد القناطر	viaduct
جسر محوري	pivot bridge
جسر مدعم بالحبال	Hooghly bridge, cable-stayed bridge
جسر معلق	suspension bridge
جسر معلق صلب	stiffened suspension bridge
جسر منصي	trestle bridge
جسر نقل	transporter bridge
جسم	body
جسوءة	stiffness
جص الباريوم	barium plaster
جلاخة ذات دواليب	wheel grinder

جلبة حشو	gland
جلفطة	caulking
جليغنايت	gelignite
جمارك	customs
جمع القمامة	refuse collection
جمعية البناء	building association
جمعية وضع الخرائط	mapping institute
جُملون (تحدب)	gable, truss
جُملون الجسر	bridge truss
جُملون السطح	roof truss
جُملون «برات»	Pratt truss
جُملون بلجيكي	Belgian truss
جُملون خشب	wood truss
جُملون طرفي	end gable
جُملون مثلثي	half-lattice girder
جُملون «هاو»	Howe truss
جناح	pavilion
جناح (مطار)	terminal
جناح المبنى	wing of a building
جناح المدراء	directors suite
جنادل	rapids
جهاز	instrument
جهاز ارسال ذو تردد عالٍ جداً	VHF transmitter
جهاز استقبال ذو تردد عالٍ جداً	VHF receiver
جهاز الاحساس الصوتي	sound sensor
جهاز الاحساس بالحرارة	temperature sensor
جهاز الضخ	pumping appliance
جهاز إنذار	warning system
جهاز تحسس الموضع	position sensor
جهاز تحكم ذاتي التشغيل	self-operated controller
جهاز تحليل أول أكسيد الكربون	carbon monoxide analyser
جهاز تسارع	accelerator
جهاز تقوية	booster
جهاز تنضيد الحروف	word processor
جهاز تنفس	breathing apparatus

ج

جهاز تنفس اصطناعي	dust respirators
جهاز تنفس تحت الماء	SCUBA
جهاز حفر	earth borer
جهاز نقطة البيع	point-of-sale terminal
جهاز نقل الحركة	transmission gear box
جهد القص	shear strain
جودة الركوب	riding quality
جودة السكك الحديدية	rail quality
جوز	walnut
جوف	recess
جول	joule
جير	lime
الجير الحي	quicklime
جيلاتين	gelatine
جيوب الرفع	jacking pockets
جيوب طينية	clay pockets
جيولوجيا	geology
الجيولوجيا التطبيقية	engineering geology
جيولوجية الماء	hydrogeology
حائل أمواج	jetty
حائل موجي من الحجارة الضخمة	rubble-mound breakwater
حات	corrosive
حاجب الريح	windshield
حاجب واق	guard screen
حاجز	partition
حاجز ارتطام	crash barrier
حاجز اصطناعي	barrage
حاجز الأمواج	storm pavement
حاجز القمامة الطافية	trash rack
حاجز المنصة	platform grating
حاجز أمان	safety barrier
حاجز أمن	security fence
حاجز أمواج	breakwater
حاجز أمواج عائم	floating breakwater
حاجز أمواج مركب	composite breakwater

حاجز انشائي	ulkhead
حاجز باطار	ramed partition
حاجز بخاري	apour barrier
حاجز ترابي	mbankment
حاجز شبكي.	rating
حاجز عائم	oating boom
حاجز كابح الشرر	park arresting screen
حاجز ماء بلاستيكي	VC water stop
حاجز واق	igger
حادث	ccident
حادث مميت	atal accident
حاسبة بالقياس	nalogue computer
حاسبة هجينية	ybrid computer
حاضن الشَّد	ension carriage
حاضن الفانوس	antern carriage
حافة الرصيف	erb, curb
حافة الطريق	erge
حافة رصيف مرفوعة	aised kerb
حافة ناتئة	erm
حافلة ركاب	ommuter coach, assenger coach
حافلة صيانة السكك الحديدية	ailway service coach
حافلة النقل السريع	apid-transit coach
حاكم بالهواء المضغوط	neumatic controller
الحالة الحدية للمبنى	imit state
حامل	addle
حامل السلسلة	hainman
حامل السلم	arriage of a stair
حامل المزراب	utter bearer
حامل ثلاثي القوائم	ripod
حامل دوار	arousel
حامل قنطري لبرج الحفر	errick tower gantry
حاملات	arriers
حانية أنابيب	ipe bender
حانية فولاذ	teel bender
حانية قضبان	im crow

ح

حاوية	container
حبل	rope
حبل الارساء	mooring line
حبل الجر	traction rope, haulage rope
حبل الدَرَجْ	stair string
حبل الرفع	hoisting rope, sling
حبل المسبار	sounding line
حبل إنقاذ	lifeline
حبل بحلقات متحدة المركز	locked-coil rope
حبل تسوية	chalk line
حبل ذو جدائل عادية	cable-laid rope
حبل رفع سلسلي	chain sling
حبل سبر الغور	lead line
حبل سلكي	wire rope
حبل قطر	tail rope
حبل للارساء	mooring wire rope
حبل مسبق التشكيل	pre-formed rope
حبل مشدود	tensioned rope
حجاب مجرى	Hallinger shield
حجاب من أسلاك إسفينية	wedge-wire screen
حجاب واق	shield
حجاب وجه بلاستيك	plastic face shield
حجارة رصف الطرق	road metal
حجب	blinding
حجر	stone
حجر إزاحة	plumb
حجر أساس صلد مضغوط	compacted hardcore
حجر البرج	derrick stone
حجر الخفاف	pumice stone
حجر الربط	bonder
حجر الرصف	cobbles, paving stones
حجر جلمودي	boulders
حجر جيري	limestone
حجر جيري سليكوني	silicaceous limestone
حجر جيري كربوني	carboniferous limestone
حجر حافة	border stone
حجر رصف	sett
حجر رصف لوحي	paving flag
حجر رصف غرانيتي	pitcher
حجر رملي	sandstone
حجر زاوية مفرغ	hollow quoin
حجر طبيعي منحوت	cut stones
حجر مسوّى	dressed stone
حجر مكسر	broken stone
حجر ملبس باليد	hand-dressed stone
حجر منحوت للبناء	ashlar
حجر واسطة العقد	keystone
حجرة	chamber
حجرة التفتيش	inspection chamber
حجرة الضغط	pressure chamber
حجرة الطعام	refectory
حجم الجسيمات	particle size
حد التحميل	endurance limit
حد التقلص	shrinkage limit
حد التناسب	limit of proportionality, proportional limit
حد الحفر	excavation limit
حد السرعة الخطية	line speed limit
حد السيولة	liquid limit
حد اللدانة	plastic limit
حد المرونة	elastic limit
حدادة	smithing
حدود البناء	building line
حدود التماسك	Atterberg limits
حدود الحفر المأجور	neat lines
حدود المُباعدة	spacing limits
حدود المستجمع	catchment boundary
حديد	iron
حديد زاوي	angle iron
حديد الزهر	cast iron
حديد زهر طروق	ductile cast iron

حديد زهر مصلَّدْ	chilled cast iron
حديد لوحي	plate iron
الحديد المطاوع	wrought iron
حديد مطروق	malleable cast iron
حديد مطروق مرن	ductile iron
حديد مغزول	spun iron
حديد مغلفن	galvanized iron
حديقة	garden
حديقة السطح	roof garden
حذاء	shoe
حذاء أمان	safety shoes
حراري	thermal
حرف	ridge
حرفة البناء	masonry
حرفي	craftsman, tradesman
حرق	burn
حرق القمامة بالموقع	on-site incineration
حرق مباشر	direct incineration
حرقة	burn
الحركة الارتدادية للرمل	saltation
حركة الماء في التربة السفلية	underflow
حركة المرور	traffic
حز	notch
حزام أمان	safety belt
حزمة	pack, pencil
حزمة قضبان خشبية	bat faggot
حساس الضوء	light sensor
حساسية	sensitivity
حشو	filling
حشو بالخرسانة العادية	mass concrete fill
حشو مفصلي	joint filler
حشوة	filler, packing, pack
حشوة خرسانية	concrete fill
حشوة معدنية لمنع التسرب	flashing
حشية	gasket, mattress
حصباء	shingle
حصباء خشنة	coarse aggregate
حصباء رملية	hoggin
حصباء مسبقة التغليف	pre-coated chippings
حصباء مُقيرة	tarmacadam
حصة	dividend
حصول	procurement
حصى	gravel
حصى رخامي	macadam
حصى الرصف	marble gravel, ballast
حصى مدهون بالقار	coated chippings
حصيرة	mat
حصيرة دعم	pad foundation
حصيرة معدنية خوائية	vacuum mat
حظائر النقل	transit sheds
حظيرة	shed, yard
حظيرة طائرات	hangar
حفار	excavator
حفار بدولاب ذو قواديس	bucket-wheel excavator
حفار كاشط	scraper excavator
حفارة آبار	well sinker
حفارة الحفر العميقة	down-the-hole drill
حفارة أنفاق	mole, tunnel mole
حفارة بالهواء المضغوط	pneumatic excavator
حفارة بعجلات	wheel excavator
حفارة بقواديس دوارة	bucket-ladder excavator
حفارة تربينية	turbo-drill
حفارة جارفة	mole plough
حفارة خنادق	trencher, trenching machine, trench excavator, ditcher
حفارة دوّارة	rotary excavator
حفارة صخور ضخمة	sinker drill
حفارة كبلية	cable drill, cableway excavator
حفارة متحركة	walking dragline
حفارة متعددة القواديس	multi-bucket excavator

ح

ح

حفارة مركبة على شاحنة	power earth auger
حفارة مزنجرة	crawler mounted excavator
حفارة هيدرولية	hydraulic excavator
حفر	excavation, boring
حفر اتجاهي	directional drilling
حفر اجترافي	wash boring
حفر أساس الجسر	bridge excavation
حفر استكشافي	exploratory boring
حفر إضافي	back cutting
حفر الأساسات	footing excavation
حفر الأنفاق بالتدريع	shield tunnelling
حفر الانفاق بالمكنة	machine tunnelling
حفر أنفاق الأنابيب	moling
حفر البئر بالهواء المضغوط	pneumatic shaft sinking
حفر التربة	earth excavation
حفر التربة الفوقية	topsoil excavation
حفر الطين	muck excavation
حفر القناة	channel excavation
حفر المهواة نزولاً	shaft sinking
حفر النفق بالكامل	full-face tunnelling
الحفر بالجرف	dredging
حفر بالدق	percussion drill
حفر بالكراءة	dredge excavation
حفر بالثقوب المتقاربة	broach channelling
الحفر بضغط الهواء	air-flush drilling
حفر بلقم ماسية	diamond drilling
حفر بمثقب مجوف	shell-and-auger boring
حفر ثقوب التفجير	cut holes
حفر حراري	thermic boring
حفر خُدي	slot excavation
الحفر خلف الدعم	poling back
حفر دوار بالدق	percussive-rotary drilling
حفر دوراني عكوس	reverse-rotary drilling
حفر رحوي	rotary drilling, rotary boring
الحفر الزائد	overbreak
حفر ضغطي	pressure drilling
حفر قلب دوار	rotary core drilling
حفر قناة الصرف	drainage excavation
حفر كبلي	rope drilling
حفر كبلي بالدق	churn drill
حفر مائل	sidetracking
حفر مرطب	wet drilling
حفر مقدم للركائز	pre-boring for piles
حفر مكشوف	open cut
حفر مواد الردم	borrow excavation
حفر نفثي	jet drilling
حفر هيدرولي	hydraulic excavation
حفر يدوي	hand boring
حفرة اختبارية	trial pit, test pit
حفرة الارتشاح	seepage pit
حفرة أقذار المجاري	cesspool
حفرة البئر	well hole
حفرة التشرب	soakaway
حفرة امتصاص	absorption pit
حفرة الامداد	borrow pit
حفرة تجميع	catch pit
حفرة تفتيش المجاري	manhole
حفرة تفتيش بمدخل جانبي	side-entrance manhole
حفرة تفتيش ضحلة	shallow manhole
الحفرة السفلية	bottom cut
حفرة الطفح	overflow pit
حفرة مجاري سدودة للماء	watertight cesspool
حفرة المجاري	sump
حقن	injection
حقن الاسمنت	injection of cement
حقن بالاسمنت	grouting
حقن كيماوي	chemical grouting
حقن المحلول	solution injection
حقن المستحلب	emulsion injection
حقن مستحلب القار	Shell-perm process (TM)

ح

حك	scrub
حك : إحتكاك	rubbing
حلزوني	volute, spiral
حلقة	ring, thimble, link, grommet
حلقة اختبارية	proving ring
حلقة الماء	water ring
حلقة تعليق المرفاع	crane slinger
حلقة دعم	supporting ring
حلقة ركيزة	pile ring
حلقة فولاذية	steel ring
حلقة متحركة	traveller
حلقة نمو	growth ring
حلية معمارية مموجة	ogee
حمأة	sludge
حمالة	loader, cradle
حمالة ثلاثية الأرجل	three-leg sling
حمالة ثنائية القائم	two-leg sling
حمالة رباعية الأرجل	four-leg sling
حمالة شبه صلبة	semi-rigid cradle
حمالة قابلة للطي	collapsible cradle
حمالة كاشطة	scraper loader
حمالة متحركة بالسير	mobile belt loader
حمالة مفردة	single sling
حماية الشواطئ	shore protection
حماية الضفة	bank protection
حماية ضد حت الرياح	protection against wind erosion
حمل	load, burden
حمل إضافي	surcharge
الحمل الأقصى	maximum load
حمل التحديب	buckling load
حمل الثلج	snow load
حمل الدولاب	wheel load
حمل دون الذروة	off-peak load
حمل الرياح	wind load
حمل زائد	overload

حمل جانبي	lateral load
حمل جسر السكك الحديدية	railway bridge load
الحمل الحرج	crippling load, Euler crippling stress
حمل حركي	dynamic load
حمل الدرب	lane loading
حمل زلزالي	seismic load
حمل ساكن	dead load, static load
حمل سكيني الحد	knife-edge loading
حمل شاقولي	vertical load
حمل صدمي	impact load
الحمل الطولي الأفقي	horizontal longitudinal load
حمل العمليات المتعاقبة	process load
حمل القص	shear loading
حمل الكسر	breaking load
حمل لا تمركزي	eccentric load
حمل اللي	torsional load
حمل مأمون	safe load
حمل متحرك	live load, moving load
الحمل المتحرك للمركبة	vehicular live load
الحمل المتراكب	superimposed load
حمل متنقل	rolling load
حمل المحور الساكن	static axle load
حمل محوري	axial load
الحمل المحول	sue load
حمل مركز	point load
حمل مركَّز	concentrated load
حمل مسبق الدعم	pre-consolidated load
حمل مستعرض أفقي	horizontal transverse load
الحمل المسموح	admissible load
الحمل المصمم	design load
الحمل المضاف	superload
حمل منتظم التوزيع	uniform load
حمل المواد المترسبة	bed load

حمل موزع	distributed load
حمم بركانية خلوية	cellular lava
الحمولة الإجمالية بالطن	gross tonnage
حمولة القطار بالطن	train tonnage
الحمولة بالطن	freight tonnage
حمولة جسر الطريق العام	highway bridge load
حمولة سائبة	bulk cargo
حناية الأنابيب	pipe bedding
حنفية	tap
حنفية مطافئ	fire hydrant
حواجز	hurdle work
حواجز أمواج بجدار رأسي	vertical-wall breakwaters
حوز الهويس	lock bay
حوز أمامي	head bay
حوض اغتسال	lavatory basin
حوض الترسيب	settling basin
حوض الجنوح	tidal dock
حوض الصرف	drainage basin, watershed
حوض الضخ	pump sump
حوض بعيد عن الشاطئ	off-shore dock
حوض تجميع	catch basin
حوض تغطيس	plunge pool
حوض تنظيف	scouring sluice
حوض تهدئة	stilling basin
حوض جاف لتنظيف السفن	graving dock
حوض جاف للسفن	dry dock
حوض جمع الماء	garland drain
حوض دوار	turning basin
حوض عائم	floating dock
حوض لرسو السفن	marina
حوض ماء الشرب	drinking trough
حوض مائي	wet dock
حوض محطة الضخ	wet well
حيز	clearance
خارج الطور	out of phase

خ

خارطة للمياه	hydrograph
خاصية حرارية	thermal property
الخاصية الصوتية	acoustical property
الخاصية الكهربائية	electrical property
خافض التوتر السطحي	surface-tension depressant, surfactant
خبث	slag
خبث الفحم أو المعادن	clinker
خبث الفرن العالي	blast-furnace slag
خبث مصانع الفولاذ	steelworks slag
خبث مفتت	crushed slag
خبير	expert
خبير زراعي	agriculturalist
خدة رباعية	quadripods
خدمات الانقاذ	rescue service
خدمات تصميم اختصاصية	specialized design services
خدمات تطوير اختصاصية	specialized development services
خدمات الانقاذ	crash services
خدمات محطة الشحن	shipping-terminal services
خرج المضخم	amplifier output
الخرج الاسمي	nominal output
خرج «لوُمن»	lumen output
خردق الرصاص	lead shot
خرسانة	concrete
خرسانة أسفلتية	asphaltic concrete
خرسانة إسمنت	cement concrete
خرسانة إسمنت بورتلاند	Portland-cement concrete
خرسانة إسمنت تمددية	expanding-cement concrete
خرسانة الخشب	wood concrete
خرسانة الصب في الموقع	cast-in-place concrete
خرسانة الكوك	breeze concrete

خ

خرسانة إنشاءات	structural concrete
خرسانة إنشائية خاصة	special structural concrete
خرسانة ثقيلة الوزن	heavy-weight concrete
خرسانة جاهزة الخلط	ready-mix concrete
خرسانة جيرية	lime concrete
خرسانة حرارية	refractory concrete
خرسانة حقن	colloidal concrete
خرسانة خبث مهواة	foamed-slag concrete
خرسانة خفيفة الوزن	lightweight concrete
خرسانة خلوية	cellular concrete
خرسانة خلوية غازية	gas concrete
خرسانة خوائية	vacuum concrete
خرسانة دبشية	rubble concrete
خرسانة ردم جافة	dry-packed concrete
خرسانة رقيقة جافة	dry lean concrete
خرسانة ركام إنشاءات خفيفة	structural lightweight-aggregate concrete
خرسانة سائبة	bulk concrete
خرسانة سابقة الاجهاد	prestressed concrete
خرسانة سريعة التصلب	rapid-hardening concrete
خرسانة سيكلوبية	cyclopean concrete
خرسانة صامدة للحرارة	heat-resistant concrete
خرسانة صامدة للماء	waterproofed concrete
خرسانة صامدة للنار	fire-resistant concrete
خرسانة صوف الخشب	wood-wool concrete
خرسانة ضعيفة	lean concrete
خرسانة عادية	mass concrete, plain concrete
خرسانة عازلة للحرارة	heat-insulating concrete
خرسانة غرانوليتية	granolithic concrete
خرسانة غير مسلحة	unreinforced concrete
خرسانة كتيمة	impervious concrete

خرسانة كثيفة	dense concrete
خرسانة مخلوطة بالمصنع	plant-mixed concrete
خرسانة مدمجة بالاهتزاز	vibrated concrete
خرسانة مرشوشة	sprayed concrete
خرسانة مسبقة الصب	precast concrete
خرسانة مستوفرة	rich concrete
خرسانة مسحوبة الهواء	air-entrained concrete
خرسانة مسلحة	reinforced concrete
خرسانة مسلحة بالألياف	fibre-reinforced concrete
خرسانة مسيكة	watertight concrete
خرسانة مصلدة	hardened concrete
خرسانة مضخوخة	pumped concrete
خرسانة مغزولة	spun concrete
خرسانة مقواة بالألياف الزجاجية	glassfibre-reinforced concrete
خرسانة مقاومة للكيماويات	chemical-resistant concrete
خرسانة مقاومة للماء	water-resistant concrete
خرسانة مهواة	aerated concrete
خرطوم الاطفاء	fire hose
خرطوم التصريف	delivery hose
خرطوم تصريف الخرسانة	concrete delivery hose
خرطوم مرن	flexible hose
خروج عن الخط	derailment
خريطة	map
خريطة جيولوجية	solid map, geological map
خريطة طبيعية	layered map
خريطة طوبوغرافية	topographic map
خريطة مجسمة	stereometric map
خط حديد مزدوج السكة	double-track railway
خزان	reservoir, storage reservoir, tank
خزان استقطار	skimming tank
خزان الترسيب	sedimentation tank

خ

خزان الجرعات الكيماوية	dosing tank
خزان الدبال	humus tank
خزان الزيت	oil storage tank
خزان إمداد	service reservoir
خزان المص	suction tank
خزان الوقود	fuel tank
خزان تجميع	collecting tank
خزان تخفيف الضغط	break-pressure tank
خزان تخمير الحمأة	Imhoff tank
خزان ترسيب	grit chamber
خزان تعفين	septic tank
خزان تغذية ضغطي	pressure feed tank
خزان تهوية	aeration tank
خزان تهوية مضغوطة	contact aerator
خزان سفلي	lower reservoir
خزان ضغطي	pressure tank
خزان علوي	upper reservoir
خزان قادوسي الشكل	hopper tank
خزان قطاعي	section tank
خزان ماء	water tank
خزان ماء ضخم	impounding reservoir
خزان ماء نقي	clear-water reservoir
خزان مسبق الاجهاد	pre-load tank
خزان موازنة	balancing reservoir
خزان هضم الحمأة	sludge-digestion tank
خزان هيدرولي	hydraulic reservoir
خزانات	lockers
خزانات مياه الأمطار	stormwater tanks
خشب	timber, wood
خشب أبيض	spruce
خشب أحمر	redwood
خشب البتولا	birch
خشب البلزا	balsa
خشب الزان	beech
خشب الساج	teak
خشب الماهوغوني	mahogany
خشب رخو	sapwood
خشب رقائقي	plywood
خشب رقائقي مغرّى	glued-laminated timber
خشب صلد	hardwood
خشب لين	softwood
خشب مجفف	seasoned wood
خشب نخر	rotten wood
خشبان (تمليط)	brandering
خشبان استواء	skids
خشن	rough
خشونة	roughness
خصائص المقطع	section properties
خصائص المقطع العرضي	cross-section features
خصائص توزيع الحمل	load spreading properties
خصائص دليلية	index properties
خضب	pigment
خضخاضة	jig
خط أفقي	horizontal line
خط أنابيب الحفر	dredge pipeline
خط أنابيب عائم	floating pipeline
خط التحارر	isotherm
خط التحرز	guard railing
خط الحمل	load line
خط الدفع	line of thrust
خط الزوال	meridian
خط العرض	latitude
خط القاعدة	base line
خط اللا انحراف	agonic line
خط المركز للجسر	bridge centre line
خط المقاومة الدنيا	line of least resistance
خط الموازنة	line of balance
خط النقل	transmission line
خط بياني	graph
خط تدفق الوقود	fuel flow line

خط ترام عكوس	jig back
خط تثليث قطري	tie line
خط جهد عالٍ	high-voltage line
خط خلوص الحواجز	obstruction-clearance line
خط رأسي	vertical line
خط رئيسي	main
خط رئيسي مضخوخ	pumped main
خط عادي	plain line
خط فرعي	branch line
خط مرتكز جدار الدعم	tip grade
خط مستقيم	through line
خط مستو	level line
خط مناسيب	contour line
خط مياه رئيسي	hydraulic main
خطأ	error
خطأ عفوي	accidental error
خطأ كبير	gross error
خطأ محتمل	probable error
خطأ معياري	standard error
خطأ متكافئ	compensating error
خطاف	grab
خطاف التقوية	reinforcement hook
خطة	plan
خطة إجمالية	outline plan
خطة الركائز	pile plan
خطة أعمال السكة	trackwork plan
خطة التنمية	development plan
خطة رئيسية	master plan
خطر الانهيار الأرضي	landslide hazard
خطر التحديب الجانبي	lateral buckling hazard
خطر مهني	occupational hazard
خطوط التدفق	flow lines
خطوط معلّقة	overhead lines
خطوط الميل المغنطيسي	isoclinic lines
خطوط النقل السريع	rapid-transit lines
خطوط أيسوكروماتية	isochromatic lines
خطوط ركاب الضواحي	commuter lines
خطوط متساوية الجهد	equipotential lines
خفض منسوب الماء	drawdown
خلاط : خلاطة	mixer
خلاط سحق	pulverizing mixer
خلاطة التربة	soil mixer
خلاطة خرسانة	batch mixer, concrete mixer
خلاطة غير قلابة	non-tilting mixer
خلاطة قلابة	tilting mixer
خلاطة متحركة	travel-mixer
خلاطة متواصلة العمل	continuous mixer
خلاطة ملاط	pan mixer
خلاطة نقل	transit mixer
خلاف كاذب	dummy argument
خلط	mix
خلط دعم التربة	plant mix
خلط في الموقع	mix-in-place
خلط مصمم	designed mix
خلل	defects
خلوص الباع الأوسط	midspan clearance
خلوص فضائي	air-space clearance
خلوص منطقة الاقتراب	approach-zone clearance
خلية ذات موصلية ضوئية	photoconductive cell
خلية ضوئية	photocell
خلية فلطائية ضوئية	photovoltaic cell
خلية كهرضوئية	photoelectric cell
خليط	admixture
الخليط الأسمي	nominal mix
خليط زائد الرطوبة	wet mix
خمسة	five
خمسة عشر	fifteen
خمسة وعشرون	twenty-five
خمسون	fifty

خ

خندق	trench, ditch
خندق الصرف	drain trench
خندق حائل للحريق	fire-break
خندق حقلي	field ditch
خندق حمأة	slurry trench
خنادق غير مبطنة	unlined ditches
خنفساء الخشب	death watch beetle
خوازيق تعشيق	interlocking piles
خواص طبيعية	physical properties
خوذة الركيزة	pile helmet
خور	creek
خيط الشاقول	plumb line
دائر	revolving
دائرة	circle
دائرة : دارة كهربائية	circuit
دائرة التفاف صغيرة	mini-roundabout
دائرة الرياح	wind rose
دائرة ترانزستور مقاوم منطقية	resistor transistor logic
دائرة ترانزستورية	transistor circuit
دائرة تغطية الترددات الصوتية	audio-frequency overlay circuit
دائرة ثرميونية	thermionic circuit
دائرة دليلية	pilot circuit
دائرة رأسية	vertical circle
دائرة المطافئ	fire department
دائرة مرور	roundabout
دائرة منطقية	diode transistor logic
دائرة منطقية مقرونة الباعث	emitter coupled logic
دائرة «موهر»	Mohr's circle
دائرة «موهر» للاجهاد	Mohr's circle of stress
دائري	round
دار البلدية	town hall
دارات متكاملة	integrated circuits
داخل المقر	in-house
دايود ضوئي	photo diode
دبال	humus
دبش	rubble
دبش خشن	rough rubble
دحاريج التمدد	expansion rollers
دخل	input
درابزين	balustrade, handrail, railing
درابزين الدرج	stair handrail
درابزين حاجز	barrier railings
دراسات سابقة للاستثمار	pre-investment studies
دراسة الاستيراد	import study
دراسة البيئة	environmental study
دراسة التضاريس	field survey
دراسة الجدوى	feasibility study
دراسة السوق	market study
دراسة المياه الجوفية	hydrological study
دراسة الوقت والحركة	time and motion study
دراسة تأثير الصدم	impact study
دراسة حركة العامل	motion study
درب	lane
درّج	calibrate
درج السلم	flight of stairs
درج ببئر مفتوح	open-well stair
درج بحافة مزخرفة	bracketed stairs
درج حلزوني	helical stairs, winding stairs
درج مفتوح	open stairway
درجات معلقة	hanging steps
درجة	step, degree
درجة سُلم	stair
درجة الانحدار الأقصى	ruling gradient
درجة التشبع	degree of saturation
درجة الميل	gradient
درجة انحدار السطح	roof pitch
درجة حرارة التشغيل	operating temperature
درجة حرارة المعايرة	standardization temperature

درجة حرارة الهواء	air temperature
درع	armour
درع واق من الاشعاع	biological shield
دعائم خشبية	timber supports
دعائم زافرة	flying shores
دعائم مائلة	raking shore
دعامة	stay, post
دعامة (كتف)	buttress
دعامة (عمود)	strut, pillar
دعامة أفقية	byatt
دعامة تثبيت	kneeler
دعامة حلقية	ring support
دعامة خرسانية	concrete pier
دعامة خشبية أفقية	soldier beam
دعامة خشبية عمودية	soldier
دعامة خلفية	back prop
دعامة ذراع المرفاع	jib support
دعامة رئيسية	main support, king post
دعامة قصيرة	jack rib
دعم	support
دعم : دعائم	shoring
دعم الأساس	underpin
دعم الانحدار	slope staking
دعم جانبي	lateral support
دعم مؤقت	temporary support
دفاعة مروحية	impeller
دفع	payment, thrust, branding
دفع بالتقسيط	payment by instalments
دفع طرفي	end thrust
دفع نفثي	jet propulsion
دفق خزان التهوية	aeration-tank effluent
دفق داخلي	internal flow
دفق دوامي	turbulent flow
دفق طباقي	stratified flow
دفق متعدد المراحل	multi-phase flow
دق	percussion, pound
دق الأوتاد ومحاذاتها	boning
دق الخوازيق	spiling
دق الخوازيق هيدرولياً	hydraulic pile driving
دق الركائز	pile driving
دقائق الخام	fines
دقيق	precise
دقة	accuracy, precision
دك	tamping
دكة ترابية	fill
دكة حجارة	rip-rap
دلفنة على البارد	cold rolling
دلفنة على الساخن	hot rolling
دلفين	dolphin
دلفين باطارات مطاطية	rubber-tyred rollers
دلفين باطارات هوائية	pneumatic-tyred roller
دلفين بعجلات هوائية	multi-wheel roller
دلفين ترادفي ثلاثي المحاور	three-axle tandem roller
دلو	bucket
دلو دوار	rotary bucket
دلو محاري	clamshell
دليل	directory
دليل إرشادات التشغيل	operation manual
دليل إرشادات الصيانة	service manual, maintenance manual
دليل التدفق	flow index
دليل الثخانة	screed
دليل التماسك	consistency index
دليل السيولة	index of liquidity
دليل الشاشة	cursor
دليل الصلابة	toughness index
دليل الضوضاء والرقم	Noise and Number Index
دليل اللدانة	index of plasticity, plasticity index
دليل المستوى	telltale

دليل إمرار الحبل	fairlead
دليل بوابة	gate guide
دمج بالتفجير العميق	explosive compaction
دموج	compaction
دموج سطحي	superficial compaction
دهّان	painter
دهان الجدران	wall paint
دهان صامد للحرارة	heat-resistant paint
دهان طمس	obliterating paint
دهان قاري	bituminous paint
دهان لدن	plastic paint
دهان لماع	gloss paint
دهان مائي	water paint
دهان مبيد للفطريات	fungicidal paint
دهليز	corridor
دوائر منطقية	logic circuits
دواة المصباح	lamp holder
دوار	revolving
دوارة الريح	weathercocking
دوامة	vortex
دوران بالجاذبية	gravity circulation
دورة الآلة	machine cycle
دورة التدريب	instruction cycle
دورة المياه	water closet
دورة المياه الجوفية	hydrological cycle
دولاب احتياطي	spare wheel
دولاب إدارة	bull wheel
دولاب انزال بالمنجم	wheel drop pit
دولاب التشغيل اليدوي	operating hand wheel
دولاب صقل	polishing wheel
دولاب محزز	indented wheel
دولوميت	dolomite
ديزل كهربائي	diesel-electric
ديناميت	dynamite
ذاكرة	memory
الذاكرة الافتراضية	virtual memory

ذاكرة الدخول العشوائي	random access memory
ذاكرة رئيسية	main memory
ذاكرة قراءة فقط	read only memory
ذاكرة مرقمة	paged memory
ذاكرة مساندة	backing store
ذراع	arm
ذراع امتداد	outrigger
ذراع امتداد هيدرولي	hydraulic outrigger
ذراع التركيب	erector arm
ذراع الرافعة	lever arm
ذراع اللحام	manipulator
ذراع المرفاع	jib
ذراع تحميل	loading boom
ذراع تطويل	boom
ذراع توجيه	control lever
ذراع توصيل	connecting rod
ذراع حر السقوط	free-falling arm
ذراع دافع	pusher arm
ذراع رفع الاسمنت	placing boom
ذراع كابولي	cantilever arm
ذراع مرفاع الخرسانة	concreting boom
ذراع مرفاع شبكي	lattice jib
ذراع (وحدة قياس)	cubit
ذروة	vertex
راتينج اصطناعي	synthetic resins
راتينج الايبوكسي	epoxy resin
رادار	radar
راديو	radio
رأس	head
رأس التسوية السريعة	quick-levelling head
رأس الجسر	bridge cap
رأس الركيزة	pile head
رأس العمود	architrave, column head
رأس المطرقة	beetle head
رأس حفر	drive head
رأس ركيزة الجسر	bridge pier cap

ر

رأس عمود بوقي	flared column head
رأس مخروطي	pan head
رأس مدق الخوازيق	monkey
راسب	sediment
راشح	filtrate
راصفة الخرسانة	concrete paver
رافد	tributary
رافدة	joist, baulk, rib, balk
رافدة أفقية	purlin, templet
رافدة خرسانية مزدوجة الكتلة	two-block concrete sleeper
رافدة خرسانية نافذة	through concrete sleeper
رافدة خشبية	timber joist
رافدة فولاذ مدلفن	rolled-steel joist
رافدة قصيرة	jack rafter
رافدة مائلة	rafter
رافدة مستعرضة	traverse
رافدة معترضة	crossing tie
رافدة نصفية	half-joist
رافدة وركية	hip rafter
رافعة	lever
رافعة بالقواديس	bucket elevator, paternoster
رافعة هويسية	canal lift
راقدة	sleeper
راقدة خشبية	timber sleeper
راقدة فولاذية	steel sleeper
راقدة فولاذية تشابكية	open-web steel joist
رئيس الرسامين	chief draughtsman
رئيس فريق المساحة	survey team leader
رئيس قسم	section leader
رباط	binder, stirrup
رباط هيدروكربوني	hydrocarbon binder
رباعي الأضلاع	quadrilateral
ربح	profit
ربط	lacing, tie
ربطة خفيفة	tie in
ربط مباشر للقضبان	direct rail fastening
ربعي رصف	quadrant
رجوعية	resilience
رحض	flushing
رحض المجاري	sewer flushing
رحوية	capstans
رحى صقل	buffing wheel
رخام : مرمر	marble
رد فعل	reaction
ردم بين حفرتين	balk
ردم حبيبي	granular filling
ردم رملي	sand fill
ردهة	foyer, lounge
رزم كثيف	high density baling
رسام	draughtsman
رسام الهندسة المدنية	civil engineering draughtsman
رسام إنشاءات	structural draughtsman
رسام خرائط	cartographer
رسام رئيسي	leading draughtsman
رسام مثمن	estimating draughtsman
رسام مساحة	surveyor's draughtsman
رسم	toll
رسم إجمالي	outline drawing
رسم الماء للمزرعة	net duty
رسم بالحجم الكامل	full-size drawing
رسم بمقياس نسبي	scale drawing
رسم بياني لعزم الثني	bending-moment diagram
رسم بياني للقص	shear diagram
رسم تخطيطي للمراحل الوظيفية	functional block diagram
رسم تشغيلي	working drawing
رسم تفصيلي	detail drawing

ر

رسم تناسبي	proportional drawing
رسم خرائط	mapping
رسم خرائط بالطائرات	airplane mapping
رسم عبور الهويس	lockage
رسم مائي وحدي	unit hydrograph
رسم مساحي باللوحة المستوية	plane-tabling
رسم مسقط علوي	plan view drawing
رسو	landing
رسوم التأجير	hire charges
رسوم العقد	contract drawings
رش الحصباء على القار	blinding
رش الملاط بالهواء المضغوط	pneumatically-applied mortar
رش حار	hot spraying
رصاص	lead
رصة الجدل	lay
رصد خلفي	back observation
الرصف	paving
رصف بالبلاط الحجري	stone-block paving
رصف بالقار الغليظ	dense tar surfacing
رصف بقوالب خشبية	wood-block paving
رصف بحجارة صوانية	peg top paving
رصف حديدي	iron paving
رصف قاري	bituminous surfacing, tar paving
رصف محوري	radial-sett paving
رصف ناعم	sheet pavement
رصيف	pavement, platform
رصيف الطريق	road pavement
رصيف المحطة	station platform
رصيف المرفأ	wharf
رصيف الميناء	quay
رصيف إنزال	slipway
رصيف إنزال مستعرض	traversing slipway
رصيف بحري على ركائز	pile pier
رصيف بحري مفتوح	open wharf
رصيف حاجز	bulkhead wharf
رصيف حامل	loadbearing pavement
رصيف خرساني	concrete pavement
رصيف صلب	solid wharf
رصيف ضيق	banquette
رصيف قاري	bituminous pavement
رصيف مركب	composite pavement
رصيف مرن	flexible pavement
رصيف معبرة السيارات	car ferry apron
رصيف مكشوف	open jetty
رصيف من قوالب الخرسانة	rigid pavement
رطم	bump
رطوبة	humidity
رطوبة الهواء	humidity of air
الرطوبة المطلقة	absolute humidity
الرطوبة النسبية	relative humidity
الرطوبة النسبية في الهواء	hygroscopic moisture
رفادة	shim
رفض	refusal
رفع	uplift
رفع المص الساكن	static suction lift
رفع الوحل	mud jacking
رفع بجهد	heave
رفع ثقيل	heavy lift
رفع خوائي	vacuum lifting
رق تعويض	compensating diaphragm
رقعة	tally
رقعة ماء	expanse of water
رقم الاسناد	reference number
رقم التسلسل	serial number
رقم تصنيف الحمل	Load Classification Number
رقم «رينولدز»	Reynolds number
ركائز احتجاز	sheet piling
ركائز أساسات قصيرة	short bored piles

ر

ركائز حماية الجسر	starling
ركائز رملية	sand piles
ركائز مستعرضة	sheet piles
ركائز ملولبة	Braithwate piles
ركام	aggregate
ركام ثقيل	heavy aggregate
ركام خاص	special aggregates
ركام خفيف	Aglite (TM)
ركام خفيف الوزن	bloated clay, lightweight aggregate
ركام دقيق	fine aggregate
ركام متوسط الحجم	half-sized aggregate
ركام مصنف	graded aggregate
ركام موحد الحجم	single-sized aggregate
ركمُ الرياح	windrow
ركن	corner
ركيزة	corbel, pier, pile
ركيزة أحادية العمود	single shaft pier
ركيز ارساء	anchor pile
ركيزة إزاحة الخرسانة	concrete displacement pile
ركيزة إزاحية	displacement pile
ركيزة الجدار	pier of a wall
ركيزة بزاوية قائمة	L-shaped pier
ركيزة بشكل T	T-head pier, T-shaped pier
ركيزة تحميل	bearing pile
ركيزة تحميل طرفي	end-bearing pile
ركيزة جسر	bridge pier
ركيزة حاجز الموج	breakwater pier
ركيزة حاجز خشب	wood-fender pile
ركيزة حجر	massive masonry pier
ركيزة خرسانية	concrete pile
ركيزة خرسانية جدارية	wall concrete pier
ركيزة خرسانية مسبقة الصب	precast-concrete pile
ركيزة خرسانية مصبوبة بالموقع	driven cast-in-place pile
ركيزة خرسانية مغلفة	cased pile
ركيزة خشبية	timber pile
ركيزة دعم	stay pile, supporting pier
ركيزة دلفين	rolled beam
ركيزة دليلية	guide pile
ركيزة رئيسية	king pile
ركيزة شد	tension pile
ركيزة صد	fender pile
ركيزة صندوقية	box pile
ركيزة عامة	universal beam
ركيزة عميقة	deep beam
ركيزة فولاذية	steel pile
ركيزة فولاذية بشكل H	steel H-pile
ركيزة فولاذية مستعرضة	steel sheet piling
ريكزة قائمة	standing pier
ركيزة قيسون	caisson pile
ركيزة مائلة	batter pile, raking pile
ركيزة متوازية الحافتين	parallel-flanged beam
ركيزة مدقوقة	driven pile
ركيزة مدقوقة بالصدم	impact driven pile
ركيزة مرفوعة	jacked pile
ركيزة مركبة	composite pile
ركيزة مصبوبة في الموقع	bored pile
ركيزة ملولبة	screw pile
ركيزة من الطوب	brick pier
ركيزة منصبية	trestle pier
ركيزة يدعمها الاحتكاك	friction pile
رمث	raft
رمز	symbol
رمز أمر التشغيل	OP CODE
رمل	sand
رمل السبك الطفالي	moulding sand
رمل سريع الانهيار	quicksand
رمل متضام	compacted sand
رمل مصنف	graded sand

ر

رمل منتظم الحبيبات	uniform sand
رمي فضلات المطبخ	kitchen waste disposal
رنان	buzzer
رنين	resonance
رواسب رملية وحصباوية	bar
رواسب مائية	waterbar
روافد السقف	ceiling joists
رواق	gallery
رود : مقياس للأراضي	rood
الريّ	irrigation
ري الطبقة تحت السطحية	subsurface irrigation
ريّ بالتقطير	drip irrigation
ري بالرحض	flush irrigation
ريّ بالغمر	flood irrigation
ريّ بشبكة مرشات	sprinkler irrigation
ريّ تحتي	sub-irrigation
الريّ الرحيب	broad irrigation
ريّ سطحي	surface irrigation
ريّ فلاحي أخدودي	furrow irrigation
الرياح السائدة	prevailing winds
ريح عاصفة	gale
ريحي	aeolian
زاوية	corner
الزاوية الاتجاهية التسامتية	grid bearing
زاوية الاحتكاك	angle of friction
زاوية الاحتكاك الداخلي	angle of internal friction
زاوية الاستقرار	angle of repose
زاوية الانحدار	angle of incline
زاوية التقاطع	intersection angle
زاوية ربعية	quadrantal angle
زاوية الرصيف الصخري	shelf angle
زاوية الميل	angle of inclination
زاوية تقوية	stiffener
زاوية غير متساوية الضلعين	unequal angle
زاوية مقاومة القص	angle of shearing resistance
زاوية نصف قطرية	radian

ز

زبون	client
زجاج	glass
زجاج أمان	safety glass
زجاج رقائقي	laminated glass
زجاج شفاف	clear glass
زجاج شفاف للنوافذ	clear window glass
زجاج عائم	float glass
زجاج غير شفاف	opaque glass
زجاج لا يخرقه الرصاص	bullet-proof glass
زجاج لوحي	plate glass
زجاج ماص للحرارة	heat-absorbing glass
زجاج مزدوج محكم بالمصنع	factory-sealed double glazing
زجاج معتم	obscure glass
زجاج معتم مقوي بالأسلاك	obscure wired glass
زجاج مقسى	tempered glass, toughened glass
زجاج مقوى بالأسلاك	wired glass
زجاج ملون	coloured glass
زحف	creep
زخرفة	decoration
زردية	pliers
زرع الألغام	mining
زفت	bitumen
زقاق	alley
زلزال	earthquake
زمرة	gang
زمن الاتمام	time of completion
زمن التشغيل	running time
زمن التصلب للاسمنت	setting time of cement
زمن التصلد النهائي	final setting time
زنجير	caterpillars
زند خشب	log
زنك	zinc
زنك لحام	brazing spelter
زورق المسح	survey launch

زورق عمل	workboat
زورق مسطح	pontoon
زيت الديزل	diesel oil, diesel, derv
زيت الغاز	gas oil
زيت بزر الكتان المغلي	boiled oil
زيت تسخين	heating oil
زيحان	offset
زيوليت	zeolite
سائق	driver
سائل التبريد	refrigerant
سائل الحفر	drilling fluid
سائل هيدرولي	hydraulic fluid
سارية	mast
ساحة الاستلام	receiving yard
ساحة الخدمة	service yard
ساحة المستودع	store yard
ساحة المغادرة	departure yard
ساحة تصنيف	classification yard
ساحة تفريغ	hump yard
ساحة صب الكتل	block yard
ساحة لعب	playground
ساعات عمل	man hours
ساعات عمل المعدات	plant hours
ساق	stalk
سبائك خفيفة	light alloys
سباك	plumber
سبر	probing, sounding
سبعة	seven
سبعة عشر	seventeen
سبعة وعشرون	twenty-seven
سبعون	seventy
سبك	casting, cast
سبورة	blackboard
سبيكة	alloy
سبيكة ألومنيوم	aluminium alloy
سبيكة ألومنيوم مُشكلة	wrought aluminium alloy
سبيكة للتشكيل	ingot
ستار	screen
ستارة	blind
ستة	six
ستة عشر	sixteen
ستة وعشرون	twenty-six
ستون	sixty
سجادة مرنة	flexible carpet
سجل أغراض عامة	general-purpose register
سجل التفجير	blasting record
سجل الحفر	borehole log
سجل المسّاح	level book, chain book
سجل قطر الحفر	calliper log
سجل قياسات المساح	field book
سجلات الصيانة	maintenance records
سحب	intake
سحب الأسلاك	wire drawing
سحب بالهواء المضغوط	pneumatic suction
سحب طبيعي	natural draught
سحب على البارد	cold drawing
سحق	pulverization, pound
سد	dam
سد أجوف	hollow dam
سد الطفح	overflow dam
سد الفائض	overflow weir
سد المد والجزر	whole-tide cofferdam
سد أنقاض	debris dam
سد إنضاب بجدار مزدوج	double-wall cofferdam
سد إنضاب خلوي	cellular cofferdam
سد بألواح إنزلاقية	sliding-panel weir
سد ببوابات تفتح رأسياً	draw-door weir
سد ببوابة هويس	sluice dam
سد ببويبات	wicket dam

سد بقنطرة رقيقة	thin-arch dam
سد بقنطرة سميكة	thick-arch dam
سد بوابة	gate seal
سد بهيكل معلق	suspended-frame weir
سد تحويل ذو ألواح تبطين	liner-plate cofferdam
سد ترابي	earthen dam
سد ثقالي	gravity dam
سد ثقالي خرساني	concrete gravity dam
سد جانبي مثلثي	triangular profile weir
سد حاد الذروة	sharp-crested weir
سد خالي من الصخور المغتربة	free-nappe profile weir
سد خرسانة بدعامات متوازية	round-headed buttress dam
سد خرسانة متعدد القناطر	multiple-arch concrete dam
سد خرساني	concrete dam
سد دعم ترابي	earthen embankment
سد ذو حشوة صخرية	rock-fill dam
سد ذو ذروة	crump weir
سد شبه منحرف	trapezoidal profile weir
سد صغير	weir
سد صغير للقياس	measuring weir
سد صغير مؤقت	gabion
سد صغير محزز	notched weir
سد صندوقي	box dam
سد طيني	clay dam, clay-puddle cofferdam
سد عريض القمة	broad-crested weir
سد غاطس	submerged weir
سد قنطري	arch dam
سد قياسي (سيبوليتي)	Cipolletti weir
سد قياسي	suppressed weir
سد للمد النصفي	half-tide cofferdam
سد ليفي	fabridam
سد مبطن عمودياً	vertical-lagging cofferdam
سد متحرك	movable dam
سد متحرك عمودي	rolling-up curtain weir
سد متعدد القناطر	multiple-arch dam
سد مجنح	wing dam
سد مدعم	buttress dam
سد مركب	composite dam
سد مزدوج الجدار	Ohio cofferdam
سد مستطيل	rectangular weir
سد مفرد الجدار	single-wall cofferdam
سد مقسم إلى مناطق	zoned dam
سد من أخشاب داعمة	crib dam
سد مؤقت	cofferdam
سد هيدرولي	hydraulic fill dam
سداد	plug
سدادة تفريغ	drain plug
سدادة قفل	padlocked plug
سداسية الأرجل	hexapods
سدسية	sextant
سدسية صندوقية	box sextant
سرج	saddle
سرطان الخرسانة	concrete cancer
سرعة الاقتراب	velocity of approach
سرعة الانسحاب	velocity of retreat
سرعة الانسياب	laminar velocity
سرعة التدفق الفوهي	orifice speed
سرعة التشغيل	operating speed
سرعة السلم المتحرك	escalator speed
سرعة الهواء الدنيا	minimum air velocity
السرعة الحدية	terminal velocity
السرعة الحرجة لـ «بلانجر»	Belanger's critical velocity
سرعة حفر النفق	tunnelling speed
السرعة القصوى	maximum speed
السرعة القصوى المسموح بها	maximum permissible speed
السرعة النوعية	specific speed
سرقة	theft

س

سطح الاقتراب	approach surface
سطح الجسر	bridge deck
سطح الحفر	digging face
سطح الخزان	reservoir roof
سطح الطريق	metalling
سطح الماء الباطني	perched water table, water table
سطح بيني للدخل والخرج	input/output interface
سطح بيني متواز	parallel interface
سطح حمل المرور	traffic-carrying deck
سطح خرساني	concrete roof, concrete deck
سطح خرسانة مسلحة	reinforced concrete facing
سطح خشبي	timber decking
سطح ذو باع طولي	long-span roof
السطح السفلي	lower surface
سطح صلب	hard standing
سطح طويل الباع	lamella roof
سطح عديم الانزلاق	non-slip surface
سطح علوي	upper surface
سطح غير مرصوف	unpaved surface
سطح فولاذي	steel roof deck, steel-deck surfacing
سطح قصير الباع	short-span roof
سطح مائل	pitched roof
سطح متوسط الباع	medium-span roof
سطح مجعد	rugous surface
سطح محبب	sandpaper surface
سطح مرصوف	paved surface
سطح مزخرف	crest surfacing
سطح مستو	flat roof, level surface
سطح مضاد للانزلاق	non-skid surface
سطح معدني	metal-deck roof
سطح مقاوم للانزلاق	skid resistant surface
سطح مقبب	cupola
سطح مكافئ	hypar
سطح منزلق	slip surface
سطوح الاقلاع	take-off surfaces
سطوح ذات مستوى مرتفع	high level decks
سطوح صخرية ملساء	slickensides
سطوح فاصلة	jointing
سطحي	superficial
السعر الأقصى	ceiling price
سعة التحمل الزائد	overload capacity
سعة التحميل القصوى	ultimate bearing pressure
سعة التخزين	storage capacity
سعة الترشيح	infiltration capacity
سعة التنقية الذاتية	self-purifying capacity
سعة الدرب	lane capacity
سعة الذبذبة	amplitude
سعة الطريق	road capacity
سعة القادوس الصدمية	struck capacity
سعة المخرج	outlet capacity
سعة المدرج	apron capacity
السعة الحملية	load carrying capacity
سفع	shot firing, blasting
سفع بالحصباء	grit blasting
سفع بالخردق	shot blasting
سفع بخردق الفولاذ	wheelabrating
سفع خثي	peat blasting
السفع الدوامي	vortex shedding
سفع رملي	sand blast
سقاطة	latch
سقالات البناء	staging
سقالات عمودية	pole scaffolds
سقالة	scaffold, scaffolding, stage
سقالة أنابيب فولاذية	tubular steel scaffolding
سقالة بمنصة دوارة	swing stage scaffold
سقالة خارجية	external scaffolding
سقالة خشب	wood scaffolding

س

سقالة رفع الاسمنت المجبول	placing plant
سقالة متحركة	movable scaffold
سقالة معلقة	suspended scaffold
سقف	ceiling
سقف أردوازي	slate roof
سقف محدب	hip roof
سقوط	fall
سقوط المطر	rainfall
سقي (المعادن)	tempering
سقيفة	attic, loft
سكان	population
سكة	track
سكة أحادية	monorail
سكة إصلاح العربات	car-repair track
سكة المراقبة	check rail
سكة بقواعد خرسانية جاهزة	prefabricated slab track
سكة حديد بعوارض خرسانية	slab track
سكة حديد بقواعد خرسانية	ribbon slab track
سكة حديد جبلية	mountain railway
سكة حديد تحت الأرض	underground railway
سكة حديد ذات بعد عياري	standard gauge plain track
سكة حديد شديدة الانحدار	funicular railway
سكة حديد ضيقة	light railway
سكة حديد عريضة	wide-gauge railway
سكة حديد مسننة	rack railway
سكة حديد معلقة	cable railway
سكة حديد مفردة	single-track railway
سكة حديد نفقية	railway in tunnel
سكة حديدية	railway
سكة حديدية عريضة	broad gauge
سكة حديدية ضيقة	narrow gauge
سكة حديدية قيد الانشاء	railway under construction
سكة حديدية مرتفعة	elevated railway
سكة داخل نفق	tube railway
سكة دليلية	guide track
سكة سلمية	ladder track
سكة متحركة	running rail
سكة متوازية	parallel tracks
سكة مكسورة	broken rail
سكة ملحومة	welded track
سكة ملحومة بصورة متواصلة	continuously welded track
سكة منبسطة القاع	vignole rail
سكة منحنية	curved track
سكك جماعية	team tracks
سلامة الرجال	safety of men
سلة كبيرة	pannier
سلخ	slough
سلسلة	chain
سلسلة الارساء	anchor chain
سلسلة المهندس المسّاح	engineer's chain
سلسلة «غنتر»	Gunter's chain
سلسلة قناطر	arcade
سلسلة قياس	measuring chain
سلسلة هويسات	chain of locks
سلك تفدين بئر التهوية	shaft-plumbing wire
سلك ربط	binding wire
سلك شائك	barbed wire
سلك فولاذي	steel wire
سلك مجدول بلقمة لولبة	die-formed strand
سلك مدوّر منبسط	smooth round wire
سلك ملدن	annealed wire
سلك منسوج مغلفن	galvanized woven wire
سلك موصل	lead
سلم	staircase, ladder
سلم المعاينة	inspection ladder
سلم النجاة من الحريق	fire escape
سلم حلزوني	spiral stairway
سلم الخدمة	service stair
سلم فولاذي	steel ladder
سلم متحرك	escalator

س

سلم متحرك مستو	flat escalator
سلم مدعوم	stock ladder
سلم منصة دوارة	turntable ladder
سلم وصول	access ladder
سليكا	silica
سليكون	silicone
سماعات رأس	headphones
السمت	azimuth
سمتنة (تمليط)	cementation
سمك الخرسانة	cover
سمكرة	plumbing
سمكري	plumber
سن اللولب	chase
سن لولبة مربع	square thread
سن لولبي كتفي	buttress screw thread
سناد	bolster
سنبك	punch
سنتيمتر	centimetre
سنديان	oak
سنفرة	emery
سهم	arrow
سوء الاستعمال	misappropriation
سوائل سريعة الالتهاب	flammable liquids
سور خشبي	wooden fence
سور منخفض	parapet
سوق	market
سياج	fence
سياج حدود	boundary fence
سياج طبيعي	hedge
سيارة إطفاء	fire engine
سيارة شحن	lorry
سيبة	step ladder
سير معلّق	overhead conveyor
سير ناقل	conveyor belt
سير ناقلة ميكانيكية	mechanical conveyor belt
سيزموجراف : مرسمة الزلازل	seismograph
سيزمومتر : مقياس الزلزلة	seismometer
سيفون	siphon
سيفون الجرعات الكيماوية	dosing siphon
سيفون معكوس	inverted siphon
سيل	torrent
سيلان الهلام	thixotropy
سيولة نقدية مخصومة	discounted cash flow
شاحنة	motor truck, truck
شاحنة خلاطة	truckmixer, agitating truck
شاحنة قلابة	tipping lorry
شاحنة مرفاع	truck-mounted crane
شاخص	banderolle, staff
شاخص تباعد	ranging rod, range pole
شاخص تسوية	stadia, target rod
شادوف أرخميدس	Archimedean screw
شارع	street
شارع عريض	avenue, boulevard
شاش	lawn
شاطئي	riparian
شاطئ	shore
شاطئ حصبائي	shingle beach
شاغور	cascade
شاقول	plumb bob
شباك ذو مفصلات رأسية	casement window
شبك معدني ممد	expanded metal
شبكة	grid, mesh, system
شبكة اصطناعية	fabricated mesh
شبكة المجاري	sewerage, sewerage system
شبكة النقل السريع	rapid-transit system
شبكة النقل بالسكك الحديدية	rail-transportation system
شبكة امدادات الماء	water supply system
شبكة أنابيب	network of pipes, piping network

شبكة تجميع مياه المجاري	collecting system
شبكة تصريف الهواء	air grate
شبكة توزيع	distribution network
شبكة ثنائية الاتجاه	double-layer grid
شبكة خطوط الضواحي	commuter systems
شبكة سلكية	wire lath
شبكة طرق	road network
شبكة فراغية	space lattice
شبكة فولاذية	steel grid
شبكة كهربائية حلقية	ring main system
شبكة مزدوجة الاتجاه	two-way grid
شبكة نقل الركاب بين المدن	intercity passenger system
شبه صلب	semi-rigid
شبه موصل	semi-conductor
شبيكة	grillage, lattice
شحن	charge
شرط خاص	special provision
شد	tension, pull
شد لاحق	post-tensioning
شد مسبق	pre-tensioning
شداد علوي	collar beam
شداد ملولب الطرفين	screw shackle
شدادة	guy
شدادة العمود	pole guy
شدادة المحولة	switch tie
شدادة رئيسية	main tie
شدادة منع الارتخاء	anti-sag bar
شدة الاجهاد	intensity of stress
شدة التحميل الصافي	net loading intensity
شدة سقوط الأمطار	intensity of rainfall
شديد الانفجار	high explosive
شرائح خشبية معترضة	brander
شرط التعويض	indemnification clause
الشروع في العمل	commencement of work
شرفة	balcony
شرفة مسقوفة	verandah
شركة تأجير	hire company
شروط خاصة	special conditions
شروط عامة	general conditions
شروط العقد	conditions of contract
الشروط الطوبوغرافية	topographic terms
شريحة خشبية	lath
شريحة ربط تناكبي	butt strap
شريحة من معدنين	bi-metal strip
شريط	band, strap, tape
شريط الوصلة	waterbar
شريط دوار	rumble strip
شريط فولاذ مغلفن	galvanized steel ribbon
شريط قياس	measuring tape
شريط قياس فولاذي	band chain
شريط مغنطيسي	magnetic tape
شطف	flushing
شعاع حراري	thermic lancing
شعبة	tine
شعرتا الشبيكة	stadia hairs
شغل بالقطعة	piece work
شفة	flange
شفة الأنبوب	pipe flange
شفة الانضغاط	compression flange
شفة التوتر	tension flange
شفة الدرجة	nosing
شفة أنبوب مغلقة	blank flange
شفة دعم	support flange
شفة صلبة	rigid flange
شفة وصل	connection flange
شق الانفاق بالهواء المضغوط	compressed air tunnelling
شق الطرق	road excavation
شق القنوات	canalization
شقة ذات طابقين	maisonette
شقة سكنية	flat

شكّال	brace, shackle
شكّال تقوية ضد الرياح	wind beam
شكّال جانبي	lateral bracing
شكّال زاوي	knee brace
شكّال سقالة	brace of a scaffold
شكّال مائل	diagonal brace
شكل القضبان	shape of bars
شكل رباعي السطوح	tetrahedron
شلال	waterfall
شلال صغير	cascade
شلل الغواص	caisson disease
شهادة عدم التآمر	non-collusion affidavit
شوكة	tine
شوكة مفصلية	clevis
شياق	arbor
صابورة نفايات الحديد	kentledge
صاحب البناء	building owner
صاد للماء	water-repellent
صاري دليلي من الفولاذ	false leaders
صاري مدعم بالحبال	guyed-mast
صاقور	pick
صاقور بالهواء المضغوط	pneumatic pick
صالة العرض	exhibition hall
صامد للحرارة	heat-proof
صامد للحوامض	acid-proof
صامد للدخان	smoke proof
صامد للعوامل الجوية	weather-resistant
صامد للماء	waterproof
صامد للهواء	airproof
صب (تصفية وتفريغ)	decanting
صب	cast
صب الخرسانة في الموقع	on-site concrete, in-situ concrete
صب في الموقع	cast-in-place, cast-in-situ
صب مسبق	precast
صب مسبق في الموقع	precast on site
صبغ	stain, pigment
صحي	sanitary
صخر	rock
صخر القاعدة	bedrock
صخر بركاني	trass
صخر طبيعي	natural rock
صخر مغترب	nappe
صدأ	rust
صدع	crack
صدع (شق)	rift
صدع إجهادي	brittle fracture
صدع تشققي	cleavage fracture
صدع شعري	hair crack
صدم (قدح)	percussion
صدمة كهربائية	electric shock
صرف (نزح)	drain
صرف الطبقة تحت السطحية	subsurface drainage
صرف القاعدة التحتية	base drain
صرف زراعي	agricultural drain
صرف سطحي	surface drainage
الصرف السطحي السنوي	annual run-off
صرف مياه الأمطار	stormwater drainage
صرف مياه الطريق	road drainage
صرف مياه النفق	tunnel drainage
صغير الحجم	undersize
صفح : صفيحي	laminate
صفحة	page
صفيحة	plate
صفّى	leach
صقل	polishing
الصقل التحضيري	bodying-up
صقيع	frost
صلابة	rigidity, stiffness, toughness
صلابة الهيكل	framework stiffness

ش

صلادة	hardness
صلب	rigid
صلصال	clay
صلصال إسمنتي	sintered clay
صلصال متماسك	stiff clay
صلصال مركّب	bentonite
صلصال مفرط الاندماج	over-consolidated clay
صلصال ممدد	expanded clay
صمام	valve
صمام إبري	needle valve
صمام ابتدائي	primary valve
صمام احتجاز الضغط	pressure-retaining valve
صمام الارتفاع	altitude valve
صمام التحكم في الدفق	flow control valve
صمام العادم	exhaust valve
صمام الهواء	air valve
صمام أمان	safety valve
صمام أنبوبي	tube valve
صمام انخفاض في منسوب الماء	low-water valve
صمام بغلاف مخروطي ثابت	fixed cone-sleeve valve
صمام بوابي	gate valve
صمام بكباس غاطس	plunger valve
صمام تحرير الضغط	pressure-release valve
صمام تحكم	control valve, controlling valve
صمام تخفيف الضغط	pressure-relief valve
صمام تخفيض الضغط	pressure-reducing valve
صمام تصريف	discharge valve, drain valve, sluice valve, washout valve
صمام تفريغ هزاز	sonic bleeder valve
صمام تنفيس للأمان	safety relief valve
صمام تنظيم الضغط	pressure control valve, pressure-regulating valve
صمام ري	irrigating head
صمام سدادي	plug valve
صمام سدادي دوار	rotary plug valve
صمام سدادي كروي	spherical plug valve
صمام ضغطي	pressure valve
صمام طرفي	terminal valve
صمام غير مرجع	check valve
صمام فاصل	isolating valve
صمام قطع	stop valve
صمام قطع أو ايقاف	shut-off valve
صمام كروي	globe valve
صمام كمي مغمور	submerged sleeve valve
صمام لا رجعي	non-return valve, reflux valve
صمام لا رجعي قلاّب	flap valve
صمام مائل	oblique valve
صمام مخروطي	cone valve
صمام مروحي	butterfly valve
صمام مص	suction valve
صمام معلق	overhead valve
صمام منع الفيضان والمد والجزر	anti-flood and tidal valve
صمام نافوري	hollow jet valve
صمام هواء مزدوج	double air valve
صمام هيدروستاتي	hydrostatic valve
صمام يعمل بمفتاح	key operated valve
صمامة التفجير	detonating fuse, Primacord fuse
الصمود الغشائي للماء	membrane waterproofing
صمولة	nut
صمولة ذاتية الزنق	self-locking nut
صمولة زنق	lock nut
صنادل ذاتية القواديس	self-hopper barges
صنبور	tap, nozzle
صنبور تنظيف	cleansing hydrant
صندل	scow

ص

صندل تمديد الأنابيب	lay barge
صندل حفر	drill barge
صندوق	box
صندوق الانذار بالحريق	fire alarm box
صندوق الرسائل	letter-box
صندوق الفرش	spreading box
صندوق الماء	water box
صندوق الموازنة	balance box
صندوق تجمع الطمي	silt box
صندوق حشو	stuffing box
صندوق حقن الاسمنت	boogie box
صندوق خندقي	trench box
صنف	item
صنفرة الخشب	sanding of wood
صهريج	cistern, tank
صهريج التمَوُّر	surge tank
صهريج تخزين مضخوخ	pumped storage reservoir
صهريج ماء	water cistern
صورة بالحجم الكامل	true-to-scale print
صورة ثنائية الأبعاد لتدفق الماء الجوفي	flow net
صورة جوية	aerial photograph
صورة جوية مائلة	oblique aerial photograph
صورة عمودية	orthophoto, vertical photograph
صورة مجهرية	photomicrograph
صوف زجاجي	glass wool
صومعة تخزين	storage bin
صومعة حبوب	silo
صيانة	maintenance
صيانة السكك	track maintenance
صيانة الطرق	highway maintenance
الصيانة الدورية	periodical maintenance
صيانة دورية	regular maintenance
صيانة عادية	normal maintenance
الصيانة الوقائية	preventive maintenance
صيغة إلتواء	flexure formula
صيغة تجريبية	empirical formula
ضابط أمان	safety officer
ضابط سرعة المرفاع	hoist controller
ضاغط	compressor
ضاغط أحادي البرغي	single screw compressor
ضاغط الهواء	air compressor
ضاغط بالقوة النابذة	centrifugal compressor
ضاغط ترددي	reciprocating compressor
ضاغط لولبي	screw compressor
ضاغط نابذ متعدد المراحل	multistage centrifugal compressor
ضاغط وحيد المرحلة	single-stage compressor
ضبط	set
ضبط الثقوب	broaching
ضبط الطفو	floating control
ضبط دائم	permanent adjustment
ضبط مؤقت	temporary adjustment
ضبط يدوي	manual reset
ضجيج الصخور	rock noise
ضخ كهربائي أوتوماتي	automatic electric pumping
ضخ مياه المجارير	sewage pumping
ضخم	oversize
ضريبة	toll
ضغط	pressure
ضغط البخار	vapour pressure
ضغط التحميل	bearing pressure
ضغط التحميل الأقصى	ultimate bearing pressure
ضغط التحميل المسموح	allowable bearing pressure
ضغط التحميل المفرط	overburden pressure

ض

ضغط التربة	earth pressure
ضغط التربة الثابت	earth pressure at rest
ضغط التربة النشط	active earth pressure
ضغط التشغيل	operating pressure
ضغط التضخيم	swelling pressure
ضغط التلامس تحت الأساسات	contact pressure under foundations
ضغط الريح	wind pressure
ضغط القص	shear stress
ضغط القمامة	refuse compression
ضغط الماء في الزيت المشبع	pore-water pressure
ضغط الموجات	wave pressure
ضغط بارومتري	barometric pressure
ضغط بين الحبيبات	intergranular pressure
الضغط التناضحي	osmotic pressure
ضغط دينامي	dynamic pressure
ضغط زائد هيدروستاتي	hydrostatic excess pressure
ضغط شعري	capillary pressure
ضغط على مرحلتين	two-stage compression
ضغط فعّال	active pressure
الضغط الكلي	total pressure
ضغط متحارر	isothermal compression
ضغط متعادل	neutral pressure
ضغط هيدروديناميي	hydrodynamic pressure
ضغط هيدروستاتي	hydrostatic pressure
ضفة	bank
ضفة محتجزة	retained bank
ضفيرة	strand
ضلع القنطرة	arch rib
ضلع عمودي	leech
ضوء الازاحة	displacement light
ضوء المروحة	fanlight
ضوء النهار	daylight
ضوء متوهج	incandescent light
ضوء مستقطب	polarized light

ط

الضوضاء الملحوظة الفعالة بالديسبل	Effective Perceived Noise Decibels
طائرة	aircraft
طائرة دون سرعة الصوت	subsonic jet aircraft
طابق	floor
الطابق الأرضي	ground floor (level 1)
طابق إضافي	penthouse
الطابق الأول	first floor (level 2)
الطابق الثالث	third floor (level 4)
الطابق الثاني	second floor (level 3)
الطابق الرابع	fourth floor (level 5)
طابق سفلي	basement
طابق سفلي عميق	deep basement
طابق علوي	upper floor
طابق متوسط	mezzanine
طاحونة	mill
طاحونة هوائية	windmill
طارد	ejector
طافية سطحية	surface float
طاقة	energy, power
طاقة التوتر	strain energy
طاقة المضخة	pump capacity
طاقة حركية	kinetic energy
الطاقة السرعية	velocity head
الطاقة الضغطية المفقودة	lost head
طاقة كامنة	potential energy
طاقة كهربائية	electric power
الطاقة المتوفرة	available power
الطاقة النووية	nuclear energy
طاقة هيدرولية	hydraulic energy
طبق	saucer
طبقات حاوية للماء	water-bearing strata
طبقة التربة الوسيطة (ب)	B-horizon
طبقة التمليط النهائية	setting coat
طبقة الرصف الأسفلتية	rolled asphalt

طبقة الرصف السابقة	pervious base-course layer
طبقة الرصف السطحية	wearing course, road surface
طبقة السطح	surfacing
طبقة القاعدة التحتية	base course
طبقة بكتريا	bacteria bed
طبقة تجفيف الحمأة	sludge drying bed
طبقة تحميل	bearing stratum
طبقة ترشيح	filter bed, filter layer
طبقة تسوية	regulating course
طبقة تغطية	blinding layer
طبقة تمليط فوقية	topping
طبقة حصى الرصف	ballast bed
طبقة خرسانية تحتية	blinding concrete
طبقة خرسانية	concrete layer
طبقة رصف	racking course
طبقة رصف سطحية	surface dressing
طبقة رصف قارية	thin surfacing
طبقة سطحية	surface-course
الطبقة السفلى	bottoming
الطبقة السفلية الأولى	first underlayer
الطبقة السفلية الثانية	second underlayer
طبقة صامدة للصقيع	frost-proof layer
طبقة ختم	sealing coat
طبقة عازلة للرطوبة	damp-proof course
طبقة قيرية	tack coat
طبقة مرشحة	percolating filter
طبقة ملامسة	contact bed
طبقة نشطة	active layer
طحن	grinding, milling
طربوش الوصل	crosshead
طرد مركزي	centrifuge
طرف	limb
طرف الرصيف	kerbside
طرف مثبت	fixed end

طرف المزواة	lower plate
طرفا خط التحويل	switch points
طرفي	terminal
طرق	strike
طرق وضع الركائز	pile-placing methods
طرقة	blow
طريق	road, track
طريق أخشاب مرصوفة بالعرض	corduroy road
طريق العربات	carriageway
طريق الكبل الهوائي	aerial cableway
طريق المداس	treadway
طريق بحاجز ترابي	embanked road
طريق تجاوز	by-pass
طريق تحت الاشراف	adopted street
طريق ترابي	earth road
طريق ثابت	permanent way
طريق حبلي	Blondin, ropeway
طريق حبلي ثنائي الكبل	bi-cable ropeway
طريق حبلي متواصل	continuous ropeway
طريق حبلي مزدوج الكبل	twin-cable ropeway
طريق حبلي معلّق	overhead ropeway
طريق حبلي هوائي	aerial ropeway
طريق حرة	freeway
طريق حول البلدة	town by-pass
طريق الخدمة	service road
طريق دائري	circular road
طريق ذو ثلاث مسارات	three-lane road
طريق زراعي	farm road
طريق رئيسي	arterial road, main road, trunk road
طريق ريفي	rural roads
طريق سريع	fast road, motorway
طريق سريع حضري	urban motorway
طريق عام	highway

ط

طريق عام قيد الانشاء	highway under construction
طريق غير محسن	unimproved road
طريق غير مصنف	non-classified road
طريق فرعي	secondary road, slip road, minor road
طريق للمشاة	footway
طريق مزدوج للسيارات	dual carriageway
طريق مرصوف	metalled road
طريق مرصوف بالحصباء	macadam road
طريق مرصوف بالقار	tarred road, bitumen macadam
طريق مرصوف وكابح للماء	waterbound macadam
طريق مرور	traffic road
طريق مناسب لجميع الفصول	all-weather road
طريق منخفض التكاليف	low cost road
طريق موصل	access road
طريقة الإسفين المنزلق	sliding-wedge method
طريقة البرمجة	programming method
طريقة البناء المتداخل	jackblock method
طريقة البناء المصنّع	industrialised building
طريقة الترميم الرملي	sand patch method
طريقة التفريغ الهوائي لاختبار الرمل	vacuum method of testing sands
طريقة الرفع بالأسطوانة	drums method
طريقة القطع والردم	cut-and-cover
طريقة القوس الدائري	circular-arc method
طريقة الكسارة المطرقية	hammer-mills method
طريقة المرفاع والكرة	crane-and-ball method
طريقة منحنى السرعة	velocity-profile method
طريقة « هاردي كروس »	Hardy Cross method
طريقة « هونيجمان »	Honigmann method
طريقة علو آلة القياس	height of instrument method
طفا	float
طفح الخزان	bellmouth overflow
طفح مياه الأمطار	stormwater overflow
طفو	flotation
طقس رديئ	bad weather
طقم نحاته	chip set
طلاء	paint
طلاء اسمنتي	cement rendering
طلاء بالغمس الساخن	hot-dip coating
طلاء مانع للانزلاق	anti-slip paint
طلب البيوكيماوي للأوكسجين	bio-chemical oxygen demand
طلب الكيماويات للأكسجين	chemical oxygen demand
طلية أولى	rendering
طلية سفلية	undercoat
طلية عاتمة	mat
طلية ملساء	hard finish
طلية نهائية	finishing coat
طلية واحدة	one coat
طمي	silt
طمي مترسب	siltation
طن	ton
طن أمريكي	short ton, net ton
طن انجليزي	long ton
طُنف فوق باب أو نافذة	labels
طوب أزرق صلب	blue bricks
طوب زجاجي	glass block
طوب سد النبع	spring closers
طوب التلبيس	facing bricks
طوب الرصف	paving brick
طوب السليكا	silica brick
طوب مجوّف	cavity bricks
طوب مزجج	glazed bricks
طوب مصلد بالاحماء	hard-burnt bricks
طوب مفرغ	air bricks
طوب من الحجر المنحوت	ashlar bricks
طوبة نصفية	closer

ط

طور الجلب	fetch phase
طوربيد	torpedo
طوف	stinger, raft, barge
طوف تحت سطح الماء	subsurface float
طوف صلب	rigid raft
طوف عائم	buoyant raft
طوق الدفع	driving band
طوق الركيزة	pile hoop
طوق القنطرة	arch ring
الطول الإجمالي	overall length
طول الإنتقال	transition length
طول السد	length of dam
طول المدرج	runway length, length of runway
طول النقل	transmission length
طول معياري	gauge length
طيران	flight
طين جلمودي	boulder clay
طين صفحي	shale
طين صفحي صدعي	heaving shale
طين صفحي كربوني	burnt shale
طين صلصالي	bentonite mud
ظاهرة «باوشينجر»	Bauschinger effect
ظروف التحمل	overload conditions
ظروف العمل العادية	normal working conditions
ظروف خاصة	special conditions
ظهر الريشة	blading back
ظهور	appearance
عائق	retarder
عابر	transient
عاتق	shoulder
عارضة	batten, girder
عارضة أفقية	template
عارضة الأرضية	floor beam
العارضة الأفقية للركيزة	pier cap
عارضة السطح	roof girder
عارضة السكة	track stringer
عارضة ألواح السطح	deck-plate girder
عارضة برجية	castellated beam
عارضة بسيطة	simple beam
عارضة تشابكية	lattice girder, open-web girder
عارضة تقوية	stiffening girder
عارضة ثابتة الطرف	fixed-end beam
عارضة ثانوية	secondary beam
عارضة حلقية	ring beam
عارضة خرسانية	concrete sleeper
عارضة خرسانية أحادية	monoblock concrete sleeper
عارضة خشبية	beam, timber beam
عارضة دائرية	ring girder
عارضة رئيسية	main beam
عارضة صندوقية	box beam, box girder, hollow-web girder
عارضة طريق	road girder
عارضة طولانية	stringer beam
عارضة غير متوترة	untensioned beam
عارضة في جدار السد	flashboard
عارضة لوحية	plate girder
عارضة متواصلة	continuous girder
عارضة مثبتة	fixed beam
عارضة مركبة	compound girder
عارضة مسبقة الصب	precast beam
عارضة مستقيمة	straight beam
عارضة مصفحة	plated beam
عارضة معلّقة	overhung beam
عارضة مفتوحة	open traverse
عارضة مفتوحة الهيكل	open-frame girder
عارضة مقوّسة	bowstring girder, hogging girder
عارضة مقيدة الحركة	restrained beam

ط

عارضة ناشرة	spreader beam
عارضة وصل	fishplate
عازل	isolator
عازل للرطوبة	damp proof
عاصف	windy
عاق	trammel
عالم جيولوجي	geologist
عالم في التربة	soil scientist
عالم في الاقتصاد	economist
عامة مغمورة	submerged float
عامل	agent, labourer
عامل استحلاب	emulsifier
عامل إشارة	signalman
عامل الاحتمالية	probability factor
عامل الأمان	factor of safety, safety factor
عامل الترابط	bond form
عامل الحمل	load factor
عامل الخفض	reduction factor
عامل الزمن	age factor
عامل السرعة	speed factor
عامل الشاخص	staff man, rod man
عامل اللا إنفاذية	impermeability factor
عامل النقل	transit man
عامل أنابيب	pipe fitter
عامل تركيب الانشاءات	constructional erector
عامل تركيب الفولاذ	steel erector
عامل تشتيت	dispersing agent
عامل تطريق	striker
عامل تلميع	polishing agent
عامل حفر	navvy
عامل خرسانة	boot man
عامل ذو فاعلية سطحية	surface-active agent
عامل ماهر	skilled man
عامل مساحة	axman
عامل معوق	retarding agent
عامل نصف ماهر	semi-skilled man
عامل وقاية	protective agent
عاير	calibrate
عبور	crossing
عبور سطحي	surface access
عتبات طرفية	end frogs
عتبة	sill
عتبة أرضية	ground beam
عتبة الباب العليا	lintel
عتبة النافذة	window sill
عتبة الهويس	lock sill
عتبة بشفة مستدقة	tapered-flange beam
عتبة بشكل T	tee-beam
عتبة متواصلة	continuous beam
عتبة محززة	dentated sill
عتبة معترضة	crossbeam
عتبة مسنمة	trussed beam
عتبة مشطوبة	mitre sill
عتلة	lever, crowbar
عجرة	node
عداد	meter
عداد الدفق الكلي	total flow meter
عداد الدفق الكمي	mass flow meter
عداد الماء	water meter
عداد تكامل	integrating meter
عداد دفعات الركام	aggregate batch meter
عداد دفق بالتموجات فوق السمعية	ultrasonic flow meter
عداد دفق مغنطيسي	magnetic flow meter
عداد فنتوري	Venturi meter
عُدة	tool
عدة تعمل بالهواء المضغوط	pneumatic tool
عدة تعمل بتفاعل المساحيق	powder actuated tool
عدة جلفطة (تغليف)	caulking tool
عدد المصاعد	number of lifts
عدسة التركيز البؤري	anallatic lens

ع

عدسة عينية	astronomical eyepiece
عدم الاستقرار الايروديناميي	aerodynamic instability
عربة	vehicle
عربة إطفاء رغوي	foam tender
عربة الحفارة	jumbo
عربة الحفر	wagon drill
عربة نقل الخرسانة	buggy
عربة الشحن	freight wagon
عربة آلية	power barrow
عربات انقاذ	crash tenders
عربة بمرفاع شوكي	fork-lift truck
عربة تفريغ المجرور	gully vacuum tanker
عربة خفة مغنطيسية	magnetic levitation vehicle
عربة قلاّبة	car dumpers
عربة معلّقة	telpher
عربة نقل واحدة	single bogie
عربة يد	wheelbarrow
عرض	proposal, display
عرض الدرج	stair width
عرض السلم المتحرك	escalator width
عرض القناة	channel width
عرض المدرج	width of runway
عرض مسار الحط	width of landing strip
عرفة رأسية	pile cap
عروة	grommet
عزل	segregation
عزل (عازل)	insulation
عزل الصوت	acoustic insulation, sound insulation
عزل الفرن	furnace insulation
عزل حراري	thermal insulation
عزم الارتخاء	sagging moment
عزم التثبيت	fixing moment
عزم التقوس	hogging moment
عزم الثني	bending moment
عزم الدعم	support moment
عزم العطالة	moment of inertia
عزم العطالة القطبي	polar moment of inertia
عزم القوة	moment of a force
عزم اللي	torque
عزم المقاومة	moment of resistance, resisting moment
عزم ساكن	static moment
عشرة	ten
عشرون	twenty
عشري سداسي	hexadecimal
عضادة	alidade
عضو إنشائي	structural member
عضو مركب	composite member
عضو مشارك	corporate member
عطاء	bid
عطاء انقاذ	salvage tender
عطاء طارئ	emergency tender
عطل	fault
عقد	contract
عقد إدارة	management contract
عقد إدارة الانشاءات	construction-management agreement
عقد اختصاصي	specialty contract
عقد الانشاءات	construction contract
عقد بسعر شامل	unit-price contract
عقد بسعر محدد	fixed-price contract
عقد بمبلغ إجمالي	lump-sum contract
عقد تشجيعي	incentive-type contract
عقد تكميلي	supplementary agreement
عقد شامل	all-in contract
عقد قنطري	barrel vault
عقد مسطح	jack arch
عقد مفاوض	negotiated contract
عقد نفقي	tunnel vault

ع

عقدة	node
عقدة اللوح	panel point
عكس	invert
عكس الواجهة	change face
علاج الخشب في الخزان المفتوح	open-tank treatment
علاقة الرجل مع ماكينته	man-machine interface
العلاقات القانونية والعامة	legal and public relations
علاّقة	suspender
علامات التحركات الأرضية	ground movement signs
علامات الطرق	road markings
علامة إسناد	reference mark
علامة الاجهاد	stress notation
علامة المنسوب	bench mark
علامة « پليمسول »	Plimsoll mark
علامة « جيوديسية »	geodetic mark
علامة حدود	boundary mark
العلامة الخلفية	back mark
علم الاحصاء	statistics
علم الأرصاد الجوية	meteorology
علم الاقتصاد	economics
علم البيئة المحلية	local ecology
علم السكون (استاتيكا)	statics
علم الصوت	acoustics
علم الفلك	astronomy
علم القياس عن بعد	telemetry
علم رسم الخرائط	cartography
علم طبقات الأرض	stratigraphy
علم ميكانيكا الصخور	rock mechanics
علو الاحتكاك	friction head
علو الضغط	pressure head
علو الماء	head of water
علو الماء في السد	weir head
علو المص	suction head
علو ساكن	static head

على نطاق واسع	on a large scale
عمال الموقع	site labour
عمال مهرة	skilled labour
عمال مؤقتون	temporary labour
عمر الحمأة	sludge age
عمر السكك الحديدية	rail life
عمق الأساس	depth of foundation
عمق القطع	cut-off depth
عمق القناة	channel depth
عمق بنية السطح	surface texture depth
العمق الفعال	effective depth
العمق الكلي	overall depth
عمل	labour, work
عمل دائم	permanent labour
عمل مباشر	direct labour
عمل مؤقت	transient labour
عملية	process
عملية أحادية المصدر	single source-operand
عملية الترشيح البيولوجي المنشطة	activated-biofilter process
عملية التلامس البيولوجي	biological-contactor process
عملية الحمأة المنشَّطة	activated-sludge process
عملية الرمل الرطب	wet-sand process
عملية تحلية الماء الغشائية	membrane process
عمود	column, pillar, post, pole
عمود إدارة خارجي	power take-off
عمود إدارة مرن	flexible shaft
عمود الاتصال الرئيسي	main rod
عمود الارساء	breasting jack
عمود التقاء	meeting post
عمود التهوية	vent shaft
عمود الركن	heel post
عمود المرفاع	crane post
عمود تثبيت فولاذي	anchoring spud

ع

عمود جداري ناتئ	pilaster
عمود خرساني	concrete column
عمود خشب	wood column
عمود درابزين	baluster
عمود درابزين الدرج	bracket baluster
عمود دليلي	pilot shaft, guide runner
عمود ربط الحبال	bollard
عمود قصير	short column
عمود متطاول	long column
عمود مربوط	laced column
عمود مشطوب	mitre post
عمود منفصل	monolith
عميل	client
عنابر الصيانة	maintenance hangars
عناصر البرنامج	program elements
عناصر هيدرولية	hydraulic elements
عناقيد ركام الحطب	wood-pile clusters
عنصر تسخين	heating element
عنصر سحب الهواء	air-entraining agent
عنصر قياسي	measuring element
عنصر كشف	detecting element
عنوان رئيسي	headline
عنوان فرعي	side heading
عوارض تقوية	stiffening beams
عوارض السد	stop logs
عوارض خشبية	timber stringers
عوامة إرساء	mooring buoy
عوامة قضيبية	rod float
عوامة هوائية	pneumatic float
عيار	calibre, standard
عين كهربائية	electric eye
عينات الحفر	borehole samples
عينات جوفية	cores
عينة	sample
عينة التربة	soil sample
عينة بكر	undisturbed sample
عينة عشوائية	random sample
عينة نموذجية	representative sample
غابة	forest
غاز البيوتان	butane
غاز الحمأة	sludge gas
غاز المجاري	sewage gas
غاز المداخن	flue gas
غاز الميثان	methane
الغاز الطبيعي	natural gas
غازات العادم	exhaust gases
غازات سريعة الالتهاب	flammable gases
غاطس السفينة	draft of a ship
غالون	gallon
غثاء الخرسانة	laitance
غدير	brook
غراء بوليسلفايد	polysulphide sealant
غراسة البذور	seeding plant
غرافيت	graphite
غرانيت	granite
غربال اسطواني دوار	trommel
غربال دقيق الشبيكية	fine screen
غربال دوار	revolving screen, rotary screen
غربال متحرك	travelling screen
غربلة	screening
غرفة	chamber
غرفة احتباس الروائح	intercepting sewer
غرفة الاحتراق المتقدم	antechamber
غرفة الاستراحة	rest room
غرفة الانتظار	waiting room
غرفة الصمام	valve chamber
غرفة الطعام	dining room
غرفة العمل	working chamber
غرفة المراقبة بالكومبيوتر	computer control room
غرفة المراقبة بالمحطة	station control room
غرفة المعيشة	living room

ع

غرفة الهويس	canal lock chamber, gate chamber
غرفة تبريد	cooling chamber
غرفة تجفيف	drying room
غرفة مراقبة	control room
غرفة مكافئة	hyperbaric chamber
غرين عضوي	organic silt
غسالة	washer
غشاء	membrane
غشاء أسفلتي مسيك	tanking
غشاء الانضاج	curing membrane
غشاء پوليثين	polythene membrane
غشاء السقف	ceiling membrane
غشاء صامد للماء	waterproof membrane
غشاء كتيم	impermeable membrane
غشاء مانع للرطوبة	damp-proof membrane
غشاء مرن	flexible membrane
غطاء حفرة التفتيش	manhole cover
غطاء خرساني	concrete cover
غطاء صرف أفقي	horizontal drainage blanket
غطاء فتحة تفتيش مزدوج	double-seal manhole cover
غلاف	casing
غلاف	sheathing, jacket
غلاف رصاصي	lead sheath
غلاية	boiler
غلاية بخار	steam boiler
غلاية ذات إنتاجية عالية	high-output boiler
غلفن	galvanize
غلفنة بالزنك المذاب	wet galvanizing
غلق المدخل لحجرة مضغوطة	man-lock
غلق هوائي	air lock
غلق هوائي للمواد	materials lock
غلي (غليان)	boil
غواص	diver
غونيت : نوع من الخرسانة	gunite
غير مستقر	unstable
غير منفذ	impervious
الفائدة الآلية	mechanical advantage
فاحص الفقاعات	bubble trier
فاحص عينات التربة	soil sampler
فاحص ذو كفاءة	competent examiner
فادن خيط السبر	sounding lead
فاراد	farad
فارشة	spreader
فارشة خرسانة	concrete spread
فأس	axe
فأس صغير	hatchet
فاصل	spacer, partition
فاصل المرشح	filter separator
فاصل من الطوب المفرغ	hollow partition
فاصمة	fuse
ڤاناديوم	vanadium
فانوس	lantern
فتحة	opening, orifice
فتحة التصريف	drainage opening
فتحة التغذيف	paddle hole
فتحة الفوهة	nozzle orifice
فتحة المصباح	lamphole
فتحة تصريف	mole drain
فتحة تفتيش عميقة	deep manhole
فترات الصيانة	maintenance interval
فترة الانشاء	period of construction, construction period
فترة الانضاج	curing period
فترة الترابط	bond length
فترة الضمان	guarantee period
فتل	twist
فتيل النسف	blasting fuse
فجوة	alcove

فحص المنشآت والمعدات وإعدادها للعمل	commissioning
فحم	coal, carbon
فحم حجري مسحوق	pulverized coal
فحوص طبية	medical examination
فدان	acre
فدان قدم (وحدة سعة)	acre foot
فرازة	separator
فرازة على الساخن	hot miller
الفراغ بين حبات الرمل	voids
فرجار	compass
فرجار ذو عاتق	beam compasses
فرجار قياس القطر	callipers
فرز انتشاري كهربائي	electro-dialysis
فرش بالصخور	enrockment
فرشة	mattress
فرشة تسوية	equalizing bed
فرشة خرسانية	concrete bedding
فرشة «رينو»	Reno mattress
فرشة صلبة	hard core
فرشة قارية	bituminous carpet
فرشة من أغصان مقطوعة	brushwood fascine mattress
فرشة من الملاط	bedding
فرضة مثلثية	triangular notch
فرط الاحماء	overheating
فرقعة	chambering
فرن	furnace, kiln
فرن الصهر	blast furnace
فرن بالقوس المباشر	direct arc furnace
فرن كهربائي	electric furnace
فريق التصميم	design team
فسحة ما بين عمودين	bay
فسيفساء	mosaic
فصل بالتفريغ الهوائي	vacuum separation
فصل بالضغط الغشائي	membrane pressure separation
فصل كهربائي غشائي	membrane electric separation
فضاء جوي	air-space
فضلات التشغيل	operation waste
فطر	fungus
فعالية ميكانيكية	mechanical efficiency
فقاعات تطابق موشوري	optical coincidence bubble
فقاعة	bubble
فقد الاجهاد السابق	loss of prestress
فقد التربة المحفورة	loss of ground
فقد الضغط	loss of head
فقد الطاقة	power loss
فقد امتصاصي	absorption loss
فقد بالارتشاح	seepage loss
فقد دوّامي	eddy loss
فلطحة بالطرق	upsetting
فلطية المستهلك	consumer's voltage
فلكة	shim, washer
فلكة مائلة	bevelled washer
فلكة مخروطية	tapered washer
فلكة ميكانيكية	mechanical washer
فلكة نابضة	spring washer
فلين	cork
فنار	beacon
فندق	hotel
الفنون التخطيطية	graphics
فني الهندسة المدنية	civil engineering technician
فني هندسة إنشاءات	structural engineering technician
فواصل	joints
فواصل لمنع تكسر الخرسانة	crack inducers
فوتوستات	photostat

ف

فولاذ	steel
فولاذ أساسي	basic steel
فولاذ الصب	cast steel
فولاذ القضبان الحديدية	rail steel
فولاذ إنشاءات	structural steelwork
فولاذ بنسبة كربون مرتفعة	high-carbon steel
فولاذ توزيع الحمل	distribution steel
فولاذ حامضي	acid steel
فولاذ سبيكي	alloy steel
فولاذ طري	mild steel
فولاذ قوي الشد	high-tensile steel
فولاذ قوي المتانة	high-strength steel
فولاذ كربوني	carbon steel
فولاذ كربوني للانشاءات	structural carbon steel
فولاذ كربوني مغلف	clad steel
فولاذ كربوني منخفض الخليط المعدني	low-alloy carbon steel
فولاذ لا يصدأ	stainless steel
فولاذ «ماراجين»	maraging steel
فولاذ مبروم	round steel
فولاذ مدلفن	rolled steel
فولاذ مسحوب على البارد	cold-formed steel
فولاذ معالج بالحرارة	heat-treated steel
فولاذ منخفض الكربون	low-carbon steel
فولاذ منغنيزي	manganese steel
ڤولت	volt
فوهة	nozzle, orifice, mouth
فوهة إبرية	needle nozzle
فوهة تصريف	discharge nozzle
في المركز	on centre
في الموقع	in-situ
فيضان	flood
ڤيلا	villa
فيرلونج	furlong
قائم الجُملون	queen post
قائم المرفاع	gin pole

قائم بوابة	gate post
قائم دعم عمودي	stanchion
قائم دعم فولاذي	steel stanchion
قائمة الكميات	bill of quantities
قائمة المواد	list of materials
قائمة خشبية	stud
قابس بمصهر	fusible plug
قابض (كلتش)	clutch
قابض احتكاكي	block clutch, friction clutch
قابض حافة	rim clutch
قابض لا انعكاسي	non-reversible clutch
قابض قدة موقفة	sprag clutch
قابض لقم المثقب	chuck
قابض لوحي	plate clutch
قابض مخروطي	cone clutch
قابض نابذ	centrifugal clutch
قابل للغمر	submersible
قابلية الانطواء	pliability
قابلية التصليد	hardenability
قابلية التطريق	malleability
قابلية الطفو	buoyancy
قابلية اللحام	weldability
قابلية الليونة	ductility
قابلية النفاذ	permeability
قادوس	bowk, hopper
قادوس المهملات	skip
قادوس التعبئة	charging hopper
قادوس السمنتة تحت الماء	treamie
قادوس بفتحة سفلية	drop-bottom bucket, bottom-opening skip
قادوس قلاّب	flip bucket
قادوس كاشط	scraper bucket
قادوس كبل السحب	dragline bucket
قادوس وزن دفعة الركام	aggregate weigh batcher skip

قاذف بالهواء المضغوط	pneumatic ejector
قاذف مياه المجارير	sewage ejector
قار	bitumen, tar
قار متخلف	straight-run bitumen
قار يحتوي على حشوة	filled bitumen
قارب إنقاذ	rescue boat
قارب مسطح	barge
قارن معشق	geared coupling
قارنات متغيرة السرعة	variable speed couplings
قارنة خرطوم الاطفاء	fire-hose coupling
قارنة كُمية	socket joint
قارنة مرنة	flexible coupling
قارورة	vial
قاطرة	locomotive
قاطرة ديزل كهربائية	diesel-electric locomotive
قاطرة كهربائية	electric locomotive
قاطع التيار	breaker
قاطع الدائرة	circuit breaker
قاطع الدائرة الكهربائية	cutout
قاطع تلقائي	contactor
قاطع دوار	gyratory breaker
قاطعة كربيد التنغستين	tungsten carbide cutter
قاطعة العشب	grass cutter
قاطعة نتوءات	bump cutter
قاع	bed
قاع النهر	river bed
قاع طيني	barge bed
قاع مجروف	dredged bottom
قاعة اجتماعات	auditorium
قاعدة	base, footing, bed
قاعدة أرخميدس	Archimedes' principle
قاعدة استقرار	seat of settlement
قاعدة الأساس	foundation footing
قاعدة الالكترود	electrode stub
قاعدة الخط الحديدي	road bed
قاعدة السكة	rail base
قاعدة العمود	plinth
قاعدة الغطاء	cover seating
قاعدة المحولة	switch stand
قاعدة المصباح	lamp standard
قاعدة المطرقة	hammer base
قاعدة المقارنة	comparator base
قاعدة بئر التهوية	shaft foundation
قاعدة « بوديتش »	Bowditch's rule
قاعدة تحتية	sub-base
قاعدة تثبيت الآلة	machine foundation
قاعدة جوية	air base
قاعدة خرسانة مسلحة	reinforced concrete base
قاعدة خرسانية	concrete base
قاعدة سمپسون	Simpson's rule
قاعدة ضيقة	strip footing
قاعدة طريق جاهز	prepared roadbed
قاعدة عرضية	spread footing
قاعدة كابولية	cantilever footing
قاعدة متواصلة	continuous footing
قاعدة مشتركة	combined base
قاعدة مفردة	single base
قاعدة نضدية	stepped foundation
قاعدة هبوط	landing platform
قالب	matrix
قالب خرساني	concrete block
قالب الصب	mould
قالب صب الخرسانة	formwork
قالب فولاذي	steel formwork
قالب متحرك	travelling form
قالب من السقاط	briquette
قامة	fathom
قامة تسوية	levelling staff
قامطة	clamp

ق

قامطة ركابية	bulldog grip
قانون «أبرامز»	Abrams' law
قانون الخطأ العادي	normal law of error
قانون «بارنز»	Barnes's formula
قانون «بارنز» للجريان في المجاري الانسيابية	Barnes's formula for flow in slimy sewers
قانون «هازين»	Hazen's law
قانون «هوك»	Hooke's law
قبة	dome
قبر	grave, tomb
قبضة	knob
قبعة صلبة	hard hat
قدة تحديد	runner
قدة رفيعة معترضة	needle beam
قدرة	power
قدرة التبريد	refrigerating capacity
قدرة الذروة	peak power
قدرة الغلاية	boiler rating
قدرة حصانية	horsepower
قدرة هيدرولية	hydraulic power
قدم	foot
قراءة	reading
قُرْص	disc
قَرَصَ (قَرَضَ)	nip
قرص بيولوجي	biodisk
قرص دوار	turntable
قرص متحرك الرأس	moving head disc
قرص مثبت الرأس	fixed head disc
قرص مرن	floppy disk
قرص مطاط مرن	flexible rubber disc
قرميدة	brick
قسم	section
قسم النسيج	weaving section
قشرة خشبية	veneer
قشور خزفية	ceramic veneer
قص	shearing
قص التخريم	punching shear
قص الريح	wind shear
قص متعدد	multiple shear
قصر	palace
قصف على الساخن	hot shortness
قصور الانعطاف	cant deficiency
القصور الذاتي	inertia
قضبان التسليك للمجاري	drain rods
قضبان حديدية	iron railings
قضبان حلزونية	hooping
قضبان محززة	indented bars
قضبان مشوّهة	deformed bars
قضيب	rod, bar
قضيب انزلاق	slide rail
قضيب الموازنة	balance bar
قضيب تثبيت	tie rod
قضيب تثبيت الزجاج	glazing bar
قضيب تسليح فولاذي	steel reinforcement bar
قضيب تسوية	levelling rod
قضيب التوزيع	distribution bar
قضيب التوصيل	busbar
قضيب ثلاثي	tribar
قضيب جر	drawbar
قضيب حفر قفاز	jumper
قضيب حفر مضلع	kelly bar
قضيب دائري الطرفين	bullhead rail
قضيب دسر	dowel bar
قضيب دليلي	guide rail
قضيب رئيسي	main bar
قضيب ربط أرضي	land tie
قضيب الرش	spray bar
قضيب رصاص	lead rail
قضيب زاوي	angle bar
القضيب السفلي	bottom bar, lower bar
قضيب سكة حديد	rail
قضيب صب الخرسانة	starter bar

ق

قضيب طولي	longitudinal bar
قضيب عالي المطاوعة	high-yield bar
قضيب علوي	upper bar
قضيب فولاذ طري	mild steel bar
قضيب مائل	inclined bar
قضيب مانع للتراوح	sway rod
قضيب متصدع	cracked rail
قضيب مدبب	swinger
قضيب المراقبة	sight rail
قضيب مربع ملتوي	twisted square bar
قضيب المستقيم المقابل	subtense bar
قضيب مسطح القعر	flat-bottomed rail
قضيب مطاط مسيك	staunching rod
قضيب مقوّس	belly rod
قضيب ملحوم تناكبياً	butt-welded rail
قضيب ملحوم متواصل	continuous welded rail
قضيب وصلة سكة حديد	rail-joint bar
قطاع	section
قطاع الدمج	merging section
قطاع أنبوبي	tubular section
قطاع انحراف	diverging section
قطاع بطرف محدّب	beaded section
قطاع جانبي	profile
قطاع جانبي رأسي	vertical profile
القطاع الحقيقي	true section
قطاع رقيق	thin section
قطاع طولي	longitudinal section
قطاع عارضة الوصل	fishplate section
قطاع عياري	standard section
قطاع فولاذ مدلفن	rolled-steel section
قطاع مائل	oblique section
قطاع نفقي دائري	circular tunnel section
قطاعات أنبوبية مفرغة	hollow sections
قطاعات مشكلة بالبثق	extruded sections
قطّاعة قرصية	disc cutter
قطّاعة مسننة	toothed cutter
قطب الجنوب المغنطيسي	magnetic south pole
قطب الشمال المغنطيسي	magnetic north pole
قطر داخلي	internal diameter
قطر الحبل	rope diameter
قطران	pitch-bitumen, tar
قطع	cutting, resection
قطع : قص	shear
قطع إسفيني	wedge cut
قطع الخرسانة	concrete cutting
قطع باللهب	flame cutting
قطع تركيب الأنابيب	pipe fittings
قطع سفلي	draw cut
قطع غيار	spare parts
قطع متعرج	indentation
قطع مكافئ	parabola
قطع من الحجر المنحوت	ashlar pieces
قطع هرمي	pyramid cut
قطع وردم	cut and fill
قطعة اختبار	test piece
قطعة أرض	plot
قطعة الارتداد	kicking piece
قطعة الشفة	lip block
قطعة الكسر	breaking piece
قطعة ضيقة	strip
قطن البارود	guncotton
قفاز جلد مدبوغ بالكروم	chrome-tanned leather glove
قفازات	gloves
قفازات مطاطية	rubber gloves
قفازات معزولة كهربائياً	electrically insulated gloves
قفز هيدرولي	hydraulic jump
قفص	cage
قفل	lock
قفل مبيت	mortise lock
قلاّبة	dumper

ق

قلاّبة الجرافة	breast
قلاّبة المحراث	breast boards
قلاّبة يتحكم فيها عامل سيار	pedestrian-controlled dumper
قلب	core
قلب الركيزة	pile core
قلوب القص	shear core
قلب سائب	loose core
قلب صخري	rock core
قلب كتيم	impervious core
قلب مركزي	central core
قلب هوائي	air core
قلم رصاص	pencil
قلنسوة	cap
قلنسوة الديناميت	dynamite cap
قمة	crown
قميص البئر	well casing
قميص بئر الماء	water-well casing
قناة	canal, channel
قناة اصطناعية	aqueduct
قناة اعتراضية	intercepting channel
قناة الاقتراب	approach channel
قناة البالوعة	gully
قناة الجزر	ebb channel
قناة الخروج	outlet channel
قناة الري	irrigation channel
قناة الفيضان	flood channel
قناة الماء الفائض	wastewater channel
قناة المجرى المائل	chute channel
قناة الهويس	lock paddle
قناة أوتوماتية لتصريف الفائض بالطرد	automatic siphon spillway
قناة تحويل	bye channel
قناة تصريف	spillway
قناة تصريف الزيت	oil discharge duct
قناة تصريف الفائض	overflow spillway, channel spillway
قناة تصريف بئر التهوية	shaft spillway
قناة تصريف بمجرى مغلق	closed-conduit spillway
قناة تصريف سيفونية	siphon spillway
قناة تصريف مائلة	chute spillway
قناة تصريف ناقوسية الفم	bellmouth spillway
القناة الحيادية لأكسيد المعدن شبه الناقل	N-channel metal oxide semiconductor
قناة رئيسية	main canal
قناة صالحة للملاحة	navigable channel
قناة صرف	drainage channel
قناة فرعية	lateral canal
قناة فنتورية	Venturi flume
قناة مبطنة	lined channel
قناة مفتوحة	troughing
قناة مكشوفة	open channel
القناة الموجبة لأكسيد المعدن شبه الناقل	P-channel metal oxide semiconductor
قنب	hemp
قنبلة ملتصقة	limpet
قنطرة	arch
قنطرة احتجاز بهيكل ثابت	needle weir
قنطرة المرفاع	crane gantry
قنطرة ثابتة	rigid arch
قنطرة ثلاثية المفاصل	three-hinged arch
قنطرة ثنائية المفصل	two-hinged arch
قنطرة خشبية	timber arch
قنطرة ذات عوارض	girder bridge
قنطرة كابولية	cantilever bridge
قنطرة متحركة	travelling gantry
قنطرة مسطحة	platform gantry
قنطرة مضلعة	ribbed arch
قنطرة معيارية	gauged arch
قنطرة نصف دائرية	semi-circular arch
قواعد	footings

ق

قواعد الطرق	road foundations
قواعد الطيران البصري	visual flight rules
قواعد جهاز التبريد	refrigerator foundations
قوانين	regulations
قوانين البناء	building code
قوانين المرور	highway legislation
قوانين ممارسة العمل	code of practice
قوة	force, strength
قوة الجر	drawbar pull, tractive force
قوة الخضوع الجانبية	offset yield strength
قوة الزلزال	earthquake force
قوة الريح	wind force
قوة الشد	tensile force
قوة القص	shearing force
قوة القصور الذاتي	inertial force
قوة دفع الجسر	bridge thrust
قوة ضاغطة	compressive force
قوة نابذة	centrifugal force
قوس جملوني	trussed arch
قوس حجري	blind arch
قوس قطاعي	segmental arch
قوس كهربائي	arc
قوس مائل المحور	oblique arch
قوى التحطيم	wracking forces
قوى التيار	current forces
قوى المقاومة	drag forces
قوى حرارية	thermal forces
قياس	measurement, gauging
قياس الارتفاعات	altimetry
قياس الاضلاع الثلاثة	trilateration
قياس البعد البصري	optical distance measurement
قياس التدفق	flow metering
قياس التدفق الكيماوي	chemi-hydrometry
قياس الترددات الصوتية	audio measurement
قياس بالسلسلة	chaining
قياس الضغط	pressure measurement
قياس الكميات	measurement of quantity
قياس المساحات	planimetry
قياس المناسيب التبادلي	reciprocal levelling
قياس تكعيبي	perch
قياس كثافة الدخان	smoke density metering
القياس الكيماوي	chemical gauging
قياس المسافة الالكترونية	electronic distance measurement
قياسات العمق المسبور	lead-line soundings
قيسون	caisson
قيسون أمريكي	American caisson
قيسون « بوستون »	Boston caisson, Gow caisson
قيسون « بنتون »	Benoto caisson
قيسون بهواء مضغوط	pneumatic caisson
قيسون خرساني	concrete caisson
قيسون خشبي	potomac caisson
قيسون عائم	floating caisson, ship caisson
قيسون « شيكاغو »	Chicago caisson
قيسون صندوقي	box caisson
قيسون صندوقي مغلق	closed-box caisson
قيسون مصفح	sheeted caisson
قيسون مفتوح	open caisson
قيسون يعمل بالهواء المضغوط	compressed air caisson
القيمة الأكثر احتمالاً	most probable value
قيمة الحجر المصقول	polished-stone value
القيمة الحرارية	calorific value, heating value
القيمة الصغرى	least count
القيمة المفضلة	desired value
قيمة مقاسة	measured value
كابول : كتيفة معلقة	cantilever

ق

كاتب أعمال إنشائية	clerk of works
كاتب تكاليف	cost clerk
كاتيون : شاردة موجبة	cation
كاثود	cathode
كاربورندم	Carborundum (TM)
كاشطة باطارات مطاطية	rubber-tyred scraper
كاشطة حفر هيدرولية	hydraulic excavator scraper
كاشطة رش حلزونية	cyclonic spray scrubber
كاشطة على عجلات	bowl scraper
كاشطة فنتورية	Venturi scrubber
كانتين : مطعم	canteen
كاوية لحام	soldering iron
كِپْ (كيلو پاوند)	kip
كباس متذبذب	oscillating piston
كباش المطرقة	hammer grab
كبح	check
كبل	cable, rope
كبل اسطوانة الحفر	wire-line core barrel
كبل التعليق	suspension cable
كبل الخدمة	service cable
كبل السحب	dragline
كبل تعليق	track cable
كبل خلفي	trailing cable
كبل رئيسي	main cable
كبل ضغطي	pressure cable
كبل طاقة للجهد العالي	high-voltage power cable
كبل فولاذي	steel cable
كبل كهربائي	electric cable
كبل مدرع	armoured cable
كبل مدرع لا فلزي	non-metallic sheathed cable
كبل مساعد	auxiliary cable
كبل معزول	insulated cable
كبل من أسلاك الفولاذ	steel-wire rope
كبل نقل الطاقة	power cable
كبلات طاقة متعددة الأسلاك	multicore power cable
كبلات متعددة الأزواج	multipair cables
كبلات متعددة الأسلاك	multicore cable
كتب الاجراءات	manuals of procedures
كتب الارشادات الدائمة	manuals of standing instructions
كتف	abutment, shoulder
كتف العقد	haunch
كتف جانبية	counterfort
كتف زافرة	flying buttress
كتل الارساء	docking blocks
كتل اسمنت	cement blocks
كتل بناء	building blocks
كتل ترشيح	filter blocks
كتل حجر الزاوية	heel blocks
كتلة	block
كتلة الأساس	foundation block
كتلة حديد	bloom
كتلة خرسانية طرفية	end block
كتلة خشبية	billet
كتلة رأس الدعامة	cap block
كتلة رافدة القص	keel block
كتلة محجبة	shielded block
كتلة معدن خام مكشطة	pig
كتلة معدنية	slug
كتلة مطاطية بأسفل البلاطة الخرسانية	void former
كتلة منع الانزلاق	scotch block
كتيفات مدعومة على أعمدة	brackets on columns
كتيفة	bracket
كتيفة حوض جاف	altar
كتيفة مسنمة	bowstring truss
كتيفة من زاوية حديدية	angle cleat
كتيم	hardpan
كثافة	density

ك

كثافة الجفاف القصوى	maximum dry density
الكثافة الظاهرية	bulk density
الكثافة النسبية	relative density
الكثافة النسبية للرمل	relative density of a sand
الكراءة بالقواديس	bucket dredging
كراءة بالقواديس الدوارة	compound dredger
كراءة بمضخة رمل	sand pump dredger
كراءة تعمل ميكانيكياً	mechanically operated dredger
كراءة حفر ماصة	cutter suction dredger
كراءة ذات كباش	grab-dredger
كراءة سلسلية	chain-bucket dredger
كراءة غرف	scoop dredger
كراءة ثابتة	stationary dredger
كراءة سحب عادي	plain suction dredger
كراءة قادوسية	hopper dredger
كراءة قادوسية ماصة	hopper suction dredger
كراءة قطع ماصة	suction-cutter dredger
كراءة ماصة	suction dredger
كراءة مقطورة	draghead
كراءة ميكانيكية	mechanical dredger
كراءة هيدرولية ماصة	hydraulic dredger
كربنة بدون كربون	blank carburizing
كربون	carbon
كربيد التنغستين	tungsten carbide
كربيد السليكون	silicon carbide
كربيد ملبد	cemented carbides
كرة تسليك المجاري	sewer pill
كرسي تثبيت	rail fastening, rail chair
كريوزوت	creosote
كسارة	ripper
كسارة أولية	primary breaker
كسارة خرسانة	concrete breaker
كسارة خرسانة هيدرولية	hydraulic concrete breaker
كسارة رحوية	rotary breaker
كسارة رصف	road breaker
كسارة صخرية	rock rubble
كسارة صخور ذات فكين	jaw breaker
كسارة عشوائية	random rubble
كسارة لفافة	gyratory crusher
كسا بالاسمنت	revet
كشف الرواتب	payroll
كشف انحراف السكة	rail-defect detection
كشك	kiosk
كعب	heel
الكفاية الاجمالية	overall efficiency
الكفاية الحجمية	volumetric efficiency
كلاّب	dog, cleat
كلال	fatigue
كلال التآكل	corrosion fatigue
كلس	lime
كلس بالتحميص	calcine
كلفن	kelvin
كلمة دليلية	keyword
كلمة السر	password
كلوريد متعدد الفنيل	polyvinyl chloride
كلوريد الكالسيوم	calcium chloride
كلينومتر : مقياس الميل	clinometer
كمُ	sleeve
كمُ بوليثين أنبوبي	tubular polythene sleeve
كمُ شدّ	tension sleeve
كماشة	pinchers
كمبيوتر	computer
كمبيوتر رقمي	digital computer
كمبيوتر سريع	cache memory
كمبيوتر صغير	minicomputer
كمبيوتر للأغراض الخاصة	special purpose computer
كمُية الكبل	capel

ك

الكمية المتآثرة operand
كمية التحرك momentum
كنتور contour
كنديلا candela
الكنس brooming
كنس الشوارع street sweeping
كنيسة church
كهرباء electricity
كهربائي electric
كوارتز quartz
كوارتزيت quartzite
كوب العينات sampling cock
كوخ hut
كورندم : ياقوت corundum
كُوس set square
كوع bend, elbow
كولكريت Colcrete (TM)
كولوم : وحدة كهربائية coulomb
كومة pile
كيروسين kerosene
كيلوغرام kilogram
كيلومتر kilometre
كيلوواط ساعة kilowatt hour
لا تمركزي eccentric
لاسلكي radio
لاكيه lacquer
لا مركزية decentralised
لا يقبل المفاوضة not negotiable
لبّاد felt
لبّاد قاري bituminous felt
لباس الغوص diving cap
لباس الغوص المعياري standard diving gear
لباس واقٍ protective clothing
اللبلاب ivy
لجنة الأمان safety committee
لحام soldering, welding, weld

لحام الأنابيب pipe welding
لحام أوتوماتي automatic welding
لحام قوسي مغمور submerged-arc welding
لحام بالأكسجين MAG welding
لحام بالأكسجين والأستيلين oxy-acetylene welding
لحام بالبرونز bronze welding
لحام بالثرميت thermit weld
لحام بالحرارة والتطريق forge welding
لحام بالصهر fusion welding
لحام بالغاز gas welding
لحام بالفولاذ اللدن plastic welding
لحام بالقوس الكهربائي electric-arc welding
لحام بالقوس المعدني metal-arc welding
لحام بالكهرباء electric welding
لحام بالمقاومة resistance welding
لحام بالنحاس braze welding, copper welding
لحام بالورشة shop weld
لحام بثاني أوكسيد الكربون carbon dioxide welding
لحام بسلك معدني MIG welding
لحام بقوس الكربون carbon-arc welding
لحام تحت سطح الماء wet welding
لحام تناكبي butt weld
لحام درزي seam welding
لحام درزي بالمقاومة resistance seam welding
لحام ذاتي autogenous welding
لحام زاوي fillet weld
لحام سريع quick solder
لحام ضغطي pressure welding
لحام في جو من الهيدروجين atomic-hydrogen welding
لحام قوسي arc welding
لحام قوسي أرجوني argon-arc welding
لحام قوسي محجب shielded-arc welding
لحام كهربائي مستمر electroslag welding
لحام معدني stick welding

ل

ل

لحام نتوئي	projection welding
لحام نقطي	spot welding, tack weld
لحام نقطي بالمقاومة	resistance spot welding
لحام ومضي	flash welding
لحام ومضي بالمقاومة	resistance flash welding
لحام يدوي بالقوس المعدني	MMA welding, manual metal-arc welding
لدائن	plastics
لدائن قارية	bituminous plastics
لدانة	plasticity
لدن بالحرارة	thermoplastic
لزوجة	viscosity
لسان توصيل ذكر	pin connection
لصاق زفتي	asphalt cement
لصق ورق الجدران	paperhanging
لغات تفسيرية	interpretative languages
لغة البرمجة	programming language
لافتات دليلية على المدرج	taxiway guidance signs
لافتات غير مضاءة	non-illuminated signs
لافتة	sign board
لافتة مضاءة	illuminated sign
لافظ هيدرولي	hydraulic ejector
لفيفة فيلم	cartridge
لقمة بشكل ذيل السمكة	fishtail bit
لقمة حفر	bit
لقمة حفر الطين	clay cutter
لقمة حفر بمسننات دوارة	roller bit
لقمة حفر يمكن فصلها	detachable bit
لقمة صفر	zero bit
لقمة لولبة	die
لكس : وحدة ضوئية	lux
لهب الأكسجين والأستيلين	oxy-acetylene flame
لوائح وقوانين البناء	building regulations
لوازم الاسعاف الأولي	first aid supplies
لوازم التسييج	fencing
لوح : لوحة	board, plate, panel

لوح اعتراضي لتغيير اتجاه التيار	baffle plate
لوح أفقي	head board
ألواح الأرضية	floor boards
لوح الارساء	anchor plate
لوح الأساس	bed plate
ألواح البئر المتحجزة	well curbing
لوح التدفق الفوهي	orifice plate
لوح التشكيل	mouldboard
لوح الدخان	smoke plate
لوح الركائز	piling board
لوح القاعدة	base plate, base board
لوح المفرق	frog plate
لوح المفاتيح	switchboard
لوح تبطين	liner plate
لوحة تحكم	control panel
لوح تثبيت	middling-board
ألواح تثبيت العقد	lagging
ألواح تغطية السقف	ceiling laths
لوح تقوية	gusset plate
لوح جانبي	side board
ألواح الجدران الخشبية	siding
ألواح حماية جوانب الخندق	sheeters
لوح خارجي	exterior panel
ألواح خشب مقواة قطرياً	racked timbering
لوح خشبي	batten plate
ألواح دعم	cross poling, poling boards
لوح دعم	tucking board
لوح دعم جانبي	head tree
لوح ربط	lacing board, tie plate
لوح ربط أفقي	waling, soldier waling
ألواح السد	stop planks
ألواح السقف	roof boards
ألواح أرضية	floor boards
لوح صلد	hardboard
ألواح طرفي المحولة	switch side plates

لوح طولي	strake
لوح عازل	insulating board
لوح فولاذي	steel sheet, sheet steel
لوح فولاذي خلوي	cellular steel panel
لوح فولاذ صلب	stiffened steel plate
لوح فولاذي مثقوب	chequer plate
لوح فولاذي مثلث	benching iron
لوح فولاذي ملحوم	welded steel plate
ألواح قاعدة منحدرة	sliced blockwork
لوح ليفي	fibre board
ألواح متباعدة	open sheeting
لوح معدني	sheet metal
لوح مموج	corrugated sheet
لوح من الفولاذ المموج	corrugated steel sheet
ألواح نايلون	sheets of nylon
لوح هزاز	vibrating plate
لوح وصل تراكبي	splice bar, shin
لوحة رسم	drawing board
لوحة مسح مستوية	plane table
لوحة مفاتيح التنضيد	keyboard
لولب دودي	perpetual screw
لولب مماس	tangent screw
لون خفيف	tint
لوي	twist
ليف الجوت	jute fibre
ليف اصطناعي	manmade fibre
لينوليوم	linoleum
ماء التغذية	feedwater
ماء التربة السفلية	subsoil water
ماء الجاذبية الشعرية	water of capillarity
ماء الشرب	potable water, drinking water
ماء المطر	rainwater
الماء المهدر من المرحاض	wc wastewater
ماء جار في دائرة محصورة	circulating water
ماء عسر	hard water
ماء شرب ملوث	contaminated drinking water
ماء شعري	capillary water
ماء غير محتجز	unconfined water
ماء غير معالج	raw water
ماء فائض	wastewater
ماء «ڤادوز»	vadose water
ماء قذر	foul water
ماء لاصق	pellicular water
ماء مستهلك	spent water
ماء محجوز	bound water, back water
ماء مزال منه المعدن	de-mineralized water
ماء معالج	treated water
ماء يجري بالجاذبية	gravitational water
ماء يحوي أملاح مذابة	leachate
ماء يسر	soft water
ماء يسري بثقل الجاذبية	free water
مائل	oblique, cant
مئة	hundred
مئزر السكة	track apron
مأخذ التيار	socket outlet
مأخذ الهواء الساخن	hot-air intake
مأخذ رئيسي صاعد	rising main
المأخذ الرئيسي لمياه الاطفاء	fire mains
مادة إسمنتية	cementitious material
مادة الترابط	matrix
مادة صامدة للماء	waterproofing agent
مادة بلورية صلبة	whisker
مادة ترشيح	filter material
مادة تفتيت	release agent
مادة جيرية مقاومة للحرارة	basic refractory
مادة سائبة	loose material
مادة صامدة للنار	fireproof material
مادة عازلة	insulating material
مادة فرش الطبقة الحبيبية	gritting material

مادة قابلة للاحتراق	combustible material
مادة كتيمة	impermeable material
مادة ماصة للصوت	sound-absorbing material
مادة مانعة للتسرب	sealant
مادة مانعة للتسرب ثنائية	two-part sealant
مادة متجانسة	homogeneous material
مادة مدمجة	compact material
مادة مضافة	additive
مازوت	fuel oil
ماسورة	pipe
ماسورة الأسلاك	cable duct, conduit
ماسورة الامداد	supply pipe
ماسورة التمَوُّر	surge pipe
ماسورة الدفق	flow pipe
ماسورة الطفح	overflow pipe
ماسورة المص	suction pipe
ماسورة إمداد صاعدة	service riser
ماسورة حفر	drive pipe
ماسورة حفظ العينة	core barrel
ماسورة خرسانية	concrete pipe
ماسورة سحب	draught tube
ماسورة صاعدة	riser
ماسورة فرعية	lateral conduit
ماسورة قائمة	stand pipe
ماسورة قائمة لتصريف الفائض	overflow stand
ماسورة ماء المطر	rainwater pipe
ماسورة مبطنة بالمطاط	rubber-lined pipe
ماسورة مزدوجة	dual-conduit
ماسورة مغلفنة	galvanized pipe
ماسورة مغمورة	immersed tube
ماكينة تشطيب الخرسانة	concrete-finishing machine
ماكينة تفتيت	comminutor
ماكينة ذات برميل قلاّب	tilting-drum machine
ماكينة صب القوالب	brick moulding machine
ماكينة ضغط	compactor
مالك	owner
مانع الضجيج	noise eliminator
مانع للتفاعل الكيماوي	inhibitor
مانومتر (مقياس ضغط)	manometer
مبادئ التشغيل	operation code
مبادئ للطيران بأجهزة القياس	instrument flight rules
مبارد شحذ المناشير	saw files
مباعدة	spacing
مباعدة سلك التأريض	earthwire spacing
مبان	buildings
مبان فرعية	ancillary buildings
مبدد طاقة قادوسي	bucket energy dissipator
مبرد	file
مبرِّد الزيت	oil cooler
مبرِّد بالسوائل	liquid cooler
مبرِّد بيني	intercooler
مبرد ذيل الفأر	rat-tail file
مبرِّد لاحق	aftercooler
مبرد مستدق	taper file
مبرد مستعرض	draw-file
مبرد مفرد القطعية	single-cut file
مبرِّد هوائي	air cooler
مبنى	building
مبنى الأجهزة الالكترونية	building for electronic equipment
مبنى التهوية	ventilation building
مبنى الحرس	guardhouse
مبنى الغلاية	boiler house
مبنى المحركات	engine house
مبنى المراقبة والتحكم	control block
مبنى المستودع	store building
مبنى بهيكل كامل	simple framed building
مبنى تموين الطائرات بالطعام	aircraft catering building

م

مبنى جاهز التركيبات الهندسية	pre-engineered building
مبنى شاهق الارتفاع	tower building
مبنى صناعي	industrial building
مبنى متنقل	portable building
مبنى محقون	grouted masonry
مبنى مضخة ماء التبريد	cooling-water pumphouse
مبنى هيكل الرفع	hoistway enclosure
مباني الميناء	port buildings
مبيِّض	plasterer
متآلف	monolithic
متانة	toughness, tenacity
متانة المواد	strength of materials
المتانة المميزة	characteristic strength
متباين الخواص	anisotropic
متجر كبير	department store
متحف	museum
متخلي	disclaimer
متدرب	apprentice
متدلٍ	overhanging
متر	metre
متراس	barricade
منزعة الركائز	pile-drawer
متر مربع	square metre
متر مكعب	cubic metre
متشعب	manifold
متطلبات الانشاء	construction requirements
متطلبات التهوية	ventilation requirement
متطلبات الري	irrigation requirement
المتطلبات الزراعية	agricultural requirement
متطلبات العطاء	bidding requirements
متطلبات الماء	water requirement
متطلبات طاقة التهوية	ventilation power requirements
متطلبات مفصلة	detailed requirements
متفجرات	explosives
متفجرات جيلاتينية	gelatine explosives
متفجر الحمأة	slurry explosive
متعدد	manifold
متعدد المناطق	multizone
متعهد	contractor
متكأ حديدي	step iron
متكامل	self-contained
متناظر	isometric
متنزه	park
متنقل	portable
متوتر	tensor
متوسط	mean, average
متوسط ارتفاع الماء	mean high water
متوسط الحرارة المشعة	mean radiant temperature
متوسط العمق	mean depth
متوسط العمق الهيدرولي	hydraulic mean depth
متوسط انخفاض الماء	mean low water
متوسط مستوى سطح البحر	mean sea level
متوسط موزون	weighted average
مثاقب بحرية	marine borers
مثاقب حفر	auger drills
مثبتات الحبل	rope fastenings
مثبتات عوامة الارساء	mooring-buoy anchor
مثبت أرضي	ground anchor
مثبت السكة	rail anchor
مثبت الفولاذ	steelfixer
مثبت جزئياً	partially fixed
مثبت خط حديدي	chair
مثبت كبل التعليق	suspension-cable anchor
مثبت الطرف	end-fixed
مثقاب	auger, drill
مثقاب الثقوب الكبيرة	post-hole auger

م

مثقاب العلامات	sign bit
مثقاب صدر	breast drill
مثقاب عمودي	pole drill
مثقاب كهربائي	electric drill
مثقاب لاستخراج العينات	core drill
مثقاب مطرقي	hammer drill
مثقاب يعمل بالهواء المضغوط	pneumatic drill
مثقب توسيع	pin drill
مثقب حجارة	puncheon
مثقب منشاري	trepan
مثلث الخطأ	triangle of error
مثلث « وايزباخ »	Weisbach triangle
مثمن	estimator
مثير	exciter
مجاري المصب	receiving waterways
مجاري المياه القذرة	foul water sewerage
مجاري أنبوبية	pipe sewer
مجاري مشتركة	combined sewers
مجاري مياه المصانع	industrial sewage
مجار منعزلة	separate system
مجار منعزلة نوعاً	partially-separate system
مجاز ضيق	throat
مجال	range
مجال الرؤية	sight distance
مجراف تحميل	loading shovel
مجراف تحميل جرار	tractor shovel
مجرفة	shovel
مجرفة آلية	power shovel, navvy
مجرفة بخارية	steam shovel
مجرفة تحميل	traxcavator
مجرفة دفعية	push shovel
مجرفة عكسية	backacter
مجرفة قلابة	rocker shovel
مجرفة ميكانيكية	mechanical shovel
مجرور	sewer
مجرور الأوساخ	foul sewer
مجرور المبنى	building sewer
مجرور بقطر كبير	large-diameter sewer
مجرور بيضاوي الشكل	egg-shaped sewer
مجرور تحت للمواسير المطمورة	submain sewer
مجرور تنفيس	relief sewer
مجرور خارجي رئيسي	outfall sewer
مجرور رئيسي	trunk sewer, main sewer
مجرور صحي	sanitary sewer
مجرور ماء المطر	storm sewer
مجرور متفرع	branch sewer
مجرور مخرج التصريف	outfall culvert
مجرور مشترك	common sewer
مجرى	culvert, duct
مجرى الخروج السفلي	outlet culvert
مجرى المسمار الملولب	bolt sleeve
مجرى الصرف	wasteway
مجرى الفائض	rollway
مجرى الماء الرأسي	head race
مجرى تصريف الأمطار	storm-overflow sewer
مجرى خرساني	concrete culvert
مجرى سحب سفلي	draw-off culvert
مجرى فولاذي	steel culvert
مجرى قناة الصرف	drainage gallery
مجرى ماء	leat
مجرى ماء التبريد	cooling-water culvert
مجرى مائل	chute
مجرى مائي	watercourse
مجرى مزدوج	double duct
مجرى هواء	air flow
مجرى وحيد	single duct
مجسام	stereoscope
مجفف السائل	liquid dryer
مجفف بالهواء الساخن	hot-air dryer
مُجمع رمزي	symbolic assembler

م

مجمع مستودعات	store compound
مجموعة المفاتيح الكهربائية	electrical switchgear, switchgear
مجموعة بكرات تفاضلية	differential pulley block
مجموعة تصنيف الحمل	Load Classification Group
مجموعة ركائز	pile group
مجموعة غلايات	boiler group
مجموعة لوحية	plate section
مجموعة محولات	bank of transformers
مجموعة مفاتيح صناعية	industrial switchgear
محاذاة	alignment
محاذاة الشق	guideway alignment
محاذاة القناة	channel alignment
محاذاة المنحنى	ranging a curve
محاري الشكل	conchoidal
محامل دلفينية	roller bearings
محامل طرفية	end bearings
محاور متعامدة	perpendicular axes
محبس	stopcock
محبس المجرور	gully trap
محبس مطافئ	hydrant
محبس ملتق	junction cock
محتوى الاسمنت الأدنى	minimum cement content
محتوى الاسمنت الأقصى	maximum cement content
محتوى الرطوبة المثلى	optimum moisture content
المحتوى الرطوبي	moisture content
المحتوى الرمادي	ash content
المحتوى العضوي	organic content
المحتوى المائي	water content
المحتوى الهوائي للخرسانة الجديدة	air content of fresh concrete
محجر	quarry

محدلة اهتزازية	vibrating roller
محراك	paddle
محراك هزاز	vibrating poker
محرض	exciter
محرك	engine
محرك أساسي	prime mover
محرك المصعد	lift motor
محرك بخاري	beam engine, steam engine
محرك ترددي	reciprocating engine
محرك حثي خطي	linear induction motor
محرك داخلي الاحتراق	internal-combustion engine
محرك ذو كباس	piston engine
محرك ديزل	diesel engine
محرك ذو سرعتين	two-speed motor
محرك رفع	hoisting engine, lifting motor
محرك صامت	noiseless engine
محرك عام	universal motor
محرك ذو عزم مرتفع	high-torque motor
محرك قفص السنجاب	squirrel-cage motor
محرك كهربائي	electric motor
محرك متعدد المراحل	multiple-expansion engine
محرك مركب	compound engine
محرك هزاز	vibratory driver
محرك وحيد السرعة	single speed motor
محشو	packed
محصدة حازمة هيدرولية	hydraulic binder
محط طائرات الهليكوبتر	heliport
محطة	station
محطة إزالة الملوحة	desalination plant
محطة أقمار صناعية	satellite terminal
محطة البضائع	cargo terminal
محطة البضائع السائبة	bulk cargo terminal

م

محطة التربين	turbine house
محطة الركاب	passenger terminal
محطة الضخ	pumping station
محطة الطاقة	power house
محطة المساحة	survey station
محطة المياه	waterworks
محطة تحويل	transformer station
محطة تسخين	heating plant
محطة توليد	generating plant
محطة توليد تحت الأرض	underground power station
محطة توليد الطاقة	power station
محطة توليد بالطاقة النووية	nuclear power station
محطة توليد مائية	hydroelectric power station
محطة سكة حديدية	railway station
محطة شحن البضائع	freight terminal
محطة ضخ أوتوماتي للمجاري	automatic sewage pumping station
محطة ضخ شاطئية	pump-ashore plant
محطة طاقة تعمل بالمازوت	oil-fired power station
محطة طاقة نووية	nuclear energy plant
محطة طرفية	terminal station
محطة فرعية	substations
محطة لتوليد الطاقة تدار عن بعد	telecontrolled power station
محطة معالجة	treatment plant
محطة معالجة النفاية النشطة	active-effluent treatment plant
محطة معالجة بالكلور	chlorination plant
محطة معالجة بثاني أكسيد الكربون	carbon dioxide treatment plant
محطة معالجة مياه المجارير	sewage treatment works
محطة مناولة الحاويات	container-handling terminal
محطة هيدرولية	hydraulic works
محللة إلكترونية	electronic analyser
محمل	bearing
محمل أسيطينات إبري	needle roller bearing
محمل الجسر	bridge bearing
محمل الهزازة	rocker bearing
محمل تمدد	expansion bearing
محمل ذوكتيفات	bracketing
محمل صلب	rigid cradle
محمل كريات	ball bearing
محمل مثبت	fixed bearing
محمل معدني	metal bearing
محور	axis
محور التعادل	neutral axis
محور السد	dam axis
محور دوران	trunnion axis
محول	transformer
محول الطاقة	power transformer
محول الفلطية	voltage transformer
محول وناقل الطاقة	transducer
محولات حجز	trap points
محولات خلفية	trailing points
محولات سكة حديد نابضية	spring points
محوّلة	switch
محيط حراري	thermal environment
محيط مبتل	wetted perimeter
محيط مرصوف	paved surround
محيطي	peripheral
مختبر	laboratory
مختبر اختبار المواد	material testing laboratory
مختبر العينات	sampler
مخدشة	scarifier
مخدشة مقطورة	rooter
مخرج	exit, outlet
مخرج التصريف	outfall drain, outfall
مخرج الخرسانة	concrete outlet

م

مخرج الخزان	reservoir outlet
مخرج الرحض	flushing outlet
مخرج الزيت	oil outlet
مخرج الماء	water outlet
مخرج تصريف بحري	sea outfall
مخرج سفلي	bottom outlet
مخزن	store, magazine
مخزن المعدات	equipment-storage building
مخزن مبرد	cold storage building
مخزن وقود	bunker
مخزون الضفة	bank storage
مخطط	layout
مخطط الأرضية	floor plan
مخطط الأساس	ground plan
مخطط البناء	buildings layout
مخطط التردد	frequency diagram
مخطط التقاطع	interchange layout
مخطط الجناح	terminal area layout
مخطط المدرج	runway layout
مخطط المطار	airport layout
مخطط المعدات	plant layout
مخطط الميناء	harbour layout
مخطط أمامي	frontal layout
مخطط بياني	alignment chart
مخطط بياني قضيبي	bar chart
مخطط بياني للعمليات المتعاقبة	process chart
مخطط تمهيدي	preliminary layout
مخطط هيكلي	skeleton
مخفض الضغط	pressure reducer
مخل	crowbar
مخمد	damper
مخمد المصد	Raykin fender buffer
مخمد حريق أوتوماتي	automatic fire damper
مد الأنابيب بالدفع	pipe-pushing
مد أو جزر تام	spring tides
مد (أو جزر) شبه نهاري	semi-diurnal tides
مد أو جزر ناقص	neap tide
مد أو جزر نهاري	diurnal tides
مدارج متوازية	parallel runways
مداد أنابيب	pipelayer
مداس	tread
مدة الصيانة	maintenance period
مدحلة	tamping roller, roller
مدحلة بخارية	steam roller
مدحلة بدواليب متراوحة	wobble-wheel roller
مدحلة تحزيز	crimper, indenting roller
مدحلة ترادفية	tandem roller
مدحلة تسوية	smoothing iron
مدحلة دك	tamping roller
مدحلة طرق	road roller
مدخل	entrance, threshold, inlet
مدخل العمال	workers' entrance
مدخل الماء	water inlet
مدخل المصعد	lift entrance
مدخل الميناء	harbour entrance
مدخل جانبي	side entrance
مدخل خارجي مسقوف	porch
مدخل مقاوم للريح	wind portal
مدخل نهري	river intake
مدخنة	chimney, stack
مدخنة متعددة المحاري	multi-flue chimney
مدرج	runway, taxiway
مدرج بدون أجهزة	non-instrument runway
مدرج طيران	apron
مدرج للاقتراب بأجهزة القياس	instrument approach runway
مدرج مرصوف	paved apron
مدرج هبوط	landing strip
مدرجة	grader

م

مدرجة آلية	autopatrol, motor grader
مدرجة طريق	road grader
مدرجة ممهدة	blade grader
مدرسة	school
مدفأة	heater
مدفأة كهربائية	electrical fire
مدفوعات نقدية	cash payment
مدق الركائز	pile-driver
مدق ركائز هيدرولي	silent pile-driver
مدق ركائز هزاز	sonic pile-driver, vibrating pile-driver
مدك	ram, tamper
مدك آلي	power rammer
مدك بالضغط	pressure-assisted ram
مدك مطرقة	hammer ram
مدك ميكانيكي	mechanical rammer
مدك هزاز	vibrating tamper
مدك يدوي	punner
مدى	range
مدير العقود	contract manager
مرآب (كراج)	garage
مرآة الأفق	horizon glass
مرآة دليلية	index glass
مراجعة	revision
مراسي إرساء	mooring dolphins
مرافع متسلقة	creeper cranes
مرافق المحطة البحرية	marine terminal facilities
مرافق المحطة البحرية الساحلية	onshore marine-terminal facilities
مرافق الميناء	port services
مرافق خدمة السفن	ship servicing facilities
المرافق العامة	public utilities
مرافئ «ستول»	STOL ports

مراقب	chargehand, checker
مراقب إنقاذ	life linesman
مراقبة	control
مراقبة الترصيف	stack monitor
مراقبة التكاليف	cost control
مراقبة السيولة النقدية	cash flow control
مراقبة العمليات المتعاقبة	process monitoring
مراقبة المدرج	apron control
مراقبة الميزانية	budgetary control
مراقبة بالشيفرة	coded control
مراقبة حركة المرور الجوي	air traffic control
مراقبة مالية	financial control
مراقبة مركزية للمرور	centralised traffic control
مربط	holdfast, cleat
مربط وصل البطارية	battery terminal
مرتبة الفولاذ	grade of steel
مرتكز جدار الدعم	toe
مرتكزات المحولة	switch toes
مرجة	lawn
مرجع إسناد	datum
مرجل بأنابيب مائية	water-tube boiler
مرحاض ذو حفرة	pit type toilet
مرحاض كيماوي	chemical toilet
مرحاض مؤقت	temporary toilet
مُرحِّل	relay
مرحلة	stage
مرحلة التنفيذ	execution phase
مرساة	anchor
مرسمة	tracer
مرسمة الميل	clinograph
مرسى	deadman, berth
مرسى ثابت للسفن	fixed mooring berth
مرسى جرسي الشكل	bell dolphin, Baker bell dolphin
مرش يدوي	hand sprayer

مرشة إطفاء	sprinkler
مرشة صهريجية	tank sprayer
مرشح	filter
مرشح أحيائي	trickling filter
مرشح بيولوجي	biological filter
مرشح ذو سعة عالية	high-rate filter
مرشح ذو حمام زيتي	oil bath filter
مرشح رملي	sand filter
مرشح رملي بطئ	slow sand filter
مرشح رملي سريع	rapid sand filter
مرشح متدرج الطبقات	graded filter, toe filter
مرشح مثقل	loaded filter
مرشح محبب	granular filter
مرشح مستمر	continuous filter
مرشح معكوس	reversed filter
مرصد	observatory
مرصن	branding iron
مرض التحني	bends
مرطم أمواج	baffle pier, groyne
مرفأ	dock, port
مرفأ اصطناعي	artificial harbour
مرفأ انزلاقي	slip dock
مرفاع	hoist, crane
مرفاع أحادي البرج	monotower crane
مرفاع السيارة	jack
مرفاع المواد	material hoist
مرفاع بالهواء المضغوط	pneumatic hoist
مرفاع بذراع دوار	swing-jib crane
مرفاع بذراع سفلي	luffing jib crane
مرفاع برج الحفر	derrick crane, Scotch derrick
مرفاع برجي	tower crane
مرفاع بقائم صلب	stiff-leg derrick
مرفاع بمنصة	platform hoist
مرفاع تسوية	level-luffing crane
مرفاع ثابت	stationary crane
مرفاع جبار	Titan crane
مرفاع جداري دوار	rotary wall crane
مرفاع دوار	revolver crane
مرفاع ذراعي	jib crane
مرفاع سحب	pull-lift
مرفاع سلمي	ladder jack
مرفاع عائم	floating crane
مرفاع علوي متحرك	overhead travelling crane
مرفاع عمودي	pillar crane
مرفاع قنطري	portal crane
مرفاع قنطري متحرك	travelling gantry crane, gantry crane
مرفاع كابولي	cantilever crane
مرفاع متحرك	travelling crane
مرفاع متسلق	climbing crane
مرفاع القاطرة	locomotive crane
مرفاع متنقل	mobile hoist, portable crane
مرفاع معلّق	overhead crane
مرفاع مقصي	shearlegs
مرفاع ملفافي يدوي	jack roll
مرفاع نقال هيدرولي	hydraulic jack
مرفاع نقالي ضخم	Goliath crane
مرفاع نقل	transporter crane
مرفق	elbow
مرفق بالوعي	sink
مرفق تمدد	expansion bend
مركّب	built up
مركب استخلاص بالفصل	parting agent
مركب الزجاج	glazier
مركب انضاج	curing compound
مركب على عجلات	wheel mounted
مركبة على دلفين	roller mounted
مركبة هوائية برية (مجنزرة)	tracked air-cushion vehicle

م

مركز الثقل	centre of gravity
مركز الشرطة	police station
مركز الضغط	centre of pressure
مركز القص	shear centre
مركز متوسط	centroid
مركزي	centralised
مرمد	incinerator
مرملة	sander
مرن	flexible, elastic
مروحة	fan
مروحة داسرة	propeller fan
مروحة ذات دفق محوري	axial-flow fan
مروحة طرد مساعدة	auxiliary exhaust fan
مرور باتجاه واحد	one way traffic
مرونة ضوئية	photoelasticity
مزامنة البيانات	data synchronisation
مزج بالهواء	aeration
مزدوج الفعل	double acting
مزدوجة حرارية	thermocouple
مزراب مفتوح	open gutter
مزرعة	farm
مزرعة مياه المجاري	sewage farm
مزلاج	latch
مزواة	theodolite
مزواة (تيودوليت)	microptic theodolite
مزاوة بقراءة بصرية	optical-reading theodolite
مزيج	admixture
مزيج جاهز	prescribed mix
مسّاح	surveyor
مسّاح أرضي	land surveyor
مسّاح إداري	managing surveyor
مسّاح طوبوغرافي	topographical surveyor
المساحة	surveying
مساحة القاعدة	base area
مساحة المقطع العرضي	cross-sectional area
المساحة الجوية	aerial surveying
المساحة الداخلية	internal area
المساحة السطحية النوعية	specific surface
مساحة طوبوغرافية	topographical surveying
المساحة المروية	irrigable area
المساحة المسطحة	plane surveying
مسار	path
مسار مزدوج	double-track
مسار السكة	track road
مسار الطيران الأدنى العادي	normal minimum flight path
مسار مرتفع	elevated track
مسار مؤقت	construction way
مساعد عامل المرفاع	banksman
مساعد مهندس مدني	civil engineering assistant
مسافة الارساء	anchorage distance
مسافة الرؤية	visibility distance
مسافة المماس	tangent distance
المسافة بين خطي السكة	track gauge
مسافة كنتورية	contour interval
مسامية	porosity
مسامير كبيرة عاكسة	reflecting studs
مساهم	shareholder
مسبار بالصدى	echo sounder
مسبار رصد	sonde
مسبار يدوي	hand lead
مسبق الاجهاد جزئياً	partial prestressing
مسبق الصب بالمصنع	factory precast
مسبك	foundry
مستتر	blind
مستجمع	catchment area
مستجمع الصرف	drainage area
مستجمع المصب	source catchment
مستجمع النبع	spring catchment
مستحلب	emulsion

م

مستحلب قاري	bituminous emulsion
مستحلب قطراني	tar emulsion
مستدق الطرف	spudding
مستشار	consultant
مستشفى	hospital
مستعمد	orthotropic
مستقر القاعدة	seat
مستقطب	polarizer
مستقيم	straight
مستندات العقد	contract documents
مستندات العطاء	tender documents, bidding documents
المستندات القانونية	legal papers
مستنقع	swamp
مستو	level, flat
مستودع	warehouse, magazine
مستودع غاز	gasometer
مستودع متعهد شحن السفن	stevedores' warehouse
مستوى أفقي	horizontal plane
مستوى الارتفاع	altitude level
مستوى الانزلاق	slip plane
مستوى الانعكاس	invert level
مستوى التحكم أو المراقبة	level of control
مستوى التشبع	saturation line, plane of saturation
مستوى التصدع	plane of rupture
مستوى الضبط الآلي	autoset level
مستوى الضجيج	noise level
مستوى الطريق	road level
مستوى الماء الباطني	piezometric surface
مستوى النصوع	luminance level
مستوى انبعاث الضوضاء	level of noise emission
مستوى أوتوماتي	automatic level
مستوى ذاتي الاستواء	self-levelling level
مستوى ذاتي المحاذاة	self-aligning level
مستوى طرف الركيزة	toe line
مستوى عالٍ	high level
مستوى منخفض	low level
مستويات أمان السكك	track safety standards
مسجل الدوران	rotation recorder
مسجل الوقت	timekeeper
مسجل ثاني أكسيد الكربون	carbon dioxide recorder
مسجل شرائط	tape recorder
مسجل المنسوب	level recorder
مسح	survey
مسح اجتيازي بالبوصلة	needle traverse
مسح أرضي	land surveying, ground survey
مسح الأبعاد	tacheometric, stadia work
مسح الانشاءات	construction survey
مسح التربة	soil survey
مسح الحدود	boundary survey
مسح الحفر	borehole surveying
مسح شمسي	solar surveying
مسح الطرق	route survey
مسح الموقع	site survey
مسح بالقمر الصناعي	satellite surveying
مسح بالسلسلة	chain survey
مسح باللوحة المستوية	plane-table surveying
مسح بحري	marine surveying, sea surveying
مسح تصويري	photogrammetric survey
مسح جوي	aerial survey, air survey
مسح جيوديسي	geodetic surveying
مسح جيوفيزيائي	geophysical surveying
مسح طوبوغرافي	topographic survey
مسح مبدئي	preliminary survey
مسح مثلثي	trigonometrical survey
مسح نهائي للبناء	as-built survey
مسح هيدروغرافي	hydrographic survey

م

مسحاج	jack plane
مسحبة دعائم	prop drawer
مسحوق الاشعال	ignition powder
مسخن	heater
مسخن بالأشعة تحت الحمراء	infra-red heater
مسخن بيني	reheater
مسخن سطح الطريق	road heater
مسخن غازي	gas circulator
مسدس إسمنت	cement gun
مسدس رش الدهان	paint spray gun
مسرح	theatre
مسطرة بشكل T	T-square, tee-square
مسطرة قياس منزلقة	slide rule
مسطرين	trowel
مسطرين زاوي	angle float
مسقط	projection
مسقط رأسي	elevation
مسقط رأسي قطاعي	sectional elevation
مسقط قنطري	Chinaman chute
مسقط مياه السد	overfall
مسلة	needle
مسلك	duct, access
مسلك الهواء	air duct
مسمار أشوك	rag bolt
مسمار المفصلة	gudgeon pin
مسمار تثبيت احتكاكي	interference-body bolt
مسمار (تثبيت) السكة	track spike
مسمار تثبيت محزز	indented bolt
مسمار تمددي ملولب	expansion bolt
مسمار خطافي الشكل	hook bolt
مسمار ذو عروة ملولب	eye bolt
مسمار ربط	holding-down bolt, tie bolt
مسمار ربط احتكاكي	Torshear bolt
مسمار قمط ملولب	clamping bolt
مسمار كبير	spike
مسمار كبير ملولب	screw spike
مسمار لمثبت الخط الحديدي	chair bolt
مسمار «لويس»	Lewis bolt
مسمار ملولب	Rawlbolt, bolt
مسمار ملولب أسود	black bolt
مسمار ملولب مخروط	bright bolt
مسمار ملولب مربع المقطع	dead bolt
مسمار ملولب للتثبيت	anchor bolt
مسمار نابضي	spring spike
مسند	bolster, outrigger
مسند سحب	suction pad, vacuum pad
مسنن عال	high gear
مسواة المساح الدائرية	circular level
مسواة المنحدر	batter level
مسواة المهندس	engineer's level
مسواة بفتحة لزيادة التهوية	chambered-level tube
مسؤولية	responsibility
مسؤولية الأضرار عن الممتلكات	property damage liability
مسؤولية المقاول العامة الشاملة	contractor's comprehensive general liability
المسؤولية التعاقدية	contractual liability
مسؤولية عن الحوادث	body injury liability
المسؤولية عن السيارات	automobile liability
مسؤولية قانونية عامة	comprehensive general liability
مسيل مراقبة	rating flume
مشبك	barrette, clamp
مشحمة السكة	track lubricator
مشرف عام	superintendent
مشروع استصلاح الأراضي	reclamation scheme
مشروع المياه السطحية	surface water scheme
مشروع تحت الأرض	underground scheme
مشروع كهرومائي	hydroelectric scheme

م

مشفّه	flanged
المشكلة الثلاثية النقط	three-point problem
مصارف أنبوبية	pipe drains
مصارف السكة	track drains
مصارف مرفق الاحتباس	interceptor drains
مصب مد وجزر	tidal outfall
مصب النهر	river mouth
مصباح	lamp
مصباح إضاءة قادوسي	hopper light
مصباح بخار الصوديوم	sodium-vapour lamp
مصباح بخار صوديوم عالي الضغط	high-pressure sodium vapour lamp
مصباح بخار صوديوم منخفض الضغط	low-pressure sodium vapour lamp
مصباح دليلي	pilot lamp
مصباح زئبقي	mercury-vapour lamp
مصباح الشارع	street lamp
مصباح فلوري زئبقي	mercury-vapour fluorescent lamp
مصد	fender
مصد أمواج ركامي	rock-mounded breakwater, mound breakwater
مصد الباب	check of door
مصد عربات القطار	buffer stop
مصد مطاطي أسطواني	cylindrical rubber fender
مصد مطاطي متدلٍ	draped rubber fender
مصدات الحريق	fire stops
مصدات السفن بالمرفأ	dock fenders
مصدر	origin
مصدر الماء	water resource
مصدم رباعي القوائم	tetrapod
مصعد كبلي بحبل مرتخ	slack-line cableway
مصرف أرضي	land-drain
مصرف التربة السفلية	subsoil drain

مصرف الحجرة	chamber drain
مصرف الزيت	oil drain
مصرف المدخنة	chimney drain
مصرف المنزل	house drain
مصرف المياه السطحية	surface-water drain
مصرف بئري	well drain
مصرف حجري	blind drain, stone drain
مصرف حقلي	field drain
مصرف حقلي مثقب	trammel drain
مصرف خندقي	trench drain
مصرف رئيسي	main drain
مصرف رملي	sand drain, sandwick
مصرف رملي عمودي	vertical sand drain
مصرف صندوقي	box drain
مصرف لولبي	herringbone drain
مصرف لولبي مزدوج	buttress drain, chevron drain
مصرف معترض	intercepting drain
مصرف مطمور	buried drain
مصرف مفتوح	open drain
مصرف هواء	air flue
مصروفات عامة	overheads
مصطبة	benching, terrace
مصطبة خرسانية	concrete benching
مصطبة درج السلم	half landing
مصطكاء	mastic
مصعد	elevator, lift
مصعد أفراد	personnel hoist
مصعد البضائع	goods lift, service lift
مصعد بالهواء المضغوط	pneumatic lift
مصعد حفر كبلي	excavating cableway
مصعد رجل المطافئ	fireman's lift
مصعد ركاب	passenger lift
مصعد صغير بين طابقين	dumbwaiter
مصعد عادي	normal lift
مصعد عياري	standard lift

مصعد كبلي	cableway
مصعد كبلي مائل	inclined cableway
مصعد كهربائي	electric lift
مصعد نقل البضائع	freight elevator
مصعد هيدرولي	hydraulic lift
مصعد هيدرولي كهربائي	electro-hydraulic lift
مصفاة	strainer
مصفاة أسطوانية	drum screen
مصفاة دقيقة	micro-strainer
مصفاة شريطية	band screen
مصفاة مياه للطوارئ	emergency water strainer
مصلحة المساحة	ordnance survey
مصلحة المياه	water authority
مصمم	designer
مصمم إنشاءات	structural designer
مصنع خرسانة	concrete plant
مصنِّف	classifier
مصنوعات حجرية	stoneware
مصنوعات خزفية مزججة	glazed stoneware
مصهر	fuse, foundry
مُصوّت	sounding
مصوّرة لتلفزيونات الدوائر المغلقة	closed-circuit television camera
مصيدة	trap
مصيدة البنزين	petrol interceptor
مصيدة الرمل	sand trap
مصيدة الوحل	mud trap
مصيدة قلاّبية	flap trap
مضاد للتصدع	anti-crack
مضبط ذو خطوتين	two-step control
مضبط مدرّج	multi-step control
مضبط الوصل والقطع	on-off control
مضخة	pump
مضخة إزاحة متنوعة	variable displacement pump

مضخة إزاحية	displacement pump
مضخة إزاحية هوائية للمجارير	pneumatic sewer ejector
مضخة البئر العميقة	deep well pump
مضخة التصريف	drainage pump
مضخة التغذية	feedpump
مضخة الحفر	borehole pump
مضخة الحمأة	sludger
مضخة الوقود	fuelling machine
مضخة بصمام مرتكز	seated-valve pump
مضخة بكباس	plunger pump
مضخة بكباس ذو دفق محوري	axial flow piston pump
مضخة ترددية	ram pump, reciprocating pump
مضخة تصريف متدفق	spout-delivery pump
مضخة تفريغ	vacuum pump
مضخة حفر آبار المناجم	sinking pump
مضخة حفر بالقوة النابذة	centrifugal dredge pump
مضخة حصى	gravel pump
مضخة حقن الاسمنت	boojee pump
مضخة خرسانة	concrete pump
مضخة خرسانة متحركة	mobile concrete pump
مضخة دافعة	force pump
مضخة دافعة بالهواء	air-lift pump
مضخة ذات أرياش	vane pump
مضخة ذات رق	diaphragm pump
مضخة ذات مسننات	gear pump
مضخة رافعة	lift pump
مضخة رحوية	rotary pump
مضخة رمل	sand pump
مضخة سلسلية	chain pump
مضخة ضخمة	mammoth pump
مضخة مغمورة	submersible pump
مضخة نابذة	centrifugal pump

م

مضخة نابذة عديمة الانسداد	non-clogging centrifugal pump
مضخة نزح	shell pump
مضخة هواء	air pump
مضخة هوائية مركبة	compound air lift
مضخة هيدرولية	hydraulic pump
مضخة وحيدة المرحلة	single-stage pump
مضخم الطاقة	power amplifier
مضخم إلكتروني	electronic amplifier
مضخم ترددات لاسلكية	radio frequency amplifier
مضخم سمعي	audio amplifier
مضخم سمعي ترانزستوري	transistor audio amplifier
مضخم متقدم	pre-amplifier
مضغط	press
مضغط بمكبس حر	free-piston compressor
مُضمن	modulator
مطار	airport, aerodrome
مطار ذو كثافة منخفضة	low density airport
مطار ذو كثافة عالية	high density airport
مطار عسكري	military airport
مطار متوسط الكثافة	medium density airport
مطار مدني	civil airport
مطاط	elastic
مطالبة بالتعويض عن الأضرار	damage claim
مطاوعة	yield
مطبخ	kitchen
مطبوعات	printout
مطحنة كروية	ball mill
مطرقة	hammer
مطرقة ارتدادية	rebound hammer
مطرقة بالهواء المضغوط	pneumatic hammer
مطرقة بمحرك ديزل	diesel hammer
مطرقة ثقيلة	sledge hammer
مطرقة ركائز	pile hammer

مطرقة ساقطة	tup
مطرقة ساقطة آلية	drop hammer
مطرقة كروية	ball hammer
مطرقة مزدوجة الفعل	double-acting hammer
مطرقة نحاتة	chipping hammer
مطرقة وحيدة الفعل	single-acting hammer
مطعم	restaurant
مطفأة بالكيماويات الجافة	dry chemical extinguisher
مطفأة بثاني أكسيد الكربون	carbon dioxide extinguisher
مطفأة حامض الصودا	soda-acid extinguisher
مطفأة حريق	fire extinguisher
مطفأة رغوية	foamite extinguisher
مطفأة مائية	water extinguisher
مظهر	appearance
معابر السمك	fish passes
معادلة الانحناء	bending formula
معادلة الحدود	equalization of boundaries
معادلة «بنكويد»	Pencoyd formula for impact
معادلة «بوسينزك»	Boussinesq equation
معادلة شخصية	personal equation
معادلة «هيلي»	Hiley's formula
معادن لا فلزية	non-metallic minerals
معالجة الأرض	land treatment
معالجة التصميد	proofing
معالجة الحمأة	sludge treatment
معالجة الماء	water treatment
معالجة الماء المهدر	wastewater treatment
معالجة المعادن على البارد	cold working
معالجة المواد	materials progressing
معالجة النفايات الصناعية	industrial waste treatment
معالجة أولية	primary treatment

معالجة بالحرارة	normalizing
معالجة بالطريقة الأنودية	anodizing
معالجة بالكريوزوت المضغوط	pressure creosoting
معالجة بالكلور	chlorination
المعالجة بمرشح توشيل	percolating-filter treatment
معالجة ثانوية	secondary treatment
معالجة ثلاثية	tertiary treatment
معالجة حرارية	heat treatment
معالجة لمنع الرطوبة	damp-proofing
معالجة مبدئية	preliminary treatment
معالجة متعاقبة	processing
المعالجة المتقدمة للماء المهدر	advanced waste water treatment
المعالجة المتقدمة للنفايات	advanced waste treatment
معالجة مياه المجارير	sewage treatment
معالم	parameters
مُعامل الاحتكاك	coefficient of friction
مُعامل الاسترطاب	hygroscopic coefficient
مُعامل الانزلاق	slip factor
مُعامل الانضغاط	coefficient of compressibility
مُعامل الانكماش	coefficient of contraction
مُعامل التصادم	impact factor
مُعامل التصدع	modulus of rupture
مُعامل التماس	tangent modulus
مُعامل التمدد	coefficient of expansion
مُعامل الخشونة	roughness coefficient
مُعامل الذبول	wilting coefficient
مُعامل الرجوعية	modulus of resilience
مُعامل الصلابة	modulus of rigidity
مُعامل القاطع	secant modulus
مُعامل القص	shear modulus

مُعامل اللاانضغاطية	nodulus of ncompressibility
مُعامل اللدانة	plastic modulus
مُعامل اللدونة للقاطع	secant modulus of elasticity
مُعامل المرونة	bulk modulus, elastic modulus, modulus of elasticity
مُعامل المقطع	modulus of section
مُعامل النفاذية	permeability coefficient
مُعامل تحويل	conversion factor
مُعامل «ينج» لقياس المرونة	Young's modulus
معايرة	setting
معايير الحواجز	obstruction criteria
معبأ	packed
معبأ بالكريات المعدنية	shot-loaded
معبر ضحل	ford
معبر مشترك	common crossing
معبر وقتي مدرج متداخل	telescoping gangplank
معبرة	ferry
معجون لتثبيت الزجاج	putty
معجونة الزجاج	glazier's putty
معدات	plant
معدات الإنقاذ	life saving equipment
معدات البناء	building equipment
معدات التسوية	skimmer equipment
معدات الحفر	borehole equipment, excavating equipment
معدات الخرسانة	concreting equipment
معدات الرفع	lifting tackle
معدات السكة الحديدية	on-track equipment
معدات الضخ	pumping equipment
معدات المساحة	surveying equipment
معدات أمان خاصة	special safety equipment

م

معدات إنشاء الطرق	road construction plant, road making plant
معدات بحركة متوازية	parallel-motion equipment
معدات بحرية	maritime plants
معدات تأريض	earthing equipment
معدات تغليظ الحمأة	sludge-thickening equipment
معدات تقيير الحصباء	tarmacadam plant
معدات جر	hauling plant
معدات حفر ونقل التربة	earth-moving plant
معدات رفع	elevating plant
معدات فرش التربة السطحية	top soiling plant
معدات هزازة	vibrating machinery
معدات واقية	protective equipment
معدل	average
معدل الاستهلاك اليومي	average daily consumption
المعدل الأقصى لهطول الأمطار	maximum intensity of rainfall
معدل الانحدار الأقصى	maximum gradient
معدل الانحدار الهيدرولي	hydraulic gradient
معدل الانشغال المعياري	Standard Busy Rate
معدل التبخر السنوي	average annual evaporation
معدل التردد	frequency rate
معدل الحفر	drilling rate
معدل الحوادث	incidence rate
معدل الخطورة	severity rate
معدل السرعة	mean velocity
معدل السرعة في الأنابيب	velocities in pipes
معدل انحدار المدرج	runway gradient
معدل حوادث مرتفع	high accident rate
معدل حوادث منخفض	low accident rate
معدل سرعة الارسال البرقي (بود)	baud rate
معدل سقوط الأمطار السنوي	average annual rainfall
معدلات التهوية	ventilation rates
معدن	metal
معدن أبيض	white metal
معدن مقاوم للاحتكاك	anti-friction metal
معزز المادة	promoter
معزز دوار	rotary booster
معزقة خلفية	backhoe
معطيات	data
معلومات ميدانية إضافية	additional site information
معوق العربة	wagon retarder
معيار الاسناد	reference standard
معيار التصميم	design criteria
معيار المسافة الخلفية	back gauge
معينات الاتصال السلكي واللاسلكي	telecommunication aids
معينات ملاحية	navigation aids
مغدف	paddle
مغذٍ	feeder
مغرفة	dipper
مغلظ ثقلي	gravity thickener
مغلظ طفو	flotation thickener
مغنسيت	magnesite
مغنطيس رافع	lifting magnet
مغنطيسي كهربائي	electromagnet
مغنيط : مولد قلطية	magneto
مغنيطومتر	magnetometer
مفاعل خافض للتوتر السطحي	bathotonic reagent
مفاعل ذري يبرد بالماء المضغوط	pressurised water reactor
مفاعل ذري يلطف بالماء	boiling water reactor
مفاعل غازي متقدم	advanced gas reactor
مفاعل ماء ثقيل مولد للبخار	steam generating heavy water reactor
مفاعل متقدم يبرد بالغاز	advanced gas cooled reactor

مفاعل يبرد بالغاز	gas-cooled reactor
مفاعل « ماچنوكس »	magnox reactor
مفتاح	key
مفتاح المصطلحات	legend
مفتاح تحكم ناءٍ	remote control switch
مفتاح ربط	podger, torque wrench, wrench
مفتاح ربط الصمولات	nut wrench
مفتاح ربط الهياكل الانشائية	construction spanner
مفتاح ربط آلي	impact spanner, power wrench
مفتاح ربط بسقاطة	ratchet spanner
مفتاح ربط مجنب	offset spanner
مفتاح ضغطي	pressure switch
مفتاح كهربائي	switch
مفتاح كهربائي حدي	limit switch
مفتاح كهربائي رئيسي	main switch
مفتش	inspector
مفتش الأعمال	inspector of works
مفجر	detonator
مفجر متأخر الفعل	delay-action detonator
مفجر هيدرولي	hydraulic burster
مفرق خطوط حديدية	frog
مفصلة	hinge
مفصلة بلاستيك	plastic hinge
مفصل كروي حُقّي	ball and socket joint
مفصلة معدنية لغلق البوابة	anchor and collar
مفك براغي	screwdriver
مفكك الترابط	bond breaker
مقارن	comparator
المقارنة بالعمود	column analogy
مقاس الدولاب	wheel size
مقاطعات	interrupts
مقاطعات حث	solicited interrupts
مقاطعات غير حاثة	unsolicited interrupts
مقاول	contractor
مقاول بناء	building contractor
مقاول رئيسي	main contractor, principal contrator
مقاول عام	general contractor
مقاول فرعي	subcontractor
مقاوم	resistor
مقاوم للحوامض	acid-resisting
مقاومة	resistance
مقاومة الاحتراق	combustion control
مقاومة الاختراق المخروطي	cone penetration resistance
مقاومة الاختراق المعيارية	standard penetration resistance
مقاومة الانحناء	bending strength
مقاومة الانزلاق	skidding resistance
مقاومة الانشاءات للحريق	structural fire resistance
مقاومة الانضغاط	compressive strength
مقاومة الانضغاط القصوى	ultimate compressive strength
مقاومة الانكسار	breaking strength
مقاومة التدفق الهيدرولي	hydraulic friction
مقاومة التفتت	crushing strength
مقاومة الثني	flexural strength
مقاومة الجر	tractive resistance
مقاومة الخرسانة	cube strength
مقاومة الدروج	rolling resistance
مقاومة الريح	wind resistance
مقاومة السطح البيني	interface strength
مقاومة الشد	tensile strength
مقاومة الشد القصوى	ultimate tensile strength
مقاومة القص	shear strength
مقاومة سلبية	passive resistance
المقاومة النهائية	ultimate strength
مقاييس الرطوبة	moisture meter
مقبرة	cemetery
مقبس	socket

م

مقبس الطاقة	power socket
مقبض	grip
مقبض الباب	doorhandle
مقبض المشبك	clamp handle
مقتصد	economizer
مقدار الحمل	magnitude of loading
مقر الجامعة	university building
مقرضة (للألواح المعدنية)	nibbling machine
مقرن	yoke
مقسّم الماء (في باطن الأرض)	watershed
مقشطة	planer
مقصورة السائق	driver's cabin
مقصورة المراقبة	control cabin
مقصورة مراقبة المدرج	apron control cabin
مقطع القضيب	bar section
مقطع زاوي	angle section
مقطع زجاج	glass cutter
مقطع عرضي	cross-section
مقطع فولاذي ثخين	bulb angle
مقطورة	trailer
مقعد النافذة	window seat
مقياس	gauge, scale
مقياس أبعاد	tacheometer
مقياس «ارمسترونغ»	Armstrong scale
مقياس أسلاك	wire gauge
مقياس أسلاك أمريكي	American wire gage
مقياس أعماق	fathometer
مقياس الاختراقية	penetrometer
مقياس الأسلاك البرمنجهامي	Birmingham wire gauge
مقياس الأسلاك الأمريكي	Brown and Sharp wire gauge
مقياس الانتشار	spread recorder
مقياس الانحدار	slope gauge
مقياس الانحراف	deflectometer
مقياس الانفاذية	permeameter
مقياس بإبرة مؤشرة	needle instrument
مقياس «بوردون» لضغط السوائل	Bourdon pressure gauge
مقياس «بوفورت» لسرعة الريح	Beaufort scale
مقياس البعد	telemeter
مقياس التحميل	loading gauge
مقياس التخطيط البياني	plotting scale
مقياس التدفق	flow meter
مقياس التدفق الفوهي	orifice meter
مقياس التدفق لمساحة متغيرة	variable area flow meter
مقياس الترنح	nutating-disk meter
مقياس تسجيل	recording gauge
مقياس تناسبي	proportional scale
مقياس التيار	current meter
مقياس الجهد	strain gauge
مقياس الحبيبات المحمولة جواً	vel
مقياس حدة الصوت	acoustic strain gauge
مقياس خطافي	hook gauge
مقياس رسم طبيعي	natural scale
مقياس الخطوط الحديدية	rail gauge
مقياس الرسم	scale of a drawing
مقياس الرمل	sand catcher, sand-grain meter
مقياس الضغط	pressure gauge
مقياس ضغط التربة	pressuremeter
مقياس الضغط الجوي	barometer
مقياس الضغط العالي	piezometer tube
مقياس العجلات	wheel gauge
مقياس كمية المطر	rain gauge
مقياس اللزوجة	viscometer
مقياس لمنسوب الماء	point gauge
مقياس مائل	inclined gauge
مقياس مدرج	dial gauge
مقياس مدرج قصير	offset scale
مقياس المد والجزر	marigraph, tide gauge
مقياس مسافات	mekometer

م

مقياس مسافات إلكتروني	Tellurometer
مقياس مستوى الماء	water gauge
مقياس المسطحات	planimeter
مقياس مفتوح التدريج	open-divided scale
مقياس منسوب الماء	staff gauge
مقياس ميكرومتري	micrometer gauge
مقياس ميل ضوئي	photoclinometer
مقيِّم	estimator
مكافحة التأكل	corrosion control
مكافحة الحريق	fire fighting
مكافحة تلوث الماء	water pollution control
مكافحة تلوث الهواء	air-pollution control
مكافحة القوارض	rodent control
مكان العبور	passing place
مكان المعطيات المشتركة	common data area
مكب	spool
مكب المفاعل	reactor core
مكباس هيدرولي	hydraulic ram
مكبح العربة المقطورة	overrun brake
مكبح نابذ	centrifugal brake
مكبر الصوت	loudspeaker
مكبس تشكيل	moulding press
مكبس هيدروستاتي	hydrostatic press
مكتب	office
مكتب البريد	post office
مكتب تصميم	design agency
المكتب التنفيذي	executive office
مكتب عمومي	general office
مكتب فني	technical bureau
مكتب المساح	surveyor's office
مكتبات خرج الكمبيوتر	object libraries
مكتبة	library
مكثف	capacitor, condenser
مكثف تبخري	evaporative condenser
مكثف يبرد بالهواء	air-cooled condenser
مكشاف الاستقطاب	polariscope

مكشاف الدخان	smoke detector
مكشاف ثابت الحرارة	fixed-temperature detector
مكشاف الرموز	condition code
مكشاف كهرضوئي	photoelectric detector
مكشاف لهب	flame detector
مكشطة	scraper, skimmer
مكشطة ذات دواليب	wheel scraper
مكشطة ذات كبل	dragline scraper
مكشطة معزقية	hoe scraper
مُكعب اختبار الخرسانة	test cube
مكنة	machine
مكنة اختبار	testing machine
مكنة إجهاد هيدرولية للسكة	rail hydro-stressor
مكنة تشكيل	moulding machine
مكنة تفجير	blasting machine
مكنة تفريز	milling machine
مكنة تهذيب حواف الطريق	siding machine, verge trimmer
مكنة حفر التربة الطرية	soft-ground boring machine
مكنة حفر الصخور	rock-boring machine
مكنة حني الروافد	beam bender
مكنة حني القضبان	rail bender
مكنة دوارة غير قلابة	non-tilting drum machine
مكنة رصف وتسوية الطرق	Barber Greene tamping levelling finisher
مكنة سنفرة بالهواء المضغوط	pneumatic sander
مكنة فرش وإنهاء	laying-and-finishing machine
مكنة لرصف الطرق	macadam spreader
مكنة ليّ القضبان	bar bender
مكنة نقر وتحزيز	notcher
مكنسة كهربائية	vacuum sweeper
مكونات الاجهاد	stress components

م

مكونات حثية	inductive component
ملاءمة التصميم	fitness of design
ملابس صامدة للماء	waterproof clothing
ملابس واقية	protective garment
ملاحة	navigation
ملاحظ عمال	foreman
ملاحظ عمال عام	general foreman
ملاحق	addenda
ملاذ	refuge
ملاط	mortar
ملاط أسود	black mortar
ملاط الاسمنت	cement mortar
ملاط الجير والاسمنت	lime-cement mortar
ملاط جيري	lime mortar
ملاط خشن	rough-cast
ملاط رقيق القوام	slurry
ملاط رملي	Shotcrete (TM)
ملاط سائل	grout
ملاط سيئ	bad mortar
ملاط «كولجروت»	Colgrout (TM)
ملاط معياري	gauged mortar
ملاط يرش بالهواء المضغوط	pneumatic mortar
ملتقى	junction
ملتقى خاضع للمراقبة	controlled junction
ملجأ	shelter, refuge
ملحق	supplement
ملحق للمبنى	annex to a building
ملحوم	welded
مُلدن	plasticizer, water reducer
ملعب رياضي	sports stadium
ملعب مدرج (استاد)	stadium
ملعقة العينات	sampling spoon
ملف إعتاق	trip coil
ملف المسخن	heating coil
ملف لولبي	solenoid
مليّمتر	millimetre
مليّميكرون	millimicron
مليون	million
ممثل الأمان	safety representative
ممر	corridor, passage
ممر إبطاء السرعة	deceleration lane
ممر الدرج	stairway
ممر رأسي	shaft
ممر السير	traffic lane
ممر سفلي	underpass
ممر ضيق (على جسر)	catwalk
ممر مائي صالح للملاحة	navigable waterway
ممر المشاة	footpath
ممر مشاة مائل	inclined footway
ممسك العدة	mandrel
ممشى	walkway
ممشى تحت الأرض	subway
ممشى جانبي	sidewalk
ممنوع الانتظار	clearway
ممهدة	blade grader
ممهدة رافعة	elevating grader
ممهدة طريق	road grader
منارة	lighthouse
منارة المطار	airport beacon
منارة عائمة	lightship
منارة لتحديد الموقع	locator beacons
المناطق الخالية على المدرج	runway clear zones
مناولة البضائع	handling cargo
مناولة البضائع العامة	general cargo handling
مناولة السلع	handling goods
مناولة المواد	materials handling
مناولة لا مركزية	decentralised handling
مناولة مركزية	centralised handling
منبع النهر	headwater
منتوجات بترولية	petroleum products
منجم	pit

م

منجم رمل	sand pit
منح العقد	award of contract
منحات مسنن	bush hammer
منحدر	ramp, batter
منحدر ذاتي التنظيف	self-cleansing gradient
منحدر عشبي	grassed slope
منحدرات مدرعة	armoured slopes
منحنى	bent, curved
منحنى إجمالي	mass curve
منحنى الاجهاد والتوتر	stress-strain curve
منحنى الأداء	performance curve
منحنى بسيط	simple curve
منحنى بيان كمية الحفر	mass-haul curve
منحنى التحول	transition curve
منحنى التدفق	flow curve
منحنى التردد	frequency curve
منحنى تصنيف الركام	aggregate grading curve
منحنى التغير لعزم الثني	bending-moment envelope
منحنى التوزيع	distribution curve
منحنى حلزوني	spiral curve
منحنى الحمل والامتداد	load-extension curve
منحنى دائري	circular curve
منحنى الدوام	duration curve
منحنى السعة	capacity curve
منحنى سلسلي هيدروستاتي	hydrostatic catenary
منحنى الماء المحجوز	backwater curve
منحنى متعاكس	reverse curve
منحنى مركب	compound curve
منحنى المرونة	elastic curve
منزحة	bailer
منزل	house
منساخ	pantograph
منسِّق	classifier
منسوب الماء	level of water, water level

منسوب الماء الأقصى	maximum water level
منسوب الماء القياسي	standard-water level
المنسوب المتوسط	average level
منسوب النقطة	spot level
منشأ	origin
منشآت الجناح	terminal buildings
منشآت الصيانة	maintenance buildings
منشآت المرافق	utility buildings
منشار	saw
منشار إطاري	frame saw
منشار آلي	power saw
منشار آلي سلسلي	chain saw
منشار دائري لقطع الأحجار	diamond saw
منشار شريطي	band saw
منشار منضدي	bench saw
منشار معادن	hacksaw
منشار يدوي	handsaw
منشار يدوي متنقل	portable hand saw
منشاق	snorkel
منصب	trestle
منصب ثلاثي القوائم	tribrach
منصة	platform
منصة امتداد	outrigger scaffold
منصة تحكم	control platform
منصة توزيع الحمل	relieving platform
منصة رافع هيدرولي	hydraulic lift platform
منصة صغيرة نقالة	dolly
منصة نقالة	pallet
منصة هبوط	landing stage
منضدة عمل	bench
منطق الحقن المتكامل	integrated injection logic
منطقة	zone
منطقة الارساء	anchorage area
منطقة انتظار السيارات	parking area
منطقة الانحدار	sloping zone

م

منطقة تجمع الركاب	passenger catchment area
منطقة تحول	transition zone
منطقة ترشيح	filter zone
منطقة ترفيه	recreation area
منطقة التشبع	zone of saturation
منطقة التهوية	zone of aeration
منطقة الجناح	terminal area
المنطقة الخارجية	outer zone
منطقة الخدمة	services area
منطقة « الڤادوز »	vadose zone
منطقة عبور المشاة	pedestrian crossing
المنطقة المتوفرة	area available
منطقة مفروشة بالخرسانة	road panel
منطقة وقوف السيارات	car parking area
منظر	sight
منظر خارجي	outside view
منظر عام	bird's eye view
منظف	scour
منظم	regulator
منظم تصريف	discharge regulator
منظم الضغط	pressure regulator
منظم ضغط الزيت	oil pressure regulator
منظمة الطيران المدني الدولية	International Civil Aviation Organisation
منظور	perspective
منع التآكل	corrosion prevention
منع الحوادث	accident prevention
منعطف	curve
منعطف رأسي	vertical curve
منعطفات بشكل S	S curves
منفث إبري	needle jet
منفذ	access, port
منفذ التصريف	drainage inlet
منوِّب	alternator
مهايئ	adaptor
مهداف	sight
مهداف التسوية السالب	minus sight
مهذبات الوشيع	hedge trimmers
مهندس	engineer
مهندس استشاري	consulting engineer
مهندس إنشاءات	constructional engineer, structural engineer
مهندس إنشاءات قانوني	chartered structural engineer
مهندس الموقع	site engineer
مهندس بلدي قانوني	chartered municipal engineer
مهندس تخطيط	planning engineer
مهندس تخطيط المدن	town planner
مهندس – رسام إنشاءات	structural designer-draughtsman
مهندس فني	technician engineer
مهندس مدني	civil engineer
مهندس مدني قانوني	chartered civil engineer
مهندس معماري	architect
المهندس المقيم	resident engineer
مهندس هيدروليات	hydraulic engineer
مهندس وضع المواصفات	specification engineer
مهواة	ventilator
مهواة التشغيل	working shaft
مهواة صاعدة	rising shaft
مهواة المخرج	outlet ventilator
مهواة المدخل	inlet ventilator
مهواة المنجم	mineshaft
مؤتمر صحفي	press conference
مؤشر اتجاه الريح	wind indicator
مؤشر ضغط الزيت	oil pressure indicator
مؤشر نصف القطر والحمل المأمون	radius-and-safe-load indicator
مؤشرات انحدار الاقتراب البصري	Visual Approach Slope Indicators

م

مواد البناء	building materials
مواد الردم المستعارة	borrow
مواد إنشائية مركبة	composites
مواد حافظة للخشب	preservatives for timber
مواد حاكة	abrasives
مواد خفض سرعة السوائل	peptizing agents
مواد ربط	binding material
مواد صهورة	fluxes
مواد مغربلة	screened material
موازن	stabilizer
مواسير تصريف	drainpipe
مواسير مطمورة	sub-main
مواصفات	specification
مواصفات الأداء	performance specification
مواصفات الأمان	safety standards
مواصفات التصميم للأعمال الترابية	earthworks design standards
مواصفات الحصول على المواد	material procurement specification
مواصفات الطرق	road specification
مواصفات المشروع	project specification
مواصفات المصنعية	workmanship specification
مواصفات المواد	materials specification
مواصفات تصويرية	descriptive specification
مواصفات رئيسية	master specification
المواصفات القياسية	standard specification, standards
المواصفات القياسية المطبقة	standards in force
المواصفات الوطنية	national standards
موجات	waves
موزع يدوي	hand distributor
موسع ثقوب	reamer
موشور بصري	optical wedge
موصل	conductor
موصل معلّق	overhead conductor
الموصلات الرئيسية	service mains
موصلات القوة الرئيسية	force mains
موظف التشغيل	operator
موظفون	staff
موظفون دائمون	permanent staff
موظفون مؤقتون	temporary staff, transient staff
موظفو الموقع	site staff
موظفو الهندسة	engineering staff
مؤقت	transient
موقع	site
موقع البناء	building site
موقع السكك	track location
موقع المطار	airport location
موقف استراحة للسيارات	lay-by
مولد	generator
مولد تربيني	turbo-generator
مولد قلطية	magneto
مولد للطوارئ	emergency operator
مولد يعمل بالبخار	steam-powered generator
موهن	attenuator
مياه الأمطار	catchwater, stormwater
مياه التعويض	compensation water
مياه المجاري	sewage
مياه المجاري القادمة	incoming sewage
مياه المجاري المنزلية	domestic sewage
مياه بواليع منزلية	domestic wastewater
مياه جوفية	groundwater, underground water
مياه سطحية	surface water
مياه محجوزة	held water
ميزاب	gutter
ميزان استواء عاكس	reflecting level
ميزان تسوية	level tube

م

ميزان تسوية بفقاعة	bubble tube
ميزان تسوية راكب	striding level
ميزان تسوية قلاب	tilting level
ميزان تسوية يدوي	Abney level, hand level
ميسم	branding iron
ميكانيكا التربة	soil mechanics
ميكانيكي	mechanic
ميكانيكي تجميع	fitter
ميكروبروسسور	microprocessor
ميكروبروسسور تداخلية	interfacing microprocessor
ميكروبروسسور للبرمجة	programming microprocessor
ميكروبروسسور وحيد الشظية	single chip microprocessor
ميكروسكوب	microscope
ميكروسكوب بتروجرافي	petrographic microscope
ميكروفون	microphone
ميكروكمبيوتر	microcomputer
ميكرومتر	micrometer
ميكرون	micron
مَيْل	banking, tilt, incline, batter
ميل	mile
ميل انجليزي	statute mile
ميل بحري	nautical mile
الميل المغنطيسي	magnetic declination
الميل الهيدرولي الحرج	critical hydraulic gradient
ميناء	harbour, port
ميناء الدخول	port of entry
ميناء اللجوء	harbour of refuge
ميناء حر	free port
ميناء شبه طبيعي	semi-natural harbour
ميناء طبيعي	natural harbour
ميناء عسكري	military harbour
نابذ	centrifuge
نابض	spring
ناتج التكثيف	condensate
ناتج الحفر	bulking
نازع الركائز	pile extractor
ناشرة	diffuser, spreader
ناظر	ganger
ناظر عمال متجول	walking ganger
ناعورة	water wheel
نافذة	window
نافذة بابية	picture window, casement
نافذة بإطارين منزلقين	sash window
نافذة توازن	balance window
نافذة مزدوجة المفصلة	double-hung window
نافذة محورية	pivoted window
نافذة ناتئة	projected window
نافذة ناتئة مدورة	bow window
نافخ دوار	Rootes blower
نافورة	fountain
نافورة مائية	water jet
ناقل	conveyor, conductor
ناقل بمنصة	pallet conveyor
ناقل حلزوني	helical conveyor
ناقل هوائي	aerial tramway
ناقلات	carriers
ناقلة بالسير	belt conveyor
ناقلة بالهواء المضغوط	pneumatic conveyor
ناقلة بترس دودي	worm conveyor
ناقلة ركاب	passenger conveyor
ناقلة كبلية	cableway transporter
ناقلة لولبية	screw conveyor
ناقوس التقاط	screw bell
ناقوس غوص	diving bell
ناقوس غوص بالهيليوم	helium diving bell

نبْت	vegetation
نبع	spring
نبيطة إنذار	warning device
نتاج	output
نتح	transpiration
نترات الأمونيوم	ammonium nitrate
نترات السليولوز	cellulose nitrate
نتروسليلوز	nitrocellulose
نتروغليسرين	nitroglycerin
نتوء	spur, ridge
نتوء الصخر	outcrop
نتوء صخري	rocky outcrop
نجار	carpenter
نجار تركيب	joiner
نجارة	joinery
النحت	scabbling
نحاتة	chippings
النحاس الأصفر	brass
نحاس الطلقات	cartridge brass
نحاس مسحوب بصلابة	hard-drawn copper
نحو الغرب	westing
نداء طارئ	emergency call
نزح	bail, drain
نزح الماء	dewatering
نزع الهيكل المؤقت	stripping
نزف	bleeding
النسبة الاقتصادية	economic ratio
نسبة الثبات الحرج للرمل	critical voids ratio of sands
نسبة الحساسية	sensitivity ratio
نسبة الركام إلى الاسمنت	aggregate/cement ratio
نسبة السرعة	velocity ratio
نسبة الضغط الهيدروستاتي	hydrostatic pressure ratio
نسبة الفراغ	voids ratio
نسبة القضافة	slenderness ratio

نسبة الماء إلى الاسمنت	water/cement ratio
نسبة «بويسون»	Poisson's ratio
نسبة معيارية	modular ratio
نسخ	photostat
نسخة مصوّرة	photocopy
نسخة جافة	Xerox copy (TM)
نسخة داعمة	back-up copy
نسخة زرقاء	blueprint
نسيج	fabric
نسيج خارجي	external fabric
نشارة الخشب	sawdust
نشر	spread
نصب	setting up
نصب تذكاري	monument
نصف قطر الانحناء	radius of curvature
نصف القطر	radius
نصف باع	semi-span
نصف خشبي	half timber
نصف قطر الدوران	radius of gyration
نصف كروي	orange-peel bucket
نصف مغمور	semi-submersible
نصل (ريشة)	blade
نصل المفتاح السكيني	switch blade
نصيحة المصنع	manufacturer's advice
نظائر	isotropic
نظارات أمان	safety glasses
نظارات نحاتة	chipping goggles
نظارات واقية من الحوامض	acid goggles
نظام	system
نظام أحادي الماسورة	one-pipe system
نظام إرشاد سلكي	wire guidance system
نظام الاشارات الهادية الأوتوماتيكي	automatic-block signal system
نظام الاشارات اليدوي	manual-block system
نظام الألواح	panel system
نظام البناء	building system

ن

نظام تحكم	control system
نظام ترقيم المدرج	runway numbering system
نظام التصريف	drainage system
نظام تصريف الماء بالارتشاح	soakaway system
نظام تصريف مشترك	combined system
نظام تصريف منفصل المجاري	separate system
نظام تصنيف التربة	Ac system
نظام تكييف الهواء	air conditioning system
نظام تهوية سطحية	surface-aeration system
نظام الحط بأجهزة القياس	instrument landing system
نظام الحمأة المنشَّطة	activated-sludge system
نظام الدارة المغلقة	closed loop system
نظام الدفق	finger system
نظام الذاكرة	memory system
نظام ري سطحي	surface irrigation system
نظام الطبقات	sandwich system
نظام الطرق	road system
نظام طولاني	longitudinal system
نظام عوارض مقوى بالكبلات	cable-braced girder system
نظام قدرة الرفع	lift power system
نظام ماء التبريد	cooling-water system
نظام مادة الترابط	matrix system
نظام مترابط	interconnected system
نظام متري	metric system
نظام مراقبة التكاليف	cost control system
نظام مرشات أوتوماتي	automatic sprinkler system
نظام مزدوج السكة	duo-rail system
نظام مستعرض	transverse system
نظام مسح القصور الذاتي	inertial surveying system
نظام المنح	bonus scheme
نظام منفصل جزئياً	partially-separate system
نظام النداء العام	public address system
نظام نصف مستعرض	semi-transverse system
نظام هواء أحادي المجرى	single-duct all-air system
نظام هواء العادم	exhaust-air system
نظام هواء منتشر	diffused-air system
نظام الهواء والماء	air-water system
نظام هوائي كامل	all-air system
نظام هوائي كامل مزدوج المجرى	double-duct all-air system
نظرية الإسفين	wedge theory
نظرية الخطأ	theory of errors
نظرية «برنولي»	Bernoulli's theorem
نعل الدلو	boot
نعل الركيزة	pile shoe
نفاثة عالية السرعة	high-velocity jet
نفاخ دوراني	rotary blower
نفايات خطرة	hazardous waste
نفايات مشعة	radioactive waste
نفاية	effluent, muck
نفاية سائلة	liquid effluent
نفخة	blow
النفط	crude oil
نفقات إضافية	overhead costs
نفق	tunnel
نفق أفقي	heading
نفق أفقي صندوقي	box heading
نفق أنبوبي مغمور	sunken-tube tunnel
نفق تحت الماء	underwater tunnel
نفق التصريف	drainage tunnel
نفق دليلي	pilot tunnel
نفق سفلي	bottom heading
نفق سكة حديدية	railway tunnel
نفق صخري	rock tunnel

نفق طريق عام	highway tunnel
نفق المجاري	sewer tunnel
نفق محفور بالقطع والردم	cut-and-cover tunnel
نفق مغمور	submerged tunnel
نفق النقل السريع	rapid-transit tunnel
نفق هوائي	wind tunnel
نقالة قلاّبة	dump barge
نقالة مقطورة	dumb barge
نقر	notch
نقر المعدن	pickling of metal
نقر مستطيل	rectangular notch
نقش الحجارة بالمنحات المسنن	bush hammering
نقص المعدات المكنية	underplanting
نقط تحويل	catch points
نقطة الاتصال	junction point
نقطة الاسناد في المطار	airport reference point
نقطة الانعطاف	turning point
نقطة الانكسار	breaking point
نقطة التفجير	shot point
نقطة التقاطع	intersection point
نقطة التماس	tangent point
نقطة التزازن	balance point
نقطة الخضوع	yield point
نقطة الصفر	zero point
نقطة الكبح	arrest point
نقطة الندى	dewpoint
نقطة الوسط	midpoint
النقطة الحرجة	change point, critical point
نقطة مراقبة	control point
نقل	removal, transfer, transport
نقل الطاقة	power transmission
نقل الطاقة الكهربائية	electrical power transmission
نقل بالحبال	rope transmission
النقل بالسكك الحديدية	rail transportation
نقل بالقطارات	locomotive haulage
النقل الشخصي السريع	personal rapid transit
نقل علوي	upper transit
نماذج الموانئ	harbour models
نمط الأمر لايجاد المعلومات	addressing mode
نمط التنمية	development pattern
نمط الدفق العادي	normal flow pattern
نمط تدفق هواء التهوية	ventilating air flow pattern
نموذج	model, sample
نموذج اتفاقية العقد	contract-agreement form
نموذج عرض	proposal form
نهر	river
نوابض ورقية	leaf springs
النواحي الاقتصادية	economic aspects
النواحي البيئية	environmental aspects
نوبة العمل	working shift
نوموجرام : رسم بياني	nomogram
هاتف	telephone
هبوط	landing
هبوط (قاعدة البناء)	settlement
هبوط السكة	rail depression
هبوط بالثقل الذاتي	inherent settlement
هبوط تداخلي	interference settlement
هبوط متفاوت	differential settlement
الهبوط النسبي	relative settlement
هدْر	waste
هدف	target
هدم يدوي	hand demolition
هرب	escape
هرتز	hertz
هز الأساسات	vibration of foundations
هزاز لدمج الخرسانة	concrete-vibrating machine

ن

هزازة	vibrator
هزازة بمحراك	poker vibrator
هزازة خارجية	external vibrator
هزازة داخلية	internal vibrator
هزازة سطحية	surface vibrator
هزازة لوحية	plate vibrator
هزهزة الخرسانة	pervibration
هليكوبتر	helicopter
هضبة صغيرة	mound
هضم	digestion
هضم الحمأة	sludge digestion
هكتار	hectare
هكتو	hecto-
هليوم	helium
هندرد ويت	hundredweight
الهندسة البلدية	muncipal engineering
هندسة البناء	architecture
هندسة البيئة	environmental engineering
هندسة الصحة العامة	public health engineering
هندسة مدنية	civil engineering
هندسة المرور	traffic engineering
هندسة الموانئ	harbour engineering
هندسة النقل	transportation engineering
الهندسة الهيدرولية	hydraulic engineering
هدم	demolition
هراس التربة	sheepsfoot roller
هواء مشبع بالرطوبة	saturated air
هواء مضغوط	compressed air
هويس دخول	entrance lock
هويس مزدوج	double lock
هويس واق	guard lock
هيئة المرور	highway authority
هياكل جانبية	road forms
هياكل منزلقة	sliding forms
هيجان الضفة	outburst bank
هيجرومتر	hygrometer
هيدوديناميك	hydrodynamics
هيدروغرافي	hydrography
هيدروكربون مهلجن	halogenated hydrocarbon
هيدرولوجية (علم المياه)	hydrology
هيدرولي	hydraulic
هيدروليات (علم السوائل)	hydraulics
هيدروليات الحدود السائبة	loose-boundary hydraulics
هيدرومتر	hydrometer
هيكل	framework, body
هيكل (شاسيه)	chassis
هيكل (إطار)	frame
هيكل إطار رئيسي	mainframe
هيكل البناء	shell
هيكل السطح	deck framing, roof structure
هيكل برجي	tower shell
هيكل ثابت	rigid frame
هيكل خشب مقوى بالشكالات	braced-timber framing
هيكل خشبي	wood framing
هيكل خشبي مؤقت	timber shuttering
هيكل دائم	permanent frame
هيكل ذو جائز وبلاطة	beam-and-slab construction
هيكل ذو جائز وعارضة	beam and girder framing
هيكل ركائز	pile frame
هيكل صب الخرسانة	shuttering
هيكل صلب	stiff frame
هيكل ضعيف	unstable frame

هـ

هيكل علوي	top frame
هيكل على شكل حرف A	A-frame
هيكل غربي	platform framing
هيكل فراغي	space frame
هيكل فرعي	ancillary frame
هيكل فولاذ إنشاءات	structural-steel framing
هيكل فولاذي	steel framework
هيكل فولاذي معلق	hanging leaders
هيكل قنطري	portal frame
هيكل كابولي	cantilever formwork
هيكل كامل	simple framework
هيكل متحرك	travelling formwork
هيكل مستو	plane frame
الهيكل المفقود	lost formwork
هيكل مقوى بالشكالات	braced framework, balloon framing
هيكل ملحوم	welded frame
هيكل من الخشب الرقائقي	plywood formwork
هيكل مؤقت	temporary formwork
هيكل مؤقت منزلق	sliding formwork
هيكل ناقص	imperfect frame
هيكل وصل الانشاءات	stunt end
واجهة	facade
واجهة العرض	show window
واجهة متوازية	parallel heading
واحد	one
واحد وعشرون	twenty-one
وادٍ ضيق	clough
واط	watt
واقي القدم	foot guard
واقي الآلة	machine guard
واقي لوح الوصل	shin guard
واقية الوحل	mudguard
وتد	peg
وتد الإرساء	mooring post
وتد إسناد	recovery peg, reference peg

وتد لمقعد تثبيت	rail key
وتد متعامد	quarter peg
وتد محاذاة	boning rod
وتر	chord, tendon
وترة	web
وترة السكة	rail web
وترة مصمتة	solid web
وجه	face
وجهة نظر	view point
وحدة الاجهاد	unit stress
وحدة إزالة التلوث	decontamination unit
وحدة الانفعال	unit strain
وحدة تبريد	refrigeration plant
وحدة التحكم	control console
وحدة التحكم في المخرج	outlet control unit
وحدة تدريع ثلاثي	tribar armour unit
وحدة ترشيح	filtration plant
وحدة تركيز	concentrating plant
وحدة تسخين	heating unit
وحدة تصحيح	correcting unit
وحدة التصريف	drainage plant
وحدة تهوية	ventilation plant
وحدة حجرية جوفاء	hollow masonry unit
وحدة حرارة بريطانية	British Thermal Unit
وحدة حفر متنقلة	portable rig
وحدة الحوادث الكيماوية	chemical incident unit
وحدة خلط الخرسانة على دفعات	concrete batching/mixing plant
وحدة خلط القار	bituminous mixing plant
وحدة خلط على دفعات	batch mixing unit
وحدة دخول مباشر على الذاكرة	direct memory access interface
وحدة صب الخرسانة	concrete placer
وحدة صناعية تجريبية	pilot plant
وحدة الضخ	pumping unit
وحدة ضغط الحمأة	sludge pressing plant

هـ

وحدة العرض البصري	VDU
وحدة فرش	laying plant
وحدة القياس	unit of measurement
وحدة قياس كميات الخلط	batching plant
وحدة الكراءة	dredging plant
وحدة مدرعة لمصدم رباعي القوائم	tetrapod armour unit
وحدة مزج مكونات الخرسانة بالوزن	weigh batcher
وحدة معالجة مركزية	central processing unit
وحدة هضم الحمأة	sludge-digestion plant
وحدة الوزن	unit weight
وحيد الفعل	single-acting
ورشة	workshop
ورشة تركيب	erecting shop
ورشة حدادة	forge
ورق استشفاف	tracing paper
ورق الجدران	wallpaper
ورق عزل الصوت	building paper
ورق لف	cartridge paper
ورق مربعات	profile paper
ورنيش	varnish
ورنيش اليابان الأسود	black japan
وزن الازاحة بالطن	displacement tonnage
وزن الحمل الساكن بالطن	deadweight tonnage
وزن السكة	rail weight, weight of rail
وزن الطائرة الإجمالي	gross aircraft weight
الوزن الفارغ	tare
وزن معادل	counterweight
الوزن النوعي	specific gravity
وسائل إخماد الحريق والانقاذ	fire-crash-rescue facilities
وسائل الحفر	means of excavation
وسائل الاتصال	communication facilities
وسائل النقل العام	public transport
وسادة الركيزة	pile cushion
وسادة المطرقة	hammer cushion
وسع الثقب	ream
وسم	branding
وسيط	agent
وسيلة ارتداد	kicker
وسيلة تحميل الركاب	passenger-loading device
وسيلة حدية لمعدل الدفق	flow limiting device
وسيلة الخرج	output device
وسيلة كشف	detection device
وسيلة كهرضوئية	photoelectric device
وسيلة منع الفيضان	anti-flood device
وسيلة واقية للتنفس	respiratory protective device
وصل	connection, coupling
وصل بالنقر	notching
وصل قنطري	bridging
وصلات الحركة	movement joints
وصلات تراكبية	fishplated joints
وصلة	link
وصلة استقرار	settlement joint
وصلة اعوجاج	warping joint
وصلة إنشاء تناكبية	single-butt construction joint
وصلة إنشائية	construction joint
وصلة بجلبة	gland joint
وصلة «بوردو»	Bordeaux connection
وصلة تحويل	crossover
وصلة تراكب	lap joint
وصلة تسليك	reeving thimble
وصلة تعويض	compromise joint
وصلة تقلص	contraction joint, shrinkage joint
وصلة تمددية	expansion joint
وصلة تناكب	jump join

وصلة تنصيفية	halved joint
وصلة جافة	dry joint
وصلة حز ولسان	tongue and groove joint
وصلة ذراع التطويل	boom joint
وصلة ذكر وأنثى	spigot-and-socket joint, ogee joint
وصلة سكة حديد	rail joint
وصلة تقلص	dummy joint
وصلة طريق	road connection
وصلة عامة الحركة	universal joint
وصلة فم العصفور	bird's mouth joint
وصلة قضيب ملحومة بالصب	cast-welded rail joint
وصلة متداخلة	recessed joint
وصلة متراكبة	overlapping joint
وصلة محورية	pivot joint
وصلة مسمارية	pin joint
وصلة مسيكة	leaktight joint
وصلة مغراة	glued joint
وصلة منزلقة	sliding joint, slip joint
وصلة نقرة ولسان	mortise and tenon joint
وصلة هيدروستاتية	hydrostatic joint
وضع الحصى تحت السكة	boxing up
وضع العلامات	setting out
وطيدة	plinth
وعاء	container
وعاء استقبال	receiver
وعاء ضغط المفاعل	reactor pressure vessel
وعاء ضغطي	pressure vessel
وعاء هوائي	air vessel
وقاء	fender
وقاية الانحدار	slope protection
وقاية بالأنود الذواب	sacrificial protection

وقاية خرسانية	concrete protection
وقاية الرأس	head protection
وقاية العينين والوجه	eye-and-face protection
الوقاية الكاثودية	cathodic protection
وقاية المرفأ	harbour protection
وقاية من التسرب الأرضي	earth-leakage protection
الوقاية من التعرية	scour protection
الوقاية من الحريق	fire protection
وقت التصلد الأولي	initial setting time
وقود بترولي	fuel oil
وقود نووي	nuclear fuel
وكالة حكومية	government agency
وكيل فرعي	sub-agent
وكيل المقاول	contractor's agent
وليجة فولاذ منغنيزي صلبة	solid manganese-steel insert
ونش	winch
ونش ذاتي الدعم	self-sustaining winch
ونش متنقل	mobile crane
وهدة الهبوط	settlement crater
ياردة	yard
ياردة مربعة	square yard
يزود بمصاطب	terracing
يحاذي	aline, align
يحرك آلياً	power handling
يحمي	guard
يضرب	man handling
يعمل بالهواء المضغوط	pneumatic
يمر تحت طريق	cross-under
يوثق	joggle
يورانيوم	uranium

و